Praise for *Urban Forests*

• Named a Staff Pick for NPR Books' Guide to 2016's Great Reads •

"Next time you're outside, look up. Trees are so ubiquitous that it's easy to take them for granted. But *Urban Forests* makes you stop and pay attention to the 'living landmarks' standing tall in America's cities. From Thomas Jefferson's time to present day, Jill Jonnes explores the essential roles trees play in urban centers—filtering air, providing habitat, offering shade, calming nerves, and more. I loved this book because it's both for history lovers and for tree devotees. It's a good read—best done under the canopy of your favorite tree."

—Jeanine Herbst, NPR Books

"A fascinating slice of both urban and natural history that tree lovers and everyone interested in city life will enjoy." —*Booklist*

"In *Urban Forests,* Jill Jonnes extols the many contributions that trees make to city life . . . [and] celebrates [the] men and women who stood up for America's city trees over the past two centuries. . . . An authoritative and admirably nontechnical account of the past, present, and future of our cities' trees."

—Gerard Helferich, *The Wall Street Journal*

"Even if you can't tell a fir from a pine, you probably judge the quality of your surroundings by its trees. For city residents, trees are perhaps the most accessible form of the natural world. . . . Jonnes traces the history of America's urban trees over two centuries—they were once viewed as an economic commodity, but people later invested personal and patriotic meaning in individual trees and in the act of planting. . . . *Urban Forests* goes beyond trees, exploring a nation's changing relationship with the whole natural world."

—Jeremy B. Yoder, *Sierra Club Magazine*

"*Urban Forests* contains some of the most readable and insightful arboreal prose I have ever come across. Jonnes dives deeply into trees and their roles in American cities through various eras of history. The text is laced with facts, dates, and figures gleaned from recent scientific studies that, rather than making one's eyes glaze over, inspire a profound respect for these resilient trees and the people who champion them. . . . A spell
relates the heartbreaking stories of America's
tragedies—the annihilation of native elm, che
introduced pests and diseases. . . . Through thes

ling anecdotes, the book elucidates the powerful emotional connection humans have with trees." —Guy Sternberg, *The American Gardener*

"Perhaps the most affecting portions of Jonnes's book delve into trees as symbols of resilience . . . as much as trees can be transportive, inviting imagination to alight on the branches arcing toward the sky, they can also anchor us. Trees, with their graceful grit, embody some of the very best traits that we can hope to emulate." —Jessica Leigh Hester, *The Atlantic CityLab*

"America's cities are full of trees but despite encountering them all the time we tend to take them for granted or know little about their natural history and civic virtues. But in a new book, *Urban Forests*, author Jill Jonnes says trees play an extraordinarily important role in our cityscapes and they are the dominant component of what is now called green infrastructure."

—Diane Rehm, NPR, *The Diane Rehm Show*

"We all know that trees can make streets look prettier. But in her new book *Urban Forests*, Jill Jonnes explains how they make them safer as well. . . . It's no wonder then, that cities like New York, Denver, and Sacramento have already invested heavily in urban planting. Now Jonnes argues that others should follow their lead. It's time, she writes, 'to get serious about creating the lushest tree canopies we can nurture.'" —Sara Begley, *Time*

"The deforestation that ran rampant in the United States through the nineteenth century spurred a band of doughty dendrologists and politicians to forest the cities. Jill Jonnes's stimulating history chronicles their collective story, from William Hamilton (who reintroduced Ginkgo Biloba to North America millennia after it was glaciated out) to the many scientists struggling to control blights and beetles. Today, Jonnes shows, despite trees' measurable benefits for human well-being and microclimate regulation, urban forestation remains at risk from short-sighted redevelopment." —Barbara Kiser, *Nature*

"Sure to fascinate nature lovers and natural scientists alike . . . A lovingly written book that should appeal to most city dwellers and all tree lovers."

—*Kirkus Reviews*

"Far-ranging and deeply researched, *Urban Forests* reveals the beauty and significance of the trees around us."

—Elizabeth Kolbert, Pultizer Prize-winning author of *The Sixth Extinction*

PENGUIN BOOKS

URBAN FORESTS

Jill Jonnes is the author of *Eiffel's Tower, Conquering Gotham, Empires of Light*, and *South Bronx Rising*. Founder of the Baltimore Tree Trust, she is also a Maryland Master Naturalist. A staff member of the 2010 National Commission on the BP Deepwater Horizon Oil Spill and Offshore Drilling, she wrote the first chapter of the report to the president: *Deep Water: The Gulf Oil Disaster and the Future of Offshore Drilling*. In the fall of 2011, she studied "Trees as Green Infrastructure" at the Woodrow Wilson International Center for Scholars in Washington, D.C. Jonnes has been a National Endowment for the Humanities scholar.

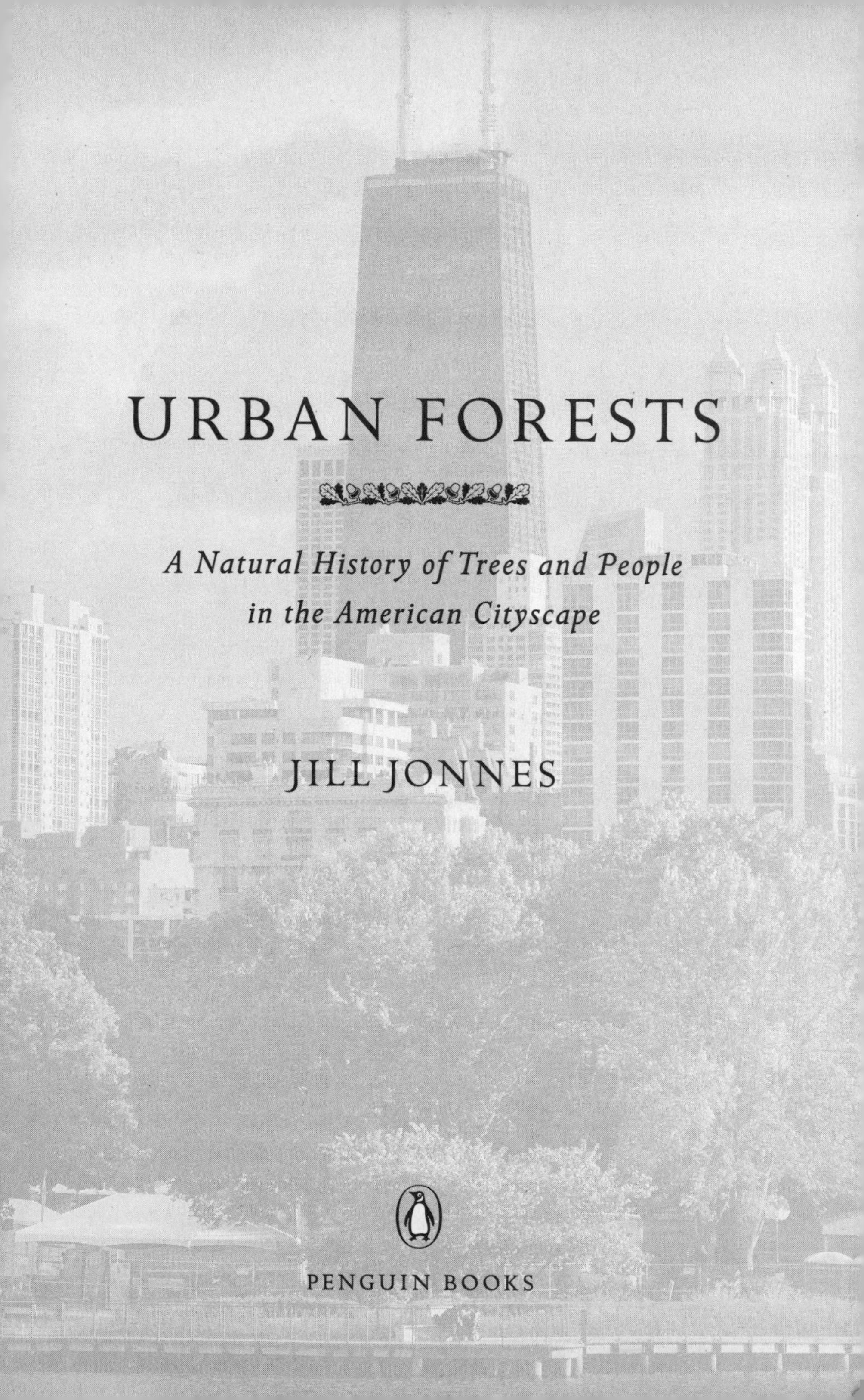

URBAN FORESTS

A Natural History of Trees and People in the American Cityscape

JILL JONNES

PENGUIN BOOKS

PENGUIN BOOKS
An imprint of Penguin Random House LLC
375 Hudson Street
New York, New York 10014
penguin.com

First published in the United States of America by Viking Penguin,
an imprint of Penguin Random House LLC, 2016
Published in Penguin Books 2017

ISBN 9780143110446 (paperback)

THE LIBRARY OF CONGRESS HAS CATALOGED THE HARDCOVER EDITION AS FOLLOWS:
Names: Jonnes, Jill, 1952– author.
Title: Urban forests : a natural history of trees and people in the American cityscape / Jill Jonnes.
Description: New York : Viking, 2016. | Includes bibliographical references and index.
Identifiers: LCCN 2016031535 (print) | LCCN 2016033155 (ebook) | ISBN 9780670015665 (hardcover) | ISBN 9781101632130 (ebook)
Subjects: LCSH: Trees in cities—United States. | Urban forestry—United States.
Classification: LCC SB435.5 .J66 2016 (print) | LCC SB435.5 (ebook) | DDC 635.9/77—dc23
LC record available at https://lccn.loc.gov/2016031535

Printed in the United States of America
1 3 5 7 9 10 8 6 4 2

Set in Albertina MT

For Amanda and Sarah,
my friends in tree adventures and so much else

He who plants a tree,
Plants a hope.

—Lucy Larcom

Contents

INTRODUCTION

"We Appreciate the Symmetry of Human and Sylvan Life"

Graft by sculptor Roxy Paine at the National Gallery of Art on the Mall in Washington, D.C. *(Photograph courtesy of Steve Lagerfeld.)*

The wonder is that we can see these trees and not wonder more.

—Ralph Waldo Emerson, "Nature"

On a May morning in Washington, D.C., a middle-aged male tourist wearing plaid shirt, tan shorts, and Birkenstock sandals wanders into the National Gallery of Art's outdoor sculpture garden by the Mall, stops, and begins angling his big-lensed Nikon camera this way and that. He backs up, steps to one side, and clicks away at artist Roxy Paine's forty-five-foot-tall silver tree, shimmering against a blue sky. *Graft* is an eight-ton polished-stainless-steel sculpture that towers above the manicured lawns and gravel paths, one bare silvered limb extending lithely up into an airy tangle of branches and twigs while the second large branch (grafted on?) reaches skyward, heavy and gnarled. A grackle lands on one of the uppermost metabranches, reconsiders, and takes off.

This faux city tree is a showstopper, its enormity and gleaming beauty eclipsing the garden's honey locust trees and other pieces of modern art. "It's like something out of a fairy tale," says a young mother standing hand in hand with her two-year-old girl, both contemplating the sculpture. "In the original Grimm brothers' version of *Cinderella,*" the mother explains, "it is not a fairy godmother who grants Cinderella her wishes but a tree that grows on her mother's grave. In my mind's eye, that tree was always silver."

A piece of art at once so close to nature and yet so metallic and shiny provokes reactions. For *Washington Post* art critic Blake Gopnik, "Paine's piece has the strangest effect: It's grander and more impressive than the live trees already in the garden, but also so much deader. The trees, in a sense, critique the human artifice, 'That's the best you can do?' But the art also seems to flatter trees we might otherwise ignore—'I wish I could be live like you,' says the stainless steel version."

Roxy Paine assembled the tree, the sixteenth in his "Dendroid" series, with his crew in November of 2009. They pulled up with three flatbed trucks carrying thirty-seven different components and soldered, fitted, and adjusted, often working like arborists from bucket trucks. The stainless-steel tree grew taller in the fall air, its thick trunk convincingly misshapen, its branches realistically windswept and wild. In a matter of days *Graft* achieved a size that in nature would require decades, and an instant celebrity bestowed upon few city trees.

Take a diagonal walk across the Mall toward the Smithsonian Castle, and before you stands a worthy rival to the gleaming *Graft*, a large and architecturally striking American elm, the iconic American tree *Ulmus americana*, a forest dweller–turned–urbanite beloved as no other tree in our history. "We had rather walk beneath an avenue of elms than inspect the noblest cathedral that art ever accomplished," proclaimed Henry Ward Beecher, nineteenth-century abolitionist and preacher. Like the young American nation, the elm tree was fast-growing and flourished in the toughest city soils and settings. In time the elms became long-lived wonders, their high, arching branches and lordly architecture perfect for urban streets and parks.

The National Mall is lined on both sides with twin elm allées, and the standout among those six hundred trees is a living sculpture, the Jefferson Elm. It hews to the apparent rule that classic American elms must have patriotic appellations, though it possesses no plaque to make itself known to passersby. Its name is fitting, for Thomas Jefferson, our third president, cherished old trees, denouncing their destruction as "a crime little short of murder." During Jefferson's tenure in the White House, poachers—"these midnight predators"—rustled a grove of seventy towering tulip poplars from District of Columbia public land to sell as firewood. Furious, the president declared at a dinner party: "The unnecessary felling of a tree, perhaps the growth of centuries . . . pains me to an unspeakable degree."

The Jefferson Elm, planted more than eighty years ago as the Depression set in, today has a massive pillarlike trunk that rises straight up thirty feet, supporting the muscular forking branches that then spiral sinuously toward the heavens, where on a sunny day its million leaves create a luminous canopy of lime green light and shadows. The arboreal embodiment of power and civic grandeur, it was witness to the departure of the nation's troops during World War II and their victorious return, to the sonorous roll of Dr. Martin Luther King Jr.'s "I have a dream" speech in 1963, and to demonstrations of every political stripe. A sentinel to the inaugurations of thirteen presidents, it could easily be present at another fifty such ceremonies.

If we had X-ray vision, or if the Jefferson Elm were sheathed not in furrowed grayish bark but in some kind of wondrous see-through glass, this tree would easily eclipse the shiny silver allure of *Graft*, for it would be a fabulous slow-motion aquasculpture. Crowds would gaze in wonder at the millions of shimmering threadlike columns of water charged with nutrients ascending the elm's trunk, each thread (known as xylem, from the Greek *xulon* for "wood") slowly transporting to the leaves high above water from the network of roots anchoring the tree.

The tree's leaves, says dendrologist Colin Tudge, suck up in the course of a day hundreds of gallons of water "from above by a combination of osmosis and evaporation." As the water transpires out each underleaf's tiny stomata (from the Greek *stoma* for "mouth") into the air, "so the sap that remains in the leaf cells becomes more concentrated—and so more water is drawn from below ... not in a crude and turbulent gush but in millions and millions of orderly threads. ... The tension within them is enormous: the threads are taut as piano wires. ... Water molecules cling tightly together. Their cohesive strength is prodigious." In short, "trees act like giant wicks." And as they respire into the air some of that water, trees serve as natural air conditioners.

But the Jefferson Elm is far more than a noble tree: It is the only elm in the entire District of Columbia with its own name, because it is that rarest of American elms—a flourishing specimen in the prime of life unaffected by Dutch elm disease. In the late 1920s the DED fungus was unwittingly introduced into several East Coast ports via French elm burl logs destined to be furniture veneer. Those shipments unleashed a slow-motion ecological and civic tragedy that would wipe out 100 million elms coast to coast, leaving streets, communities, whole cities bereft of character and beauty. The Mall elms are a poignant reminder of those now largely lost classic American cityscapes, where elms created "cathedrals of shade" along so many streets and avenues.

It is no accident that the National Mall, sometimes called the nation's front lawn, features only American elms. The native elm projects a distinct tree persona, writes Michael Pollan, "a soaring, optimistic one, eminently confident, youthful and sociable. For impressive as elms are singly, a whole street or square of them, twining their arching canopies overhead to form the most perfect roof in nature, is enough to make one feel civil, even neighborly. If some trees encourage romance, and others solitary reflection, the American elm seems almost to foster republican sentiments."

The modern American city is a great place to look at and learn about trees, because this fundamentally unnatural environment has a far bigger variety than any crowded real forest. *Urban Forests* is a celebration of the elm and other great city trees, and of the Americans—presidents, plant explorers, visionaries, citizen activists, scientists, nurserymen, and tree nerds—whose arboreal passions have shaped and ornamented the nation's municipalities from Jefferson's day to the present. In an era when four fifths of Americans live in or near cities, we are and have long been highly conscious of our built environment. Yet most of us know little of that other essential part of urban life—our grown environment. We are surrounded by millions of trees, trees of a hundred sizes, shapes, and species: London planes, Norway maples, lindens, Callery pears, ash, locusts, red maples, oaks, and flowering cherries, to name a few. Despite their ubiquity and familiarity, most of us—typical cosmopolites—take them for granted and know little of their specific natural history or many civic virtues. And yet trees, nature's largest and longest-lived creations, play an extraordinarily important role in our cityscapes. They are not only critical to public and individual health but are also the dominant component of what is now called green infrastructure, defining space, mitigating storm water, cooling the air, soothing our psyches, and connecting us to nature and to our past. "In trees, we see ourselves," says historian Thomas J. Campanella, "We appreciate the symmetry of human and sylvan life. In the seasons of a tree we find a map of our own lives."

Our affinity for trees dates far back to the origins of our species, believes tree canopy scientist Nalini M. Nadkarni. "From the first glimmers of humanity's dawn, we have evolved with trees. Our grasping hands, opposable thumbs, and binocular vision once enabled us to leap between branches with easy confidence, though we now use those endowments for other things, such as playing baseball and dodging in and out of traffic."

Although we no longer live in trees, we still intuitively appreciate them. The angular ginkgo or flowering horse chestnut seems so palpably alive in spaces dominated by concrete and buildings, and on a sweltering city day we naturally gravitate to the trees, absorbing their shade, their cooler, cleaner air and dappled light.

In the decades after the Civil War, tree evangelists of the first urban tree movement proselytized so successfully on behalf of the necessity of planting and caring for trees that the celebration of Arbor Day (invented in 1872), the founding of arboretums, and tree planting swept America. In this book we'll meet the colorful characters who lived (and sometimes died) for the trees that today populate our cities, plant hunters like Ernest "Chinese" Wilson and Frank Meyer, who braved bandits, plagues, and raging rapids in remotest Asia. We'll also delve into the tales of the trees themselves. Among the wealth of tree species in the United States are many naturalized immigrants from China, Japan, and Europe, including such classic urban examples as the ailanthus (of *A Tree Grows in Brooklyn* fame), the ginkgo and the dawn redwood (both "living fossils" unchanged since the era of the dinosaurs), the ornamental cherry, and the much-maligned Callery pear.

Like people, urban trees can have deadly enemies. Often they are human, but a tree's worst foes are generally foreign diseases and insects. Between 1860 and 2006 global trade served as the means for 455 tree-loving insect species to debark on our shores, and sixteen damaging tree diseases. For decades American chestnut blight (first detected in the Bronx in 1904) and Dutch elm disease (first detected in Ohio in 1929) were the most devastating threats to urban trees. Then came the Asian long-horned beetle (first detected in Brooklyn in 1996), an indiscriminate consumer of half of urban tree species (with a preference for maples) whose lackadaisical lifestyle has limited its presence to a handful of battlefields, where a rare victory may prevail. Far, far more devastating has been the invasion of the Asian emerald ash borer (first detected in Michigan in 2002), now destroying billions (yes, billions) of native ash trees in a swath of twenty-seven states, with further conquests likely when it breaches the Rocky Mountains and heads west.

In the 1970s, as both huge old trees and small woodlands disappeared in the wake of Dutch elm disease, developers, and old age, city dwellers once again began to experience a visible dearth of urban trees. Many officials who were dealing with other serious financial issues dismissed trees as an unaffordable amenity. In the face of that resistance arose a second wave of American urban tree evangelists, a group of enlightened leaders, citizen activists, artists, arborists, and scientists who helped create an urban-forestry movement, launch million-tree initiatives, and formulate a new body of science that employs the latest technology to illuminate how essential trees are to public and environmental health.

As we humans wrestle with how to repair the damage we have wrought on nature and how to slow climate change, urban forests can and should serve as one part of an obvious, low-tech solution. Every American metropolis, says U.S. Forest Service scientist Greg McPherson, should have a "maximally functional" urban forest. In truth, every city, where more than half of the eight billion humans on earth now choose to live, should have the lushest urban forest we can devise. "We should shoot for a performance standard," says McPherson, "like how many megawatts of air conditioning we can save, or how many pounds of nitrogen dioxide we can absorb, reducing ozone and smog." Today, as trees achieve new status as a kind of green infrastructure, scientist Rowan Rowntree suspects that "we are only 50 percent of the way to knowing what trees really do for us."

Not only scientists are looking at city trees with new eyes and questions; so are artists. When Irish artist Katie Holten created her *Tree Museum* in 2009 along the five-mile Grand Concourse in the Bronx, this conceptual piece featured one hundred of the avenue's street trees, each marked with a tiny plaque with a phone number that "museumgoers" could call to hear a forester, poet, or local activist talk about that particular elm or ginkgo. "If nothing else," Holten said, "I hope that people will realize that trees are alive, they grow and they have names: like people."

Writer Ian Frazier ventured forth on a soggy July day to attend the *Tree Museum*, reporting that "Tree No. 3 is an ancient, lofty ailanthus growing out of the base of an apartment building. The Chinese used the bark and the seeds of the ailanthus to treat baldness and mental illness, the recording says. A man in the house next to the ailanthus comes out to say how wonderful the tree's shade is on hot summer days, when he and his friends sit in folding chairs on the sidewalk."

On Frazier's walk north, listening to the tales of different trees, he encounters many urban scenes, from drug dealers to salsa gatherings. And then he arrives at the far north of the avenue and Tree No. 100, a cottonwood that is "a majestic four-trunk specimen, something you might find in Montana on the banks of the Missouri River." Frazier dials up the cellphone number on the little plaque embedded on the sidewalk and listens as "the recording tells how neighborhood activists saved this tree from destruction and created a little park around it. . . . In the breeze its leaves shiver silvery-green against the Bronx sky."

It is perhaps a solace (or just amusing) to learn that the serious student of dendrology (the scientific study of trees) finds that even "developing a precise definition of a tree is difficult and unrewarding." We must, then, admire David Allen Sibley of birding fame for attempting to solve the problem. In his guide to trees Sibley writes: "If you can walk under it, it's a tree; if you have to walk around it, it's a shrub." We'll stick with that definition.

CHAPTER ONE

"So Great a Botanical Curiosity" and "The Celestial Tree": Introducing the Ginkgo and Ailanthus

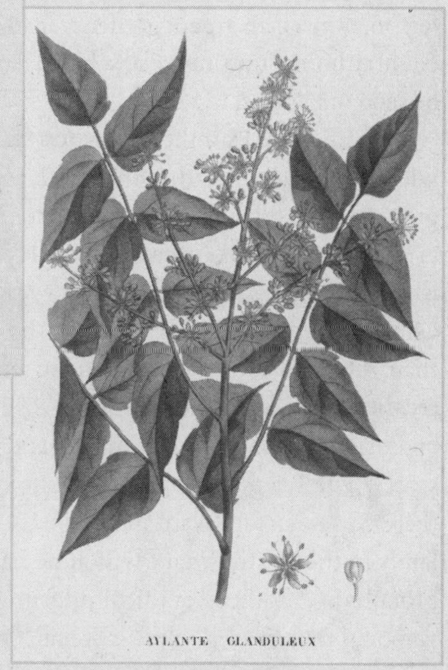

Left: *Ginkgo Biloba*. (*Courtesy of the LuEsther T. Mertz Library of the New York Botanical Garden.*)
Right: *Ailanthus altissima*. (*Courtesy of the Peter H. Raven Library of the Missouri Botanical Garden.*)

On July 7, 1806, the wealthy Philadelphia plant collector William Hamilton, sixty-one, dipped his quill pen into an inkwell and began a postscript to a letter to President Thomas Jefferson, who was summering down at Monticello. "In the autumn, I intend sending you if I live," wrote Hamilton, then in the throes of searing gout pain, "three kinds of trees which I think you will deem valuable additions to your garden." Hamilton took competitive pride in possessing every possible botanical rarity. Over the course of twenty-five years, he had transformed the Woodlands, his six-hundred-acre estate overlooking the Schuylkill River, into the young nation's premier showcase for exotic plants and trees. Jefferson, himself an ardent gardener, had pronounced his friend Hamilton's country home "the only rival which I have known in America to what may be seen in England."

Nothing gave Hamilton more joy than showing off his vast greenhouse with its ten thousand plants and his landscaped "natural" pleasure grounds, whose lawns sloped down to the river, artfully "interspersed with artificial groves . . . of trees collected from all parts of the world." He relished his visitors' amazement as they stared at strange "foreign trees from China, Italy, and Turkey," fingering the unusual leaves and bark, inhaling their "balmy odours." As Massachusetts congressman Manasseh Cutler recalled of a visit to the greenhouse: "Every part was crowded with trees and plants, from the hot climates, and such as I have never seen. All the spices. The Tea plant in full perfection. In short, [Hamilton] assured us, there was not a rare plant in Europe, Asia, Africa, from China and the islands in the South Sea, of which he had any account, which he had not procured." Another botanical pilgrim to the Woodlands favored with a personal tour peered in pleasure at "the bread-fruit tree, cinnamon, allspice, pepper, mangoes, different sorts, sago, coffee from Bengal, Arabia, and the West-Indies, tea, green and bohea, mahogany, Japan rose, rose apples. . . . The curious person views it with delight, and the naturalist quits it with regret."

Hamilton, famous as a genial host and a great talker, was not, however, generous with his green prizes. Once during a dinner party, he entered his greenhouse to pick a special camellia for the table's centerpiece and came upon a young lady with said flower in her hair. Hamilton hurled a curse, exclaiming, "Madam, I had rather have given you one hundred guineas than that you should have plucked that precious blossom." Philadelphia nurseryman Bernard McMahon complained to Jefferson that while he had bestowed upon Hamilton "a great variety of plants . . . he never offered me one in return. . . . I well know his jealousy of any person's attempt to vie with him, in a collection of plants."

And so it was no small gesture for Hamilton, dismissed by a neighbor as "interested only in his house, his hothouse and his Madeira," to be offering Jefferson among his gifts one of his rarest treasures, an offspring of his "*Ginkgo biloba* or China Maidenhair tree . . . said by Kaempfer to produce a good eatable nut." In 1784 Hamilton had again bested his plant-collecting rivals, securing before all others in America these exotic ginkgoes with their distinctive fan-shaped leaves. He had sailed to London that year to settle various debts, and while there he had ordered his private secretary back at the Woodlands to "take time by the forelock" and begin sowing seeds in large quantities of native trees such as "white flowering Locust, the sweet or aromatic Birch, the Chestnut Oak, Horsechestnuts, Chinacapins, Judas Trees, Dogwoods, . . . Magnolias," and not to fail to "put into a nursery handsome small plants of Elm, Lime, Locust, sweet Birch, white pine, ash leaved maple, sugar maple, aspen poplar, Zantoxylon or tooth ache tree, magnolia," etc. In a subsequent letter Hamilton asked his secretary about the fate of seeds he had sent home by sea: "Did not any of the seeds vegetate of a Bushel of Horsechestnuts, a peck of Spanish chestnuts, 3 pounds of pistichia, 11 quarts of Portugal Laurel, 5 pound of silver Fir . . . or have they all gone to the Dogs too?" And as winter approached in 1785, he instructed, "Secure the tender plants from the severe weather, otherwise all my pains will have been to no purpose," mentioning specifically his ginkgo.

The first ginkgo seeds were long believed to have been brought to Europe from Japan in 1693 by Engelbert Kaempfer, a physician-botanist with the Dutch East India Company. Those seeds produced a tree at the Botanic Garden in Utrecht, Holland. By 1754 England's Kew Gardens had acquired its first ginkgo, and today we presume that it was that tree's offspring that came to grace Hamilton's grounds.

For all Hamilton's considerable botanical mastery, he could not have known, as the leaves of his growing ginkgoes turned a pure yellow each autumn, that this species had been commonplace in the temperate North American forests of the Jurassic age. An abundance of fossils record that the ginkgo tree was among the many fauna and flora of North America that were glaciated out and effectively driven into extinction on the continent by the Ice Age. When Hamilton brought the ginkgo back to one of its former habitats, he was unwittingly reintroducing an evolutionary superstar: Scientists now know that the ginkgo biloba tree or its ancestors have existed on earth for 250 million years, longer than any other tree now living. (Both the eighteenth-century Utrecht and Kew Garden ginkgoes are still alive today.)

Nor, one suspects, could Hamilton have envisioned how rampantly ginkgo would eventually recolonize the New World, though not as a part of the forests it once dominated. Instead, ginkgo's modern incarnation would be as a cultivated urban tree, an elegant urban denizen valued for its distinctive fan-shaped leaves and above all its ability to grow in the smoggiest of cities, indifferent to poor soil and polluted air, a tree that had also survived whatever pests had once afflicted it in the era when dinosaurs stomped through ginkgo forests.

Such was Hamilton's reputation as a fanatic plant collector and propagator extraordinaire that on March 22, 1807, Jefferson wrote to alert him that explorer Meriwether Lewis would soon arrive at the Woodlands with packets of some of the precious seeds—"these public treasures"—gathered on the three-year Lewis and Clark expedition. Wrote Jefferson, "I am sure I make the best possible disposition of them." During his presidency Jefferson continued to trade various plants and trees with Hamilton, dispatching to the Woodlands on March 1, 1808, eight aspen trees with "paper whiteness of the body . . . and [stem and leaves] more tremulous." In a long letter the president praised Hamilton, saying, "Your collection is really a noble one, & in making & attending to it you have deserved well of your country." Some thought the less-than-patriotic Hamilton, who had faced two trials and acquittals for treason during the Revolutionary War, was lucky to have held on to his family's estates and the five hundred acres that were the site of Lancaster, Pennsylvania. That town's ground rents, while a never-ending headache to collect, were the basis of his great wealth.

After Hamilton's death in 1813, the Woodlands declined, as subsequent owners failed to maintain the expensive grounds. Factories, oil refineries, and railroads soon overwhelmed the nearby bucolic landscape, and in 1840 the estate's last owner sold it to a cemetery. Yet the ginkgoes lived on, still such a rarity that not long after the cemetery opened, a young Hudson River nurseryman named Andrew Jackson Downing showed up to wander its new paths in search of these legendary trees. Jackson, twenty-six, soon vaulted to fame as the nation's original home and garden guru, the Martha Stewart of his day, by publishing his *Treatise on the Theory and Practice of Landscape Gardening* (1841). The first such book aimed at an American audience, it would eventually go through eight editions and sixteen printings.

A slight, poetic man with intense brown eyes, dark, curling locks, and a square jaw with a dimpled chin, Downing had strong opinions about tasteful architecture, the intricacies of pomology (apple raising), and the necessity of beauty in gardens and everyday life. The youngest of five children, he had grown up in Newburgh, New York, with a beautiful view of the Hudson River and the mountains. He completed school at sixteen and went to work in the family's commercial nursery, situated on the ten acres surrounding its home and offering customers hundreds of varieties of apple, pear, peach, and cherry trees. In 1838, at age twenty-four, Downing married Caroline Elizabeth DeWint, whose mother was a niece of John Quincy Adams.

During regular visits to nearby wealthy estates as part of his work, Downing came to know and admire a number of ardent gardeners, including Henry Sargent, a wealthy New York banker who was beautifying his new estate, Wodenethe, across the Hudson. Sargent and other clients became friends, opening Downing's eyes to a world of aesthetics, culture, and knowledge that he would embrace and share with other aspiring Americans through his writings and landscape design.

After locating Hamilton's fifty-six-year-old ginkgoes amid the new Woodlands Cemetery grave sites, Downing described the tallest as being fifty-five feet high and three feet, four inches in circumference. Another famous ginkgo, not as tall, could be admired in Boston, wrote Downing, "standing on the north side of that fine public square, the Boston Common. It originally grew in the grounds of Gardiner Green, Esq. of Boston; but though of fine size, it was about three years since, carefully removed to its present site, which proves its capability for bearing transplanting. Its measurement is forty feet in elevation, and nearly four in circumference." Through Henry Sargent, the up-and-coming Downing would meet horticulturally inclined Boston Brahmans, including (we speculate) a very young second cousin of Henry's named Charles Sprague Sargent, who decades later would become America's most influential tree expert.

As Downing became famous, a steady stream of visitors began to appear to admire his faux-sandstone Elizabethan villa and parklike gardens on the family land in Newburgh. They described a slight, almost Spanish-looking man of "easy elegance," who with his lovely wife (they had no children) "seem to live for the beautiful and the agreeable in life amid a select circle of friends . . . and a cheerful and unembarrassed social intercourse seems to characterize the life of this circle." In contrast to William Hamilton, who was stingy with his plants and flowers, a guest of Downing's who exclaimed in pleasure over a creamy magnolia blossom would soon find one perfuming his place at the breakfast table each morning for the rest of his stay. One observant guest noticed that amid all this ease, lush gardens, and laden fruit trees, Downing, a passionate tree lover, was in his study very early working, and often returned there long after guests were asleep.

And so we like to imagine Downing toiling discreetly away before dawn some fine morning, writing in his *Treatise* about the ginkgo: "This fine exotic tree, which appears to be perfectly hardy in this climate, is one of the most singular in its foliage that has ever come under our observation. . . . The fruit is a drupe, about an inch in length, containing a nut. . . . They are eaten after having been roasted or boiled, and are considered excellent." Or so he had read, for when Downing wrote this, no female ginkgo in America had yet reached puberty and begun the tree's uniquely active sex life. "Abroad," he noted, the ginkgo "has fruited in the south of France, and young trees have been reared from the nuts." "The ginkgo tree," added Downing, "is so great a botanical curiosity, and is so singularly beautiful when clad with its fern-like foliage, that it is strikingly adapted to add ornament and interest to the pleasure ground." Of course, he soon installed one of these lovelies in his garden.

The ginkgo was by no means Hamilton's only big arboreal claim to fame and urban legacy. For in the same year—1784—that he sent his ginkgoes back to the Woodlands, he also secured his other great Oriental novelty: the tree of heaven (*Ailanthus altissima*) from northern China. The ailanthus was more than just an ornamental fast grower. Its wood was well suited to the kitchen steamers basic to all Chinese cuisine, while one treatment for mental illness called for an elixir of ground-up ailanthus root mixed with two liters of the urine of small boys and a paste of black soybean and various other medicinal herbs, left overnight, boiled, and then administered once daily. For balding men hoping to stimulate hair growth, Chinese pharmacists proposed a pulverized mixture of ailanthus, peach, and catalpa leaves, their extracted juice smeared on bald spots.

Like the ginkgo, the ailanthus had a colorful provenance. In the mid-eighteenth century the Chinese imperial court banned foreigners from the Celestial Kingdom, except for certain seaports. However, the emperor did accept the presence in Peking of a handful of French Jesuit priests, well schooled in science and technology. One, Father Pierre Nicolas Le Chéron d'Incarville, joined the mission in 1740 at age thirty-four. He had studied botany under Bernard de Jussieu, superintendent of the Jardin Royal des Plantes in Paris. Arriving in Macao and traveling up through South China to Peking, d'Incarville soon began sending herbarium specimens and seeds back to Jussieu as best he could. Mistaking the ailanthus for the highly valued Chinese varnish tree—their slender, pinnately compound leaves are very similar—in 1751 he sent a supply of the ailanthus's papery seeds back to Paris.

In France, Jussieu planted some and dispatched a few more across the Channel to the Royal Society. English nurseryman Philip Miller of the Chelsea Physic Garden sowed some of them and watched his little trees of heaven grow at a prodigious rate, three to five feet a year until they reached sixty feet tall, with plentiful suckers (growths from the roots) that made it easy to raise yet more trees. The tree was already popular in Europe when Hamilton dispatched one home so that he could become the proud first possessor in America of this rarity.

Also like the ginkgo, the ailanthus was destined to become an American city tree par excellence. But while the ginkgo grew at such a slow pace that it remained a precious collectible for another century, the tree of heaven would enter the nursery trade by the 1820s, heavily promoted by the Robert Prince Nursery of Flushing, New York, for its rapid growth, frothy foliage, and indifference to coal smoke and insects. Downing, in his *Treatise*, championed the exuberant ailanthus, enthusing that, in its first half dozen years of life, it "outstrips almost any other deciduous tree in vigour of growth, and we have measured leading stems which had grown twelve or fifteen feet in a single season. . . . There are, as yet, no specimens in this country more than 70 feet high; but the trunk shoots up in a fine column, and the head is massy and irregular." Unlike ginkgoes, ailanthus rarely live more than sixty years, and Downing made no mention of having seen any when he visited the Woodlands Cemetery. By then, Hamilton's original ailanthus specimens may have perished but certainly would have sired many hundreds, if not thousands, of offspring.

Was it any surprise, then, that in New York and Philadelphia, wrote Downing, who was often in these cities to meet his publishers and growing public, that "the Ailanthus is more generally known by the name of the *Celestial tree*, and is much planted in the streets and public squares. For such situations it is admirably adapted, as it will insinuate its strong roots into the most meager and barren soil, where few other trees will grow, and soon produce an abundance of foliage and fine shade. It appears to be perfectly free from insects; and the leaves, instead of dropping slowly, and for a long time, fall off almost immediately when frost commences."

Even as Downing wrote these words, the ailanthus had already effectively conquered New York City. Some five years earlier, citizens of the "uptown districts" had watched in bafflement as squads of men wielding axes began chopping down stately horse chestnuts, lindens, and other street trees that had "converted the arid pavements into refreshing and pleasant avenues, their widespread heads throwing cooling shadows on the walls and windows and porches of those New Yorkers' dwellings," making summer in the city bearable. But in this antebellum era, before the invention of window screens, "a lady has been startled with a caterpillar in her drawing-room: another lady has been very much annoyed with similar visitors; and a third has been tormented daily with perfect showers and swarms of insects of various kinds, blown into the open windows upon her elegant furniture." The old-fashioned, familiar trees with their troublesome insects were soon supplanted along the streets and avenues of Gotham by the reportedly bug-free ailanthus, "until," complained one fan of the old trees, "'uptown' has become completely orientalized."

It was during the ensuing decade, as the ailanthus became the most common street tree in New York, that Downing emerged as *the* authority on fashionable horticulture, landscaping, and small-town architecture. Passionate, well informed, and with a winning style, he wrote, edited, and authored *Cottage Residences* (1842), *Gardening for Ladies* (1843), and, with his brother Charles, his next blockbuster, *The Fruits and Fruit Trees of America* (1845). In the course of thirteen printings, the book elevated Downing to the rank of top pomologist in an era when an agricultural nation obsessed over the merits of the Newtown Pippin apple over the Belle de Boskoop.

Although Downing was ascending new heights of fame and influence, in 1846 he became engulfed by money woes and entangled in a bitter lawsuit with his father-in-law, John Peter DeWint. Downing quietly sold the nursery he had bought from his father and another piece of land before the lawsuit was resolved in his favor in April of 1849. Those money troubles may have figured in Downing's accepting a job in July 1846 as editor of *The Horticulturist*, a new journal whose popularity provided an ideal bully pulpit for what was becoming one of Downing's major preoccupations: the need for Americans to cherish, value, and plant more "trees—the noblest and proudest drapery that sets off the figure of our fair planet." He was perplexed by his countrymen's willingness to live in a world without trees, evidenced in "every desolate, leafless, and repulsive town and village in the country." Surely, he inveighed, "if our ancestors found it wise and necessary to cut down vast forests, it is all the more needful that their descendants should plant trees. . . . The first duty of an inhabitant of forlorn neighborhoods . . . is to use all possible influence to have the *streets planted with trees*."

In an 1847 editorial Downing rhapsodized about the "beautiful towns and villages of New England, where the verdure of the loveliest elms waves like great lines of giant and graceful plumes above the house tops." He further lamented that "in Philadelphia, we have seen with regret, whole rows of European Linden cut down within the last ten years, because this tree, in cities, is so infested with odious worms that it often becomes unendurable. On this account that foreign tree, the Ailanthus, the strong scented foliage of which no insect will attack, is every day becoming a greater metropolitan favorite."

By July 6, 1850, when Downing sailed for Liverpool aboard the steamship *Pacific* on his first trip overseas, he had received a Gold Medal of Science from Frederick William IV of Prussia, while Queen Anna of Holland had bestowed on him a garnet ring. English nobility welcomed and feted him in their highest horticultural circles. The Duke of Northumberland showed him the ancient trees of Syon House, the Duke of Devonshire hosted him at Chatsworth, and Sir William Hooker squired him about the Royal Botanic Gardens at Kew. Downing marveled above all at London's public parks, "laid out," to his delight, "in the main with a broad and noble feeling of natural beauty.... This makes the parks doubly refreshing to citizens tired of straight lines and formal streets, while the contrast heightens the natural charm." St. James's Park struck him as "more like a glimpse into one of the loveliest pleasure-grounds on the Hudson, than the belongings of the Great Metropolis." He observed with close professional interest the tremendous popularity of these great English public parks, conscious of how Americans flocked to the greenery of picturesque cemeteries in or near cities.

When Downing debarked back in New York that fall, he brought with him a new partner, landscape architect Calvert Vaux. What Downing had seen abroad convinced him that public parks, where city dwellers could linger among trees, gardens, and shaded walks, would be "better preachers of temperance than temperance societies, better refiners of national manners than dancing schools, and better promoters of general good feeling than any lectures on the philosophy of happiness ever delivered in the lecture room." He began advocating for what he referred to as a "central park" for New York City.

Not long after Downing settled back into his home in Newburgh, he received a letter offering the greatest commission of his career: an invitation from President Millard Fillmore to come to Washington, D.C., on November 25 for a meeting to draw up a landscaping plan for the whole of the National Mall, from the foot of the U.S. Capitol grounds all the way up to the White House. These 150 acres were described as a "bleak, unhospitable common," notable for its muddy roads, deep ditches, and barrenness, there being but a few scraggly sycamores. Downing envisioned numerous different sections, all connected by serpentine paths, carriage drives, meadows, and hundreds of "choice trees in the natural style." He supervised the grading and draining of the area and then the planting of the first section of the park around the Smithsonian with young trees in the spring of 1851, promising his supporters: "If you gentlemen who have influence in Washington will stand by me I will make the capital 'blossom like a rose.'"

Then, on July 28, 1852, a bright, blue summer day, Downing embarked from Newburgh with his wife, her mother, and her siblings on the luxury steamer *Henry Clay* for a journey down the Hudson River to New York. Starting in Albany, the captain raced another steamboat, pausing only to take on passengers at appointed stops. At 2:45 p.m., as the steamer gained speed and raced toward Yonkers, the *Henry Clay*'s boiler overheated and burst in the engine room, engulfing the craft in flames and smoke. As the passengers panicked, the steamboat ran aground, and a wall of fire left those attempting to flee only one line of escape: overboard into the river's deep waters. Downing, a strong swimmer, helped his wife and her family off the vessel as best he could, but when the chaos subsided, he, at age thirty-six, was among the seventy dead. Without his charismatic presence, the Washington mall project quickly petered out, and the antebellum United States lost its most passionate and knowledgeable advocate for trees and nature in the fast-industrializing cities.

Yet even as subscribers to *The Horticulturist* learned of Andrew Jackson Downing's untimely death, they opened the August 1852 issue to discover his final crusade. His editorial, "Shade Trees in Cities," proclaimed, "Down with the ailanthus!" He now condemned the once-favored tree of heaven as a foreign invader that respected no boundaries, inveighing, "The vices of the ailanthus [are] . . . it *smells horribly,* both in leaf and flower . . . [and] it *suckers* abominably, and thereby over runs, appropriates and reduces to beggary all the soil of every open piece of ground." With the United States already alarmed about the "wrong kind" of immigrants undermining the nation, ailanthus had become a proxy for their troublesome incursion. Downing admitted his was in part "a patriotic objection. . . . This petted Chinaman or Tartar, who has played us so falsely . . . has drawn away our attention from our own more noble native American trees, to waste it on this miserable pigtail of an Indiaman."

Why, Downing demanded of his readers, do we "fill our lawns and avenues with the cast off nuisances of the gardens of Asia and Europe"? "Take refuge, friends, in the *American Maples*. Clean, sweet, cool, and umbrageous, are the Maples. . . . No tree transplants more readily." Or, he suggested, consider that tree used far too rarely: "We mean the *Tulip Tree*. . . . What can be more beautiful than its trunk—finely proportioned, and smooth as a Grecian column? What more artistic than its leaf, cut like an arabesque in a Moorish palace—what more clean and lustrous than its tufts of foliage. . . . what more lily-like and specious than its blossoms—golden and bronze shaded? And what fairer and more queenly than its whole figure, stately and regal as Zenobia?"

Such was Downing's sway, even from the grave, that the ailanthus was soon widely denounced and scorned as a vile, malodorous foreigner, perhaps even poisonous, sickening the vulnerable with its loathsome emanations. On Capitol Hill horticultural authorities "executed capital punishment on all the unfortunate Ailanthus found in the capital grounds." With similar denunciation afoot in Gotham, one citizen warned late that year, "See how many streets and public places are every where planted with this tree. To cut them down or dig them up, what an inconvenience, what a loss! And how dismantled those streets will look without them. We shall all be old, and perhaps dead and forgotten before other trees can be brought to yield such shade, and produce such an effect as these do. . . . If the flowers are really offensive, have them cut off before they open."

For the next decade the editors of the *New York Daily Times* kept up a steady campaign against the ailanthus, proudly proclaiming themselves "Know-Nothings, decidedly, in our opposition. . . . While we have so great a variety of hardy, sweet and beautiful indigenous trees, many of which grow as rapidly as the ailanthus, it is altogether too bad that our streets and public grounds should be encumbered by this filthy and worthless foreigner." A fan jumped to the poor tree's defense, pointing out that its unpleasant-smelling flowers were in bloom for only nine days and that during the rest of the year, "it is beautiful to the eye, refreshing as a shade, of quick growth—but above all *it is always free from those jumping or measuring worms*." And so the debate went, year after year.

Indifferent to such xenophobia, the ailanthus, even as it was slowly removed as a street and official park tree, more than made itself at home in the New World, its large, papery seeds sailing easily through the air, conquering new territory, establishing new beachheads. One New York writer said in wonder, "Give it a handful of cinders, plaster dust, or crushed asphalt, and it will flourish." One suspects that the tree of heaven has become far too much of a common upstart to please William Hamilton, its original American patron. It remains the most successful introduction of a tree ever made to the American urban landscape.

In the decade before his early death, Andrew Jackson Downing had also served as the pivotal voice in the nascent movement to preserve forests and plant more trees. While Hamilton and other wealthy men like Thomas Jefferson were preoccupied with beautifying their own private estates with trees, most American settlers were cutting down the primeval forest, with its ancient trees of a size and scale unknown in Europe or England. In Hamilton's day the destruction was not far advanced, as Charles Maurice de Talleyrand-Périgord reported while in exile here in 1794. The French aristocrat found himself, not 150 miles beyond Philadelphia, guiding his horse through "wild nature in all its pristine vigor . . . ; forests old as the world itself . . . here and there the traces of former tornadoes that had carried everything before them. Enormous trees all mowed down in the same direction."

A mere fifty years later, George Emerson, a mathematician, educator, tree lover, and part-time botanist, wrote in an 1846 report to the Massachusetts legislature regarding the state's forests: "A few generations ago, an almost unbroken forest covered the continent. Now those old woods are everywhere falling. The axe has made, and is making, wanton and terrible havoc. The cunning and foresight of the Yankee seems to desert him when he takes the axe in hand. The new settler clears in a year more acres than he can cultivate in ten, and destroys at a single burning many a winter's fuel, which would be better kept in reserve for his grandchildren."

When studying the denuding of Massachusetts hills, Emerson had reported, "All along the coast of New England, numerous little streams, which were formerly fed by the forests, and often rolled a volume of water sufficient to turn a mill in summer, are now dried up at that season, and only furnish a drain for the melting snows of spring, or the occasional great rains of autumn." With the trees gone, the woods could not absorb and soften the rains and wind nor moderate the harsh heat.

The following year Downing, writing in *The Horticulturist*, likened "to murder in the fourth degree" the way "settlers, pioneers, and squatters girdle and make a *clearing*, in a centennial forest, perhaps one of the grandest that ever God planted, with no more remorse that we have in brushing away dusty cobwebs." Other contributors to the magazine begged to differ: "Our ancestors had a valid excuse for thus destroying the primeval forest. . . . Each tree afforded a secure shelter for the savage Indian, who, with tomahawk and scalping knife, darted from behind huge grey boles, to inflict a sanguinary death upon those who ventured beyond the 'clearing;' thus he who destroyed a tree brought his labor to an excellent work."

Out in California *New York Tribune* correspondent Bayard Taylor, passing through Sacramento while covering the Gold Rush, admired giant oaks and sycamores six feet in diameter, saved from the original forest to line rude streets. But "the emigrants have ruined the finest of them by camping at their bases, which, in some instances, have burned completely through, leaving a charred and blackened arch for the superb tree to rest upon. . . . The destruction of these trees is the more to be regretted as the intense heat of the summer days when the mercury stands at 120 F and renders their shade a thing of absolute necessity."

Elbert Hubbard, founder of the Arts and Crafts movement, recalled with shame the arboricides of his nineteenth-century Illinois boyhood: "I can well remember when men and boys would cut down some mighty oak, that had been over two hundred years in growing, just to catch an innocent possum that had taken refuge in its branches. When this great tree, which had defied the storms of centuries would fall to the ground with a thunderous crash, we would lift a savage yell of exultant triumph. I have seen a fire started in black walnut, birch, beech and elm trees to smoke out and kill a swarm of bees. Not only did these 'first citizens' kill the swarm of bees, but they killed the tree. . . . And so recklessly and completely did ax and torch do their work that, in certain sections, millions of acres have been rendered almost uninhabitable. . . . The miasma of ignorance, and the impulse to destroy spread like a fog over our entire land. . . . The woods were the place where lurked the enemies of man—savages, wildcats, bears, snakes, bees and all things that were supposed to strike scratch, sting, and bite."

As late as 1885, historian Francis Parkman wrote, "The early settler regarded the forest as an enemy to be overcome by any means, fair or foul, as the first condition of his prosperity and safety; and his descendants do not yet comprehend how completely the conditions are changed. The old enemy has become an indispensable friend and ally."

As these settlers and the lumbermen systematically waged war on the ancient trees and forests, they often spared the American elm, because its wood was considered inferior to that of the English elms planted by the first Puritans. The American elm's high, arching branches and light, dappled shade did not interfere much with growing crops or pastures but offered welcome respite for cattle and farmers on hot days. As those spared elms grew, they became admired local landmarks: "The American elm, is, in most parts of the state," wrote Bostonian George Emerson, "the most magnificent tree to be seen. . . . The sturdy trunk and the airy sweep of the branches are always there, and few objects of the kind are more beautiful." Like Downing, Emerson was pleased to see that in "the excellent practice, becoming every year more common, of ornamenting towns and villages and sheltering sunny roads, with rows of trees, the elm is chosen often to the exclusion of all other trees." Nothing transplanted as well as the American elm, grew so quickly, required so little attention to thrive, and lived to such a respectable old age.

To underscore this point, Emerson published pages of elm data, including one set gathered by Oliver Wendell Holmes, poet and newly minted Harvard physician. Equipped with measuring tape, Dr. Holmes headed west from Boston in September of 1837 to locate gargantuan specimens. When he encountered the Great Elm in Springfield, Massachusetts, he placed his tape five feet up its massive trunk and found its circumference to be twenty-four feet, five inches. In Medfield, at the home of Thaddeus Morse, a towering elm was thirty-seven feet in circumference. Perhaps no tree in America was more famous than the Washington Elm in Cambridge, Massachusetts, "so called because beneath its shade or near it [on July 3, 1775], Geo. Washington is said to have first drawn his sword, on taking command of the American army." Emerson made sure that its girth too was duly recorded. The honors were done by a Mr. Dixwell, who reported that one foot up from the ground, the circumference of the historic tree's trunk measured thirteen feet, two inches.

When Henry David Thoreau witnessed the felling of a huge old elm, fifteen feet in circumference, in Concord (residents worried the tree might topple upon their houses when it creaked during a bad storm and cut it down), he mourned, "Our town has lost one of its venerables." Thoreau viewed its destruction as close to a crime. "Is it not sacrilege . . . for [the tree] is of greater moment to the town than that of many a human inhabitant would be. . . . How the mighty have fallen! Methinks its fall marks an epoch in the history of the town."

Downing himself was an enthusiast of elm-bowered New England towns. By 1847, ever exhorting "persons of taste and spirit" to mobilize around civic tree planting, Downing regretted that "there is a fashion in trees, which sometimes has a sway no less rigorous than that of a Parisian *modiste*." He cited the ill-fated Lombardy poplar fad, which had ended with streets full of trees "decrepit and condemned."

With Downing's death, the woods and the city trees lost their most high-profile and passionate spokesman and advocate, a man who could write, "If our ancestors found it wise and necessary to cut down vast forests, it is all the more needful that their descendants should plant trees. We shall do our part, therefore, towards awakening again, that natural love of trees, which long warfare against them—this continual laying the axe at their roots—so common in a new country, has, in so many places, well nigh extinguished. We ought not to cease, till every man feels it to be one of his moral duties to become a planter of trees."

CHAPTER TWO

"No Man Does Anything More Visibly Useful to Posterity Than He Who Plants a Tree": Inventing Arbor Day and Cities of Trees

The first schoolchildren's Arbor Day in 1882 in Cincinnati, Ohio. (*With permission from American Forests.*)

At 2:00 p.m. on Friday, April 27, 1882, John Bradley Peaslee, superintendent of schools in Cincinnati, Ohio, stood on a ridge in that city's hilly Eden Park enjoying the sun, the breeze, and the sight of Governor Foster leading a procession of three companies of police, marching bands, the Veteran Guards, the Duckworth Club, and seven thousand blue-uniformed girls and boys from thirty school districts up into the park on Gilmore Avenue. In their wake surged thousands upon thousands of citizens, women with frilled parasols and men in derbies or top hats, all pressing in as best they could by streetcar, carriage, and foot.

Peaslee, forty-one, a handsome Dartmouth graduate and lawyer with wavy brown locks and a Van Gogh beard, had been orchestrating these festivities for months. He registered with satisfaction the discipline of his students trooping upward under blue banners, each school district then forming a human circle, square, or triangle along the mile-long park ridge. Before the Civil War these gentle slopes had been part of the Longworth family's famous "Garden of Eden" vineyard, ruined when powdery mildew struck its Catawba grapes, which were said to produce a sparkling wine rivaling French champagne.

What wondrous occasion was attracting this cheering and singing throng (reported the next day to have numbered thirty thousand) on such a gorgeous spring afternoon? For half a century tree lovers had been bemoaning America's status as "a nation of wanton destroyers of our unparalleled heritage of trees," and yet here were the people of Cincinnati gathering to cheer on something unprecedented—a vast pageant to inaugurate a new state holiday dedicated to planting and honoring trees: Arbor Day. A decade earlier, J. Sterling Morton, pioneer, farmer, newspaper editor, and Lincoln-loathing political activist, had dreamed up Arbor Day in Nebraska as a practical celebration to encourage settlers to improve the state's economy by creating woods and fruit orchards on the treeless plains.

But in the ensuing ten years, despite Arbor Day's success in Nebraska—where farmers had indeed planted millions of young trees—only Kansas and Minnesota had signed on. Now Superintendent Peaslee, when asked to arrange some appropriate event for Ohio's first Arbor Day *and* America's first American Forestry Congress (taking place here in the "Queen City"), had recast the former as an inspirational school holiday. "I will take the public-school children into the Park on that day," he declared to the forestry organizers, "and have them plant and dedicate trees to American authors."

Peaslee himself was as passionate about literature as he was about trees, and the latter love affair had started when he was very young in Atkinson, New Hampshire. Spotting a little sugar maple (*Acer saccharum*) sapling near his fishing spot, Peaslee dug it up, brought it home, and planted it. "With what exultant joy I showed that tree to my father," he recalled. As the seasons passed, he installed more small maples alongside that original, their red leaves making a brilliant show each autumn lining the road by their house. "Now, after more than forty years," he wrote, "my attachment to these maples is so strong that I have almost as great a desire to see them as I have to visit the friends of my boyhood."

Clearly the idea of honoring American literary celebrities with trees struck a chord, and while Peaslee and his seven thousand students arrayed themselves across the six-acre slope to versify and declaim at what would be known as the Author's Grove, elsewhere along this milelong former vineyard, prominent ladies, Civil War veterans, and local community leaders dedicated Peaslee's other arboreal projects: the Presidents' Grove, Citizens' Memorial Grove, Pioneers' Grove, and Battle Grove, the latter poignantly stocked with a Bunker Hill elm, a Valley Forge oak, and trees from every major Civil War battlefield from Gettysburg to Stone Mountain to Chickamauga.

A few students from each of the eighty districts stepped forward to dedicate their school's tree—a maple, an elm, or a sweet gum—to a chosen writer, such favorites as Washington Irving, Lucy Larcom ("He who plants a tree plants hope"), Nathaniel Hawthorne, and (from the colored district) the best-selling author of *Uncle Tom's Cabin*, Harriet Beecher Stowe, and poet Phillis Wheatley. Students read short personal essays and a few lines of their author's stories, recited poems ("Woodman, spare that tree!" of course), attached a wooden ID tag to each slender trunk, and then threw handfuls of dirt on the christened tree.

Over in the Presidential Grove atop the ridge, Dr. and Mrs. F. M. MacKenzie, he a Civil War veteran and she a great-great-niece of President George Washington, were planting a tree with a revolutionary pedigree: a little white oak they had reverently dug up six years earlier as a sapling from near the first president's tomb at Mount Vernon. Nurtured in their home garden in Delhi, Ohio, they were now patriotically ceding their horticultural prize to this memorial grove for the ages. Using a silver-plated steel spade with an ebony handle, Mrs. MacKenzie flung dirt onto the planted young oak, announcing her hope that "future generations may look thereon and recall the noble deeds of the immortal Washington."

Thomas Jefferson was honored with a young bur oak, a tough Midwest native, its sapling size giving no hint of its massive maturity and future offspring—acorns the size of small hen eggs, topped with wild hairy acorn caps. (Hence the tree's other name: the mossy cup oak.) The assembled ladies and handful of presidential descendants dedicated a total of twenty-one trees, one for each American president to date, including another bur oak for native son James A. Garfield, dead just seven months. By the end of the afternoon, the new Presidential Grove featured four American elms, five bur oaks, five red oaks, three pin oaks, two scarlet oaks, one swamp oak, and one sugar maple.

We can forgive Superintendent Peaslee some hyperbole when he later avowed, "No sight more beautiful, no ceremonies more touching, had ever been witnessed in Cincinnati." The spectators saw "the first memorial groves ever planted in America.... [It was] the first great celebration of a memorial tree-planting on the continent." (Peaslee was making a fine distinction, for the superintendent of the U.S. Botanic Garden in Washington, D.C., William R. Smith, had been planting memorial trees on his grounds for years, collaborating with men as varied as ardent abolitionist Pennsylvania congressman Thaddeus Stevens, who chose an Oriental sycamore, and Shakespearean tragic actor Edwin Forrest, whose tree was a weeping cypress. But it was true that these were not actually groves.)

On the following Arbor Day Peaslee returned with *seventeen thousand* students to expand the Author's Groves, this time installing eight-inch-square granite markers with the name of each school and its honored author recorded in chiseled, raised letters. By then the American Forestry Congress, the nation's first conservation organization, had formally proposed that states and students everywhere adopt Peaslee's version of Arbor Day, a charming new American holiday on which schoolchildren could learn about and plant trees. "Should this celebration become general," said Peaslee, "such a public sentiment would lead to beautifying by trees of every city, town, and village, as well as the public highways, church, and school grounds, and the homes of the people." And indeed, promoted by the Forestry Congress, Arbor Day now picked up new momentum.

J. Sterling Morton, the founder of Arbor Day, had had his epiphany as a young man in early April of 1855 while standing for the first time on the elevated ridge of his 160-acre homestead in Nebraska City in the Nebraska Territory. A strong, compact figure with a large, well-shaped head, the twenty-three-year-old newlywed from Detroit, Michigan, stared out at the vast undulating prairie before him and felt he "could not but be oppressed by the sense of treelessness. . . . No forest was visible on either side, as far as the eye could reach, and only here and there, along the banks of small creeks and in deep ravines, would a few fire-spared trees be found. Even these were mostly maimed, scarred and deformed by the surges of flame which had swept down upon them from the burning prairies during nearly every fall of their precarious lives. Thus everywhere the waves of rich land stretched bare of shade to the horizon." Suddenly Morton realized how much he missed trees.

Unlike Peaslee, an exemplary young Quaker active in every form of civic betterment from serving as director of the Humane Society to membership in the German Literary Club of Cincinnati, Morton was something of a hellion, a pugnacious bantam of a man quick to act and rare to repent. Expelled twice from the University of Michigan at Ann Arbor just before he was due to graduate, Morton wangled a degree out of Union College in Schenectady, New York. He then enraged his parents by marrying his high school sweetheart, Caroline, and departing for the Nebraska frontier, where he was soon embroiled as a Democrat in local politics.

Morton ignored his father's orders to come home and study law, instead becoming editor of the *Nebraska City News*. From this perch he vaulted to sudden power in July 1858 when President James Buchanan appointed him secretary of the new Nebraska Territory, a job he held for three years (including six months of that time as governor). When he was not busy farming and improving his land, Morton was vehemently politicking to make Nebraska a slave state, using his talents as a polemicist and orator to express his loathing for blacks, abolitionists, and, in time, President Lincoln and his "imbecile and malignant administration." While Morton ran numerous times for public office, in Republican Nebraska he made no headway.

Throughout these years Morton was a determined tree planter. Soon after building his simple frame house, he became one of the "first men in Nebraska to plant fruit trees. His neighbors all laughed at him, saying that fruit trees could not live upon the cold and bleak prairies." (This would prove to be true farther west in Nebraska, where the "Great American Desert" began.) But during those first hot pioneer summers, recalled Morton, "when the rays of a burning sun poured down all day long, the thought of cooling shades, of rustling leaves, or winds gently murmuring among trees, came unbidden, as dreams of springs haunt the mind of a fever patient." Soon his home was surrounded by American chestnuts, Osage orange trees, black walnuts, and tidy orchards.

In the postbellum years, Morton's 1872 invention of Arbor Day brought out his better angels and a gentler kind of fame. He valued trees, he explained, as "missionaries of culture and refinement" and upheld them as living symbols of "unswerving integrity and genuine democracy . . . for they refuse to be influenced by money or social position and thus trees tower morally, as well as physically, high above Congressmen and many other patriots of this dollaring age." Morton cast his new holiday as uniquely American in spirit: "All other anniversaries look backward; they speak of men and events past. But Arbor Day looks forward; it is devoted to the happiness and prosperity of the future."

Four years after Peaslee's celebrations in Cincinnati's Eden Park, the celebration of Arbor Day had reached California and its new capital: Sacramento, an early California Gold Rush town established on four square miles along the American River in 1849 by John Sutter Jr. When *New York Tribune* correspondent and author Bayard Taylor passed through in 1850, he reported that Sacramento's "original forest trees, standing in all parts of the town, give it a very picturesque appearance. Many of the streets are lined with oaks and sycamores, six feet in diameter, and spreading ample boughs on every side." In a city with brutal summer heat, "The beautiful sycamores are not merely an ornament to our city. . . . They are invaluable," editorialized the *Sacramento Transcript* that same year, incensed that the proprietor of the Sutter Hotel had nearly been murdered by a man named Ormsby, who, with a dozen men, had begun felling the gigantic sycamores fronting the hotel for their wood. When confronted, Ormsby "actually attempted the life of the proprietor . . . who escaped injury almost miraculously, several passes having been made at him with an axe." The city marshal showed up too late to save the trees.

By 1854 Sacramento had become the state capital, and roaring conflagrations had destroyed even more of its original giant sycamores. The fast-growing native cottonwoods planted in their place were unleashing drifts of irksome white silky down from their seed pods that snagged and littered, clumping up on sidewalks and streets. In 1857 James McClatchy, editor of the new *Sacramento Bee,* advised pioneers who had begun taking the ax to these "noxious" street trees to hold off until the "maple, locust, and china trees substituted in their place" achieve "a respectable size" or the fast-growing city would suffer again from "the intense heat of the summer."

When the town's residents, homesick for English and American elms and other eastern and European trees, discovered that those familiar species flourished in the local climate, they took up planting with a vengeance. As those trees grew, one visitor remarked that "in this respect Sacramento looks like a New England city." Sacramento—set in the baking plains of the Central Valley—soon proclaimed itself the "City of Trees," and its citizens valued these new oaks, elms, sycamores, camphors, and walnuts for their "positive beneficial effects" beyond shade: moderating the climate, increasing the wood supply, contributing to health, driving out malarial influences, attracting birds, and, of course, beautifying the landscape.

In 1886 California became the eighth state in the union to adopt Arbor Day, designating the last Saturday in November, after the autumn rains. Sacramento newspapers urged every family to participate by planting "a young orange, lemon, fir, walnut, elm, or other ornamental shade or fruit-bearing trees." The *Sacramento Daily Union* proposed that "children should be encouraged to name the trees singly or in small groves, and compete with each other in securing the best results."

Among the many Sacramentans who heeded the call to action was nine-year-old George H. P. Lichthardt Jr., who planted camphor trees in front of his family's store, the Sutter Fort Cash Grocery, at the corner of Eighteenth and M streets. The Lichthardts, who prided themselves on supplying the finest of Chinese and Japanese teas, from gunpowder to spider leg, lived above the store. One of those six trees would long outlast young George and the white wood clapboard store, growing ever larger above a new two-story commercial brick building as the nation fought two world wars, landed on the moon, and invented the Internet. It eventually achieved an age, girth, and majesty that would dominate the whole corner, a beloved arboreal presence whose hundredth-birthday celebration in 1986 was presided over by then-mayor Anne Rudin. It greeted the twenty-first century in fine health but began to sicken in 2012. On August 22 of that year, at age 126, the much-pruned camphor tree was declared a danger and taken down while a crowd of fifty mourners watched.

Within a decade of his reinvention of Arbor Day, John Bradley Peaslee's version of the tree fete had gone national. Marveled one Pennsylvania educator, "When I was a lad I never saw or heard of planting a tree in a school yard. Now, in the remotest parts of the State, I see growing in school yards the trees under whose ample branches the children of the next generation will play." By 1896 only Utah, Delaware, and the Indian Territory had failed to officially sign on.

In 1884 Peaslee had moved to solidify his claim to this new style of Arbor Day with a booklet promoting what he now called "the Cincinnati Plan" for celebrating the holiday titled *Trees and Tree-Planting*, something of a how-to book full of practical advice for schools and extracts children could recite, such as this from literary giant Oliver Wendell Holmes: "I have written many verses, but the best poems I have produced are the trees

I planted.... What are these maples and beeches and birches but odes and idylls and madrigals? What are these pines and firs and spruces but holy hymns, too solemn for the many-hued raiment of their gay deciduous neighbors?"

Nebraskans were not pleased with Peaslee's poaching on their ever-more-successful holiday, for J. Sterling Morton's invention was fast becoming "the scholastic festival of our times." In 1888 the state honored Morton as the originator of Arbor Day, changing the date of its own celebration to fall on his birthday, April 22. A local newspaper editor sought out hosannas from the eminent: "Mr. Morton deserves the gratitude of the whole land," obliged naturalist John Burroughs. "The birds and animals, as well as the people, profit by his wise forethought. Every tree planted upon this day will serve to keep green his memory." James Russell Lowell, retired editor of *The Atlantic* and former U.S. ambassador to Spain and then England, wrote: "I think that no man does anything more visibly useful to posterity than he who plants a tree."

When Grover Cleveland became president in 1893, J. Sterling Morton joined the twenty-second president's cabinet as the first secretary of agriculture. Still trim with broad shoulders at age sixty, Morton had become a prosperous railroad attorney (the very calling his disapproving father had urged on him). A reporter for the *Boston Daily Globe* observed that the new member of the cabinet was no midwestern bumpkin, what with his well-cut business suit, fashionable yellow shoes, costly scarf pin, and, sparkling away on his left hand, "a diamond large as the end of my thumb." Morton, a widower, mainly resided in Chicago, where his sons Joy and Mark were successful businessmen running the Morton Salt Company.

J. Sterling used his cabinet post to launch his own PR offensive, commissioning *Arbor Day: Its History and Observance*, published by the Government Printing Office in 1896. While the booklet conceded that Arbor Day had for the first time "acquired a wider interest... as it became connected in observance with public schools... in Eden Park," Peaslee merited but one brief quote praising school talks on trees as "most profitable," in the USDA's seventy-nine-page official narrative.

Peaslee, however, was determined to reinstate himself as a leading player in the Arbor Day drama. In 1900 he self-published his memoir, *Thoughts and*

Experiences, which included an entire setting-the-record-straight chapter on the "Origin of 'School Arbor-Day.'" In his account, when the good burghers of Cincinnati convened the nation's first American Forestry Congress, one Colonel De Beck proposed devoting a day to a grand public procession and speeches in Eden Park. "While the Colonel was speaking," Peaslee wrote, it was he, Peaslee, who had the brainstorm that conjoined literature and trees, his two passions.

Hoping also to inspire the rest of the state to plant trees, the brain trust behind the Forestry Congress persuaded Ohio's governor to announce on March 18, 1882, that the last Friday in April would henceforth be proclaimed Arbor Day in the Buckeye State. Again, wrote Peaslee, it was *he* who carried out this new vision of Arbor Day, one that recruited schools and their students to take part in tree planting. Moreover, it was also he, Peaslee, who dreamed up the idea of memorial groves.

Alas, no one paid him any heed. Peaslee had been successfully deleted from the historical record, until searchable historical databases plucked his memoir from obscurity and the local newspapers of the day verified his bona fide claims that it was *his* charming vision of Arbor Day—young people planting young trees that would grow up with them—that swept the nation.

Having prevailed as the sole founder of the holiday, Morton, as secretary of agriculture, initiated high-profile Arbor Day celebrations in the nation's capital. On April 22, 1895, he could be found planting an American elm for his sixty-third birthday on his agency's grounds at Twelfth Street, joining an existing grove of elms.

By 1895 Washington, D.C., was finally fulfilling its intended destiny as the nation's preeminent "City of Trees." When the capital had first come into being in the final years of the eighteenth century, George Washington, whose letters home mingled war news ("I tremble for Philadelphia") and tree worries ("it runs in my head that I have heard some objection to Sycamores"), and Major Charles Pierre L'Enfant had envisioned a stately, parklike municipality. They intended to preserve the site's existing undulating wilderness of valleys and hills, especially the lush woods with their "tall and umbrageous forest trees of every variety, among which the superb Tulip-Poplar rose conspicuous." The capital was designed so that Congress and the White House would occupy elevated terrain at opposite ends of grand avenues. All the streets that radiated out from them would be shaded and embowered by the thousands of existing mature oaks, maples, and poplars.

This plan presumed that everyone loved and respected trees as much as former president Washington, who had devoted countless happy hours to scouring his own woods for the perfect maple or dogwood to be transplanted by his slaves in fashioning his all-American landscape at Mount Vernon. (Today's current rage for planting native species dates to early in our history, and its original proselytizer was, fittingly, the father of our nation.) Instead, land speculators soon swarmed into the new District of Columbia, and their "wholesale destruction of the woodlands brought about dreary wastes and an atmosphere of devastation. The City's principal thoroughfares were bleak in their bareness."

When Thomas Jefferson, as ardent a lover of trees as Washington, became president in March of 1801, he ordered that the wide dirt road so grandly named Pennsylvania Avenue—which connected the unfinished Capitol building and the under-construction White House—be lined with double rows of newly fashionable Lombardy poplars from Italy. His planting sketch, sent over to the city's superintendent, further instructed that willow oaks, Jefferson's favorite tree, be interspersed to eventually replace the fast-growing poplars. Alas, the frugal landscaping budget allowed only for the foreign poplars, which did not flourish in the muggy capital.

Meanwhile, the city's commissioners wrote stern letters to landowners and residents warning against the "cutting of wood of any description," especially of "Ornamented Trees" on acreage around the White House and on the many plats intended for public squares and the long central mall—the latter still largely a malarial swamp. But with no means of enforcement, preserving the ancient trees on public lands "could not be done," lamented Senator Daniel Carroll from Virginia: "*The people*, the poorer inhabitants cut down these noble and beautiful trees for fuel. In one single night seventy tulip-Poplars were *girdled*, by which process life is destroyed, and afterwards cut up at their leisure by people."

President Jefferson startled Margaret Bayard Smith, wife of a dear friend and his dinner hostess one evening, when he exclaimed: "How I wish that I possessed the power of a despot. Yes, I wish I was a despot that I might save the noble, the beautiful trees that are daily falling sacrifices to the cupidity of their owners, or the necessity of the poor." When his fellow diners raised their eyebrows, Jefferson elaborated that with such "absolute power, I might enforce the preservation of these valuable groves. Washington might have boasted one of the noblest parks, and most beautiful malls, attached to any city in the world."

But surely, insisted one guest, the president had the authority to at least save the most magnificent specimens on the city's public lands. It would, replied the president, require armed guards, which he could not post at every towering tree. "In a few years, not a tree will remain, and when it is too late, the Legislature will regret that measures were not taken for their preservation." Such laments did not surprise those who knew Jefferson, for every day he set forth from the White House on horseback to take "long rides in every direction around the city, and [he] never [returned] without some branch of tree, or shrub, or bunch of flowers in his hand. He is acquainted with every tree and plant, from the oak of our forests, to the meanest flower of our valleys."

In 1833 the U.S. Congress hired one James "Jemmy" Maher, a boisterous, well-liked Irishman, as its public gardener. After landscaping the Capitol's scrubby grounds, Maher was appointed public gardener for the city and steadily expanded his green empire during the ensuing decades. No one seemed to mind when he established his own private tree nursery across the river and then sold those trees to himself as official gardener. In 1844 "Jemmy" supervised a complete arboreal renovation of Pennsylvania Avenue, lining it with multiple rows of 360 silverleaf maples and aspens. "Had Mr. Maher the sole control of the trees on the avenue for some years past," observed one fan, "many think we should have a complete shade now on that great thoroughfare." By the time of Maher's death in 1859, twenty-one of the city's very wide avenues (some cobblestoned but most still dirt) were shaded and beautified by his plantings—sycamores, Norway and sugar maples, European and American elms, along with a dozen varieties of ash.

Maher's trees, however, soon became casualties of the Civil War, as they were routinely chopped down for firewood by Union troops bivouacked all over the city, even on the White House grounds. Almost seven decades after John Adams became the first president to live in the capital, this forest of stumps only underscored the fact that Washington, D.C., was still little better than "a straggling parcel of villages," according to Senator Simon Cameron, "every one of which was unfit to represent any part of the great capital of this country."

In 1868 the local *Evening Star* described a city in "deplorable condition," where aside from a "few of the main thoroughfares . . . roughly paved with cobble-stones," the side streets were "mere dirt roads, in which pigs, goats, cattle, and geese roamed unmolested, scattering the little mounds of garbage dumped into the street by the householders." There was no drainage or sewers, just "filthy gutters . . . and a stinking open cesspool called 'the canal,' the receptacle of dead cats, rotten eggs, and garbage of all sorts . . . keeping the hand to the nose."

When boosters proposed that D.C. host the 1871 World's Fair, Nevada senator William H. Stewart, rich from litigating Comstock Lode claims, demurred: "Let us have a city before we invite anybody to see it." Fed up and out of patience, some, including Horace Greeley, the *New York Tribune* editor and political powerhouse who denounced Washington as "not a nice place to live . . . the dust is disgusting, the mud is deep," were actively lobbying to relocate the nation's capital west, to the more cosmopolitan Gateway City of St. Louis, Missouri.

At that, Washington's biggest real estate developer, the self-made Alexander Robey Shepherd, roared into action. Like a character out of Dickens, Shepherd had had to quit his private school at the age of twelve to become the sole support of his newly widowed mother and six siblings. By his twenties, he was owner of the city's largest plumbing and gas-fitting firm and had built 1,500 row homes. A large man, broad of forehead and with a masculine lantern jaw, Shepherd was, observed a contemporary, "a free liver and a free spender, a man with strong passions and appetites. . . . His large way of doing things was part of the nature of the man. . . . His imperious will recognized not merely no master, but no difficulty that stood in the way of what he wanted to do."

On September 4, 1871, President Ulysses S. Grant appointed Shepherd, his longtime Willard Hotel drinking companion, a staunch Republican, founder of the local Union Club, part owner of the *Evening* Star, director of four streetcar companies, and proud owner of an estate called Bleak House, to the vice presidency of the city's new Board of Public Works. Within weeks the man soon known as Boss Shepherd launched a massive makeover of the city. When Congress reconvened two months later, they "gazed with wonder at the change in the general topography of the city . . . at the miles of incomplete sewers, half-graded streets and half-paved sidewalks."

In an astonishing three years, Shepherd's work gangs paved 118 miles of streets and 207 miles of sidewalks; constructed 123 miles of sewers, 30 miles of water mains, and 39 miles of gas mains; filled in the fetid Tiber Canal; erected streetlights; and built numerous new schools. They ran roughshod—sometimes literally—over anyone or anything in their way, and a thousand angry citizens petitioned to Congress to investigate the Board of Public Works for a wide variety of outrages.

The overbearing Shepherd did have the redeeming quality of understanding the importance of trees, even if he freely confessed his ignorance on the topic. As he remarked to William R. Smith, superintendent of both the U.S. Botanic Garden and Shepherd's new "Parking Commission": "Now, Smith, you know I know nothing of tree planting, but I suggest oak trees if possible on some streets, but put a tree wherever there is damned room." For Shepherd, who showed "a breadth of vision unknown in that day in city improvement work . . . tree planting was made an essential part of the plan."

At the end of the following spring, Superintendent Smith, a convivial Scotsman trained at Kew Gardens who had for the past twenty years created and resided in the government's Botanic Garden at the foot of Capitol Hill, reported that Shepherd's men had planted 6,236 trees, including silverleaf maples, sugar maples, Norway maples, American elms, American and European lindens, tulip trees, American white ash, scarlet maples, various poplars, and ash-leaved maples. Wrote Smith: "This list comprises all the best trees that are available for street and avenue planting, combining rapidity and regularity of growth with a sufficient diversion of forms."

Smith's two fellow parking commissioners (all serving pro bono) were equally knowledgeable: William Saunders, forty-nine, was a fellow

Scotsman trained in London who for ten years had run the Department of Agriculture's Experiment Garden on the Mall and would later also help introduce a variety of plants from exotic locales, including the navel orange. During Saunders's early career in Baltimore and Philadelphia, he had designed the grounds of private estates, public buildings, and cemeteries, including most famously (for President Abraham Lincoln) the Gettysburg Soldiers' National Cemetery.

By 1869, historian Pamela Scott notes, Saunders was "particularly concerned with how to use street trees to promote a healthy city environment." To counter the capital's clouds of unhealthful dust, raised by horses, wandering cows, and pigs trudging through dried manure, Saunders advocated "proper varieties of deciduous and evergreen trees and shrubs introduced throughout the avenues and wide streets . . . [to] not only produce a fine landscape effect, but what is of far more importance, they would very efficiently check the progress of dust-laden siroccos."

The third Parking Commission member was John Saul, fifty-two, raised in a family of Irish horticulturists, who had come to D.C. in 1851 to work with the still-mourned Andrew Jackson Downing during his truncated effort to beautify the city's derelict public grounds. In 1854 Saul had purchased an eighty-acre farm and become the city's leading commercial grower of ornamental trees.

The Tree Troika quickly laid claim to a portion of the Almshouse grounds just east of Capitol Hill for a large nursery. After being put in "fitting condition by draining, manuring, and deep and frequent ploughings," it became home to 23,240 juvenile trees (many started from seed) destined for life on the city streets. While most were the usual hardy species, Smith also nurtured "many hundreds of rare ornamental foliaged and flowering trees. . . . These were procured when quite small at little cost." The park commissioners concluded that what their young arboreal charges required to achieve health and longevity was planting holes of "ample size," "judicious waterings" during hot months with no rain, and the protection of the all-important "cheap tree boxes." The latter—six-foot-tall wooden strips placed three inches apart and bound with iron hoops—seemed "essentially necessary, owing to the injury the [trees] were receiving from the nibbling of goats and horses, evils that still require to be closely watched."

Finally, as the young trees matured, they would need informed and timely pruning to shape proper structure.

Boss Shepherd—now elevated to governor of the Territory of the District of Columbia—had estimated his vast (and necessary) public works to cost $6 million, but within three years one of several probes and audits revealed overruns of $13 million, thus plunging the city into bankruptcy and leading to an abrupt and ignominious end to his reign. But Superintendent Smith and the Parking Commission survived, and in their 1875 report Smith listed the 5,383 trees planted that year and worried about runaway horses wrecking young trees but also predicted: "Washington will in future be known as the 'City Among Trees.'"

Ten years later, in late May of 1885, a reporter from the *New York Times* stood upon the Capitol's dome, looked out over the city, and marveled at the "vast labyrinth of leafing trees probably unequaled in extent, variety, and symmetry in any other city in the world. In a few years our capital will become preeminently the Forest City of the Nation." Wherever this journalist strolled in the course of reporting on the doings of the capital, he could not fail to notice: "Trees all around you, in platoons, in columns, in ranks, single, double, and quadruple, they shade and ornament with their luxuriant foliage not only the most magnificent avenues where fashion takes its daily airing, but equally the squalid and unimproved streets inhabited by the very poor. . . . And one needs only to notice on some hot August day the pedestrians dodging from tree to tree to escape the blistering sun, the horses picking their way where the trees shade the street, and little children continuing their play at noontide beneath their grateful shelter, to be convinced that they have an important bearing upon public health and comfort."

A few years later, New Yorker Peter Henderson declared in *Harper's New Monthly Magazine*: "The city of Washington, the capitol [*sic*] of the nation, exceeds in beauty any city in the world." While he admired the "grand conception of the plan," the wide avenues "smooth as marble, and its hundreds of palatial residences . . . above all, [the city's] magnificent trees make it without a peer. . . . Such is the effect of the wonderful growth of the street trees, seen from the Capitol or other high buildings, that it to some extent presents the appearance of a city built in a forest."

Like Saunders of the Tree Troika, Henderson appreciated other practical benefits offered by the trees: "Malaria, once such a bane to Washington, has been materially checked, and the night temperature during summer that used to be almost unendurable, has now been materially lessened. . . . Now the shaded pavement absorbs little heat, and the nights are comparatively cool." By now the Parking Commission had planted 63,014 trees, almost half being some kind of maple, followed in number by the Carolina poplar ("it will flourish even where there is a pall of coal smoke"), various elms, lindens, honey locusts, and thirty-some other species planted in small, experimental numbers, including 145 specimens of the still-exotic ginkgo. There was even the possibility that the reviled tree of heaven (or ailanthus) might make a comeback.

The reporter found William Saunders, now a fifteen-year veteran of the Tree Troika, seasoned world plant promoter, and superintendent of what Agriculture now called its Propagating Garden. Knowing much about the world of trees and plants in American and foreign climes, Saunders could say with some authority: "This city is the pioneer and most successful example of the application of arboriculture to the improvement and ornamentation of public streets." With his colleagues he was also creating a systematic record of which species flourished as street trees, what afflictions they were prone to, and the sums these incurred.

And what of the man who made all of this possible—the redoubtable Alexander Shepherd? Denounced by his enemies as a crook and a bully, he was hailed by others as "the man who redeemed and beautified Washington and came out of office a poor man." Fired as territorial governor for his prodigal ways, he was soon bankrupt but repaid his creditors and in 1880 proceeded south to Batopilas, Mexico, to remake his fortune by running a silver mine. Operating, as ever, in a large way, he departed with "his wife Mary, seven children, a cook, a nanny, seven men, four dogs, and baggage enough to require a private railroad car to San Antonio. From there, the entourage proceeded in two Army ambulances."

In 1887, when Shepherd returned for a visit home, grateful citizens by the thousands turned out to honor him with a parade up Pennsylvania Avenue, now a bower of shade with double rows of fast-growing elms and Carolina poplars lining the whole five miles. William R. Smith had honored Shepherd (the "Master Spirit") by planting a special elm in his Botanic Garden, one of three young trees coaxed to life from the remaining roots of an old American elm George Washington had (the story went) personally planted near the Capitol as it was being built. In July of 1895, just months after Secretary of Agriculture J. Sterling Morton planted the elm at Twelfth Street, Shepherd came through town again for a visit, and six thousand Washingtonians crowded into the Willard Hotel to attend a banquet in his honor.

CHAPTER THREE

"A Demi-God of Trees" and "The Tree Doctor": Charles Sprague Sargent and John Davey

Above: John Davey, "The Tree Doctor." (*Courtesy of the Davey Tree Expert Company © 2016.*)
Below: Charles Sprague Sargent, director of Harvard University's Arnold Arboretum. (*Courtesy of the Arnold Arboretum Archive Collection.*)

In June of 1893 John Muir and his editor, Robert Underwood Johnson of *Century Magazine*, stepped down from a horse and buggy at the Brookline estate of wealthy Boston Brahman Charles Sprague Sargent. Muir, a craggy, wild-bearded writer famous for his passionate defense of America's besieged wilderness, was here to meet one of the men he admired most in the world. At age fifty-two Sargent was the nation's leading dendrologist, scholar, and evangelist of trees and creator of the nation's first (and most important) "tree museum," Harvard University's Arnold Arboretum in the Jamaica Plain neighborhood of Boston.

Muir was freshly off the train from New York City, where he had been so hot that he "felt all the time as if somebody had rubbed the inside of my underclothes with very sticky molasses." Entering Sargent's 130-acre estate, Holm Lea, he beheld what he declared to be "the finest mansion and ground I ever saw." In a letter to his wife back in Martinez, California, Muir especially rhapsodized about Sargent's gardens: "fifty acres of lawns, groves, wild woods of pine, hemlock, maple, beech, hickory, etc. and all kinds of underbrush and wild flowers and cultivated flowers—acres of rhododendrons twelve feet high in full bloom, and a pond covered with lilies, etc., all the ground waving, hill and dale, and clad in full summer dress of the region, trimmed with exquisite taste."

Muir reported that while Sargent's "servants are in livery," his Boston host and his wife and family made him "in a few minutes . . . [so] at ease and at home, [I was] sauntering where I liked. . . . We had grand dinners, formal and informal." When Muir departed three days later, he and Sargent, a tall, sturdy, bearded, formal soul described by some as "cold roast Boston," were fast friends, having bonded over their mutual love of trees and devotion to saving America's forests. The two became correspondents (Muir dispatching handfuls of acorns gathered from favored oaks) and travel companions, together tromping through remote forests and hillsides, reveling in trees.

Fifty years earlier one of Andrew Jackson Downing's dearest friends was Hudson River Valley neighbor and New York banker Henry Winthrop Sargent, Charles Sprague Sargent's much older second cousin. In 1841, the year Downing became famous, the New York Sargent had semiretired to Wodenethe, his twenty-two-acre country estate across the river from Downing's villa and gardens above Fishkill Landing. At Wodenethe, Henry, who preferred plants to finance, assembled a remarkable collection of rare conifers. In 1859, seven years after Downing's untimely death, he lamented in a supplement to the sixth edition of Downing's *Landscape Treatise* that no one had claimed his late friend's "decided and marked influence."

And yet all these years later it seemed Downing's tree-loving spirit had found austere reincarnation in the unlikely person of the famously reserved Charles Sargent, a relentless workaholic whose acerbic persona could quickly veer into grumpiness when confronted with those who did not see things his way. As Sargent confessed to Muir in one letter a few years after they met, "You have a way of saying things which give me the greatest pleasure because I know that you understand the subject [trees] and are sincere—qualifications possessed by few of my correspondents." Still, Sargent was much admired, even beloved, by many who worked for and with him.

It is possible that Andrew Jackson Downing knew (or perhaps simply met) Sargent when Charles was very young, for Downing visited Wodenethe, and Downing came to Brookline. However, nothing presaged that Sargent would in time surpass the revered Downing in his arboreal achievements and influence. In 1864 Charles graduated eighty-eighth in a class of ninety at Harvard, served in the Union army during the Civil War (as did a third of his classmates), and then spent three years traveling in Europe, returning at age twenty-seven to run the family's estate in Brookline.

In 1872 Sargent signed on to work under Asa Gray, the great Harvard botanist. Within eighteen months he received the small-bore appointment as first director of the Arnold Arboretum, an institution to be created on 125 acres that Sargent described as "a worn-out farm, partly covered with natural plantations of native trees nearly ruined by excessive pasturage."

Moreover, Sargent noted, this unpromising farm was "to be developed into a scientific garden with less than three thousand dollars a year."

When Sargent learned that the City of Boston had engaged the celebrated landscape architect Frederick Law Olmsted to design a park system, he ingeniously enlisted Olmsted to incorporate the arboretum lands into that system. Charles revealed great political and financial moxie, triumphing over both a hostile Boston City Council and Harvard's President Eliot to create a unique hybrid: The Arnold Arboretum would be both a public park (all roads and paths to be built and maintained by the city) *and* America's first living tree museum, its grounds belonging to Harvard at one dollar a year rent for a thousand years. As 1882 ended, Sargent began creating the arboretum, whose size would double in time thanks to his further machinations and skills as a fund-raiser. His plan was to create a "pleasure park" that would also showcase in evolutionary order every tree or shrub hardy to New England.

Sargent was a networker par excellence. From the start he exchanged seeds and plants with botanic gardens all over the world, including Kew Gardens in England, Jardin des Plantes in Paris, and the Imperial Botanical Garden in St. Petersburg (to name but a few), while cultivating American botanists, estate owners, and those all-important engines of green commerce, the nurserymen. By April of 1887, when Sargent celebrated his forty-sixth birthday, 120,000 carefully "selected and artfully placed young plants," including many rare foreign specimens, were growing, including all the species of oaks. Even with so many trees, though, "the Arboretum still left a lot to the imagination."

"The arboretum is a museum," Sargent wrote, "and like other museums, its first and real duty is to increase knowledge. The fact that the grounds where the museum objects (trees and shrubs) are displayed are beautiful and well arranged is a valuable asset of course, for it draws persons here who would never come if the arrangement was not attractive. The Arboretum is sometimes spoken of as a School of Forestry or of Landscape Gardening. It is neither, but a place for the study of trees and shrubs as species, their character, distribution, nomenclature, uses and their commercial and aesthetic value." Every tree and bush had an ID tag and accession number that corresponded to the catalog card in the botanical archive that listed further details of the tree's origin, species, and planting date.

Although heir to a merchant-banking fortune, Sargent was not Rockefeller rich, and to raise just the fifteen thousand dollars needed to run the arboretum each year required constant begging appeals to Boston friends, relatives, and movers and shakers. When the Bussey administration building opened in 1892, Sargent's characteristic frugality meant the offices and library had no form of lighting, for he "regarded electric lamps as extravagances, and candles or kerosene lanterns as fire hazards. On short winter days, no one could see to work after four o'clock."

Unlike Downing and Muir, Sargent would never become famous, for even as he became an ever-more-renowned Harvard professor, he "was so averse to public speaking" that he was never known to give a single lecture at the university or anywhere else. But over the next decade he would secure his reputation by virtue of his prodigious scholarship (which required much study and travel) and his creation of the arboretum.

In 1884 he had published for the tenth U.S. census *Report on the Forests of North America*, a major study that was a mere prelude to his magnum opus, the fourteen-volume definitive guide to American trees, *The Silva of North America* (1891–1902). "Sargent is king in the U.S. as regards Arboriculture," wrote the influential German forester Dietrich Brandis, inspector general of forests in colonial India and a mentor to many young foresters. "Owing to his marvelous knowledge of American trees he is looked upon as a Demi God." Amazingly, we don't know why, how, or when Sargent came to so love trees. No journalist or biographer shed any light, and certainly Sargent himself, despite writing thousands of pages about the wonders of trees, never offered a clue.

When the final volume of *Silva* came out in 1903, John Muir rhapsodized in a fourteen-page review in the *Atlantic Monthly* over this "truly great book on a great subject by a master . . . a description of all the trees that are known to grow naturally in North America." Muir described how at a time when the nation was awakening to its need for trees, Sargent had crisscrossed the continent, traveling through its fast-disappearing forests and its new booming cities, "thousands of miles every year, mainly by rail of course, but long distances by canoe or sailboat . . . exploring untrodden wildernesses, like Charity, enduring all things,—weather, hunger, squalor, hardships . . . jolting in wagons or on horseback over plains and deserts. . . . While trees were waving and fluttering about him, telling their stories, all else was forgotten. Love made everything light."

Sargent also understood the need for a periodical along the lines of Downing's magazines aimed at the public and in 1888 had helped launch and underwrite the weekly *Garden and Forest*. While admired, it never achieved the same popularity as Downing's own publications, for its articles lacked the charm and verve of his pen and personality. Certain topics remained perennial favorites, such as one featured in a December 29, 1897, article urging "American Trees for America," lamenting that U.S. gardeners suffered such "ignorance with regard to the true beauty and value of native trees, which appears to be peculiar to us as a nation."

Out in Kent, Ohio, one John Davey, another tree lover with a very different provenance and temperament from Sargent's, was about to become famous as the "Tree Doctor." In 1901, at age fifty-four, Davey self-published a seminal book by that title, which methodically instructed with photos and text how to properly prune and care for trees. Marveled one admirer, "John Davey worked out the science of tree surgery alone. . . . All the tree surgery in America traces a pedigree to this one man, five feet six, weight one hundred and thirty in the shade." Davey and his book would launch the modern field of arboriculture, or the commercial care of trees in and around American cities. Looking at his book today, one is struck by how bloglike it is: lots of eye-catching images that perfectly illustrate the punchy text.

Unlike Sargent, Davey wore his passion for trees on his sleeve and could—and did—wax rhapsodic about any aspect of his arboreal loves: "Examine a leaf," he would exhort. "How many will see its beauty, saying nothing of its utility? Did you ever stop to think that all the millions of tons of timber ever grown were made in the leaf?" He hated to witness the routine disfiguring of trees. "Nature does not form those beautiful and health-giving tops of shade trees to be cut to pieces to furnish 'beer money' for a lot of Tree Fools," he wrote in *The Tree Doctor*.

Davey, an English immigrant, was a man of uncommon curiosity, high standards, and self-discipline. He was born in 1846 on a farm in Somersetshire, where at age four he asked if he could help plant potatoes in front of the family's cottage. His father cut a potato in half, offered the small lad a spoon, showed how the halves should be put in the earth and covered, and then said, "Listen to me carefully. Do it right or not at all." Never would John Davey forget those words. They drove his whole life. At age thirteen, when his mother died, his father sent him to work on a nearby farm. By twenty-one John was ready for more training and moved to Torquay, with its famous gardens and greenhouses, to apprentice himself, reveling during those six years in mastering horticulture, floriculture, and landscaping.

He moved with his wife, Bertha, and their first child to Kent, Ohio, in 1882. As the new superintendent of Standing Rock Cemetery, he made it a living laboratory for "his revolutionary idea that trees would be saved by a practical curative process." Over the next three years Davey could often be found suspended from ropes high up in the branches, pruning deadwood or applying special salves. His skillful ministrations rejuvenated the neglected mature trees in this overgrown burial ground, and the public flocked to enjoy the now-lovely memorial park. Sought out as the local "tree man," Davey also opened his own nursery business. He discovered a taste for writing pamphlets and lecturing on subjects as disparate as the need for women's suffrage and the stars and planets, but above all the beauty and importance of birds and trees to humankind and the interdependence of man and nature.

As a man of the soil, Davey had nurtured trees for his entire life and harbored the deepest respect for these fellow "living creatures." He also loved the birds that made trees their homes. As a new American citizen, John Davey was "appalled at the senseless waste of trees that went on in his adopted country. They were treated almost like an enemy that had to be destroyed. All they meant to many was cheap building material or an inexpensive source of fuel. Primeval forests were ripped out to provide farmland. . . . As for sick trees, the attitude was, 'All plants die so what's so different about a tree?' Trees that were trimmed were made to look like hat racks by local 'tree butchers.'"

Year after year in Ohio, John Davey remained a close student of trees and their health, and he avidly embraced the new medium of photography to illustrate such lectures as "The Salvation of Trees and Birds." In his role as local tree man, he had observed that when trees were properly pruned—limbs large or small sawed off close to the branch or trunk—the "wound" healed nicely. He advocated cleaning out decay and employing rods and chains to support and maintain structurally weak mature trees, extending their lives sometimes by decades. He also understood that tree roots required sufficient room and proper soil and air to thrive. He railed at the "tree butchers," know-nothings employed by the new gas and electric companies who hacked away and maimed trees rather than artfully making room for the new electrical wires and gas pipes that crews were installing in fast-growing American towns and cities.

In 1900 Davey combined his hundreds of photographs of good and bad tree care with simple explanatory text to create ninety pages of the book manuscript he titled *The Tree Doctor.* In its introduction he declared that he had had

> the care of trees and plants for more than thirty-five years and [was] an ardent lover of nature. The ghastly wounds of his friends, the trees, and their various suffering (if you will allow the expression) cry aloud and pierce his inmost soul and bid him arise and plead their cause. The author is not so conceited to suppose he "knows it all." Whatever knowledge he possesses he has learned from others, or gained it from observation. . . .
>
> I cover, in THE TREE DOCTOR, practically all that has ever been written on tree culture. . . . You pay a dollar for a tree, shrub or plant, then lose it. You try again and lose it, because you know not how to proceed. . . . THE TREE DOCTOR [price: one dollar] will prevent all this waste of money. . . . All scientific terms are avoided. The language used is chosen so that it will convey a knowledge of the facts to the ablest scholars of the land, or to the merchant, farmer, mechanic, laborer, man or woman, boy or girl."

When Davey discovered that no publisher was willing to bring forth this long-aborning labor, he boldly borrowed $7,000 (the equivalent of $280,000 today) and had the book with its red linen cover published in 1901. With a great deal of promotion from Davey and his four grown sons, the book began to find its audience.

John Davey was not one to mince words, and in a soon-to-be infamous chapter titled "The Calamity of Cleveland," he took on a sore subject: "'Why can't we grow decent trees on the Public Square' has been asked by thousands of Cleveland's best citizens," he wrote. "One says the failure is caused by *gas leakage*, another assigns it to *electric currents*, a third attributes it to *smoke*, another *this*, another *that*. Good citizens of Cleveland, there is just *one* cause of all your failures with shade trees.... Every *delicate, succulent little fibre* that gathered moisture has become exhausted and *died of thirst*. What shall I say? 'Shame!' No, because you knew not what you were doing." The problem, explained Davey, was simply that Cleveland had planted these trees in "purest and barrenest sand. Do you expect to grow a majestic elm or a gorgeous maple in such material? If you do you will wait til Doom's Day and a million years after and *then fail*. This *sand* will not retain enough moisture for large trees."

As Davey and his methods became known, he and his two oldest sons were soon overwhelmed with work. Eager to expand and spread his gospel of tree care (and pay off his big debt), in 1905 Davey established the Davey School of Practical Forestry to staff his business with young men trained to his exacting standards. The Davey tree men made important converts among wealthy families with large estates and huge old trees: the Goodrich family of Akron, Ohio; George Eastman of Kodak fame in Rochester, New York; and Henry Flagler of Standard Oil, whose Orienta Point mansion and grounds overlooked Long Island Sound. In one case John Davey was called in to save a rich man's great elm that was "rather sad looking. It seemed to be dying from root to crest... but only on one side." The Tree Doctor diagnosed the trouble: "thick sod on one side that prevented natural nutrients and moisture from reaching the roots." He returned with his crew and forty feet from the trunk dug a semicircular trench through the sod. He filled the trench with rotted manure, dug out all the rest of the sod, and worked in "several wagon loads of manure into the soil." When spring came round the following year, "the tree was beautiful and green on all sides," and John Davey had another customer for life.

His third son, Martin, had a flair for trade and had been dispatched to sell books and tree work. As Martin was proving a natural leader, John Davey, now in his midsixties, was happy to concentrate on further editions of his book. On February 4, 1909, Martin took charge of the company, incorporated as the Davey Tree Expert Company in Kent, Ohio, the very town where his father had first demonstrated the power of arboriculture at Standing Rock Cemetery. The company's motto? "Do it right or not at all." That first winter Martin—not wanting to lose his trained men when the snowy weather ended the outdoor working season—opened the

Davey Institute of Tree Surgery, offering higher pay to all who enrolled for the winter. Martin also expanded the company's sales team and crews into the South, where they could find work in all seasons. By 1911 the Davey Tree Expert Company was operating in thirty-one states and Canada.

In 1901, the same year Davey's influential book first appeared, Theodore Roosevelt ascended to the presidency in the wake of William McKinley's assassination. Not since Thomas Jefferson had the United States had such a proselytizing tree lover in the White House. At Sagamore Hill, Roosevelt's Oyster Bay home on Long Island, the Roosevelt family motto, *Qui plantavit curabit* ("He who has planted will preserve"), was "carved under the north entrance door," reported the *Washington Post*. "On his return there each season, the President visits each tree on the place, noting the progress each has made since he last saw it. The peculiarities of each tree has as distinct a place in his mind as the features of his children." The family "planting" motto was also reportedly tattooed somewhere on his person.

Sargent came to view Roosevelt with a jaundiced eye, and when the president established the new U.S. Forest Service, Sargent felt this agency viewed trees and forests as mere board feet to be managed, rather than treasures to be preserved. Sargent dismissed the president in a letter to John Muir as taking a "sloppy, unintelligent interest in forests." Sargent lamented to Muir in 1902, "There is no one but you and I who really love the North American trees."

By the time Teddy Roosevelt and his family moved into the White House, its eighteen acres of fenced grounds were well shaded and wooded and boasted several celebrity trees that had been gifts of his predecessors. Most famous was a towering "stately American elm (*Ulmus americana*) said to have been planted by President John Quincy Adams" some eighty years earlier on a mound southeast of the White House, just one of the two hundred trees of twenty varieties he contributed to beautifying the grounds. Rutherford B. Hayes's best-known legacy was actually his wife's—the White House Easter Egg Roll. While his children filled the grounds with dogs, a goat, and the nation's first Siamese cat, Hayes left his mark by planting another presidential American elm near the west entrance. In April 1892, President Benjamin Harrison planted a sweet gum tree, the favorite tree of Alexander Hamilton, on the northeast lawn.

President Theodore Roosevelt planting a tree in Fort Worth, Texas, in 1905. (*Courtesy of the Theodore Roosevelt Collection, Houghton Library, Harvard University [Roosevelt 560.52/1905-059].*)

When President William McKinley retook office from Harrison in March of 1893, he celebrated by planting a scarlet oak on the west lawn. During his first term (1885–89), McKinley, forty-nine, had married twenty one-year-old Frances Folsom Cleveland (who remains the youngest First Lady ever), and shortly after their 1886 White House wedding, she had planted two Japanese threadleaf maple trees. In short, a tradition was emerging that presidents and their wives should leave their arboreal mark on the White House grounds as a way to gradually educate and inspire their fellow citizens and leaders to appreciate the oaks and maples and lindens of their parks and streets. (Contrast their evangelism with the attitude of our fortieth president, Ronald "When you've seen one tree you've seen them all" Reagan.)

No surprise, then, that Theodore Roosevelt, ardent birder and conservationist, reveled in his power to sponsor 150 new or enlarged national forests, mainly via presidential fiat, nor that he embraced Arbor Day as zealously as he did. April 8, 1905, found the ever-vigorous president, attired in dark suit and top hat, at a Fort Worth, Texas, school demonstrating his hands-on love of tree planting. While another man held a tall sapling in place, the president shoveled dirt all around its large root ball. Two years later, on April 15, 1907, Roosevelt issued an "Arbor Day Letter to the School Children of the United States," the first such from a president, proclaiming "the importance of trees to us as a Nation, and of what they yield in adornment, comfort, and useful products," declaring that the well-being of humans was entwined with trees, and warning: "To exist as a nation, to prosper as a state, and to live as a people, we must have trees." It is a shame that the Russo-American oak (*Quercus dentata*) planted by Roosevelt on April 6, 1904, south of the east wing of the White House grounds, died in 1957. No other tree planted by him survives elsewhere in the nation's capital.

More recently Senator Ted Kennedy mourned the passing of a particular English elm in 1978, calling it "a cantilevered miracle of nature that stretched out across the sidewalk and over the roadway. . . . President [John F.] Kennedy when he was a Senator, liked to call it the Humility Tree, because Senators instinctively ducked or bowed their heads as they approached the limb and passed under it." When President Barack Obama opened his bedroom window the first spring he lived at the White House in 2009, he could smell the perfume of the flowering *Magnolia grandiflora* that President Andrew Jackson had planted back in 1829 in memory of his wife, Rachel, "a round, fat dumpling" of a woman hounded to death, many believed, by the cruel press attacks on the legality of their marriage.

While these tree lovers and evangelists inveighed against the careless destruction of old forests and the vital importance of trees in the urban landscape, no one in the early twentieth century could have anticipated the loss of an entire species of tree, extirpated in a way never before witnessed in America.

CHAPTER FOUR

"This Fungus Is the Most Rapid and Destructive Known": A Plague Strikes the American Chestnut

Philadelphia's Fairmount Park was the scene for *Gathering Chestnuts* by J. W. Lauderbach. This engraving appeared in the *Art Journal*, 1878. *(Courtesy of the American Chestnut Foundation.)*

In early summer of 1904, all across New York City American chestnuts (*Castanea dentata*) were in full flower, their creamy, fireworklike bursts of catkins swaying in the winds, abuzz with bees and beetles harvesting the trees' rich pollen. Up at the Bronx Zoological Park, chief forester Hermann Merkel stood in its prized woods, a cool oasis of native chestnuts, hemlocks, oaks, birches, locusts, and poplars. As the sunlight filtered through in thin shafts, Merkel, a German landscape architect, was running his fingers over the trunk of a young American chestnut. Why, he wondered, did a number of the young chestnuts appear to be ailing, their bark dead in patches and some of their branches full of withered brown leaves?

Yet the real stab of concern came when Merkel shifted his gaze aloft, high up into the canopy: Even towering old chestnuts with twelve-foot girths showed twigs and branches whose foliage was brown and brittle long before autumn. These largest of native trees, eastern monarchs of the tree kingdom, were valued for their statuesque beauty, lumber, and delicious fall nuts. Merkel had never seen them afflicted in this way. Founders of the five-year-old zoo cherished its parklike grounds, a verdant respite for the masses from nearby Manhattan who streamed in to see the nation's few remaining bison and such exotica as giraffes and chimpanzees, before enjoying picnics and naps under the trees. Perhaps, Merkel thought, the brutal winter just past was to blame.

The following spring Merkel again peered closely at the stricken specimens. Whether aged giant or recently planted sapling, every American chestnut he examined now appeared afflicted in the same way as the trees had been the previous summer. Merkel observed on each tiny dots of a raw sienna color speckling the rough bark. Some of these dots had already grown and hardened into warty, dark umber protuberances, spreading until they encircled and seemingly strangled a branch or even a whole tree trunk. When Merkel tapped his fingers on the patches of lifeless bark, they sounded hollow; on some affected trees the dead bark was cracking and falling off. Seeking expert advice, Merkel dispatched samples of the diseased bark to the U.S. Department of Agriculture in Washington, D.C., where scientist Flora W. Patterson identified the problem as the common canker fungus *Cytospora* and recommended "the cutting out and immediate burning of all affected branches and limbs, and the spraying of all the trees."

Merkel quickly secured two thousand dollars in emergency funds and launched an all-out assault. Throughout June and July he supervised dozens of hired tree climbers who clambered about the chestnuts, sawing off infected limbs, while a power sprayer mounted on a horse-drawn cart maneuvered through the dense undergrowth, stopping here and there while men climbed ladders to direct hoses and nozzles to spray ghostly "Bordeaux mixture" on the chestnuts to kill the fungus. Months into this campaign, however, the 438 treated trees were still visibly dying; the chestnuts' normally vibrant slender six-inch-long green leaves with their coarse serrated edges had become desiccated, while the handsome bark with its vertical wavelike pattern sported the telltale orange dots that soon girdled limbs and trunks, blocking off all food and water to the tree.

In the late summer of 1905, Merkel, now deeply alarmed, turned for help to a new staff mycologist across the way at the New York Botanical Garden. Like the zoo, the botanical garden had only recently opened, its 250 acres of woods and gardens set on a former farm chosen for "its graceful hills and glacial landscape, with a 1790s manmade waterfall, an old mill, rocky outcroppings and uncut native forest, the freshwater Bronx River flowing through the middle of it." William Alphonso Murrill, thirty-six, a PhD botanist from Cornell, was a tall, athletic Virginian who had grown up on farms and knew well the beloved American chestnut. Since boyhood Murrill had been a close observer of trees, and he recalled fondly how "while in the woods we gathered chestnuts, chinquapins, walnuts, hickory-nuts, black haws, persimmons, and pawpaws. . . . I attended my first picnic in a grove of maples. The hole where I learned to swim was under a big willow. . . . The church stood in an oak-chestnut grove, where we had a merry-go-round on an oak stump." An ardent scholar of the natural world, Murrill delighted in being out among trees of any kind and enjoyed nothing more than roaming the countryside with his butterfly net made of green mosquito netting and broomstick, his haversack stocked with geological hammer, cyanide bottle, cord, and paper, and hanging from his belt a lined cigar box for specimens, large tin plant case, knife, and trowel—all the tools he needed to add to his fossil, fungi, insect, and rock collections.

As soon as Merkel explained his plight, Murrill hurried with the forester back to the zoo, where the two walked the grounds studying the sad spectacle of the dying American chestnuts. Murrill carefully broke off infected twigs, leaves, and bark from a few of the trees for his specimen bag. Back at the botanical garden he could not help but notice that here too, even in its forty-acre virgin hemlock forest, American chestnuts were displaying similar symptoms. In his laboratory Murrill used the twigs and bark from the zoo, writes Susan Freinkel in her history *American Chestnut,* to grow "specimens of the fungus until he had satisfied himself that he had a pure culture. He then transferred the cultures onto various media—agar, bean stems, and sterilized chestnut twigs—and placed them in glass tubes sealed with wads of cotton. Sure enough, the fungus grew." Murrill had never seen such a fungus with its "beautiful yellow" pustules, but he believed it was a new, virulent mutation of the large and common fungi genus *Diaporthe,* native to the eastern United States. He named what was killing the local American chestnuts *Diaporthe parasitica Murrill.* By November he had infected living chestnut twigs and watched the fungus blossom and swiftly kill.

Come the spring of 1906, Murrill saw that the young American chestnut trees housed in his sealed-off greenhouse had emerged from their winter's dormancy. He applied his pure culture and waited. Very quickly, "the Naturalist," as Murrill thought of himself, reported the "fungus enters through a wound or dead limb and works beneath of the cortex in the layers of the inner bark and cambium. The bark soon dies and changes color and later becomes rough and warty from the presence of numerous yellowish-brown fruiting pustules. . . . [Those] send out peculiar twisted spore-masses containing millions of minute summer spores." From then until autumn, he later observed, those spores proliferated, spread by squirrels, birds, insects, or just the wind to other chestnuts, and when the fungus found the minutest opening in the tree's bark, it attacked "twigs, branches and trunks of chestnut trees, irrespective of size or position . . . [and] proceeds in a circle about the affected portion until it is completely girdled. . . . [Then] the life of the tree is measured by a few years at best."

In the Bronx Zoo's 1906 annual report, Merkel reported, "It is safe to predict that not a live specimen of the American chestnut (*Castanea dentata*) will be found two years hence in the neighborhood of the Zoological Park." Murrill concurred, writing that same month that "My observations in the Bronx this season have led me to take a gloomy view. . . . The disease seems destined to run its course, as epidemics usually do, and it will hardly be safe to plant young trees while the danger of infection is so great."

Beyond the New York parks the American chestnut was a stalwart of the nation's forests, a fast-growing keystone species in some regions ranked by foresters as "the most valuable tree in the country," prized for its wood and its copious crops of nuts. But it was also a common and beloved shade tree found in city parks, urban woods, and yards everywhere east of the Mississippi. Henry David Thoreau was surprised to "have seen more chestnuts in the streets of New York than any where else this year" and amused "that the citizens made as much of the nuts of the wild wood as the squirrels. . . . All New York goes a-nutting: Chestnuts for cabmen and newsboys."

"Chestnutting" parties were a jolly autumn tradition. As soon as the first hard frost hit, the tree's prickly bur nut covers burst open, revealing velvety interiors cosseting shiny brown nuts the size of large marbles. "It was a great day," wrote Henry Ward Beecher, "when, with bag and basket, the whole family was summoned to go 'a chest-nutting!'" Generally, the nut-gathering season lasted three weeks after that first hard frost, and city parks and woodlands would come alive with young people and families traipsing about looking for the tawny-leaved trees. "Now and then the sound of falling nuts is heard as they drop from the trees. This is music in the ears . . . and [people] are soon busy gathering the glossy brown chestnuts. . . . [They] are brought home, where in the evening some are eaten raw, others have the shells slit and are then roasted or boiled, making some sort of chestnut festival."

Long before the American chestnut became stricken with the deadly canker, tree lovers were often pained to watch what Murrill decried as "savage hordes of small boys" assaulting the trees with projectiles to dislodge nuts, or ascending the huge branches and whacking with clubs at the foliage.

By 1908 Murrill was the reigning expert on this devastating chestnut canker. Two years earlier the Naturalist had settled near the zoo with his wife, Edna Lee Luttrell, in a large house with extensive grounds. There he "entertained visitors, enjoyed his many interesting neighbors, gave lawn parties, grew dahlias and peaches, took part in the local sports and walked for miles through the adjoining fields and woodlands, often accompanied by a friend or his faithful dog." At the botanical garden itself, a combination sylvan public pleasure ground and serious research institution (bankrolled by such wealthy plutocrats as Andrew Carnegie and J. P. Morgan), Murrill would ascend to assistant director the following year, a promotion he ascribed to the fact that he "spoke in public with ease, wrote a good letter, was accustomed to meeting people socially, made friends readily and could represent the institution on almost any occasion. Moreover, [he] was already well known both in this country and abroad as a fairly good scientist with a promising future."

At first, Murrill had advised establishing a half-mile buffer zone beyond the infection zone in which every chestnut tree (diseased or not) would be removed, believing this would slow the rapid advance of the canker. But as he walked the garden grounds in 1908, Murrill had no illusions left about the fate of the American chestnut. "This fungus is the most rapid and destructive known," he told a reporter. Mincing no words, he predicted the complete loss of the nation's billions of American chestnut trees. Here was modern America's first and worst natural ecodisaster, a forerunner of other tree plagues to come. Murrill thought the disease was a mutation, but where it originated no one truly knew, and he did not believe there was there any plausible cure.

Forest Park in Queens was soon filled with twenty thousand dead and dying chestnuts, and New Yorkers and those in nearby states, heartsick at losing their beloved trees, wrote in desperation to Murrill, the expert. "Some [letters] are from men, others from women; they are from owners of large estates, or occasionally from some one who had a single fine chestnut tree which he is distressed to lose." Henry Ward Beecher presumably spoke for many when he observed: "Nature was in a good mood when the chestnut tree came forth. It is, when well grown, a stately tree, wide-spreading, and of great size. . . . This darling old fellow is a very grandfather among trees. What a great open bosom it has! Its boughs are arranged with express reference to ease in climbing." But Murrill offered no hope: "The best we can advise is to have the trees cut immediately for the timber, which is valuable." He had no patience for the religiously inclined who blamed the nation's "sinfulness and extravagant living" for the demise of the chestnuts, nor for tree owners resorting to such quack remedies as boring holes in the trees and filling them with sulfur.

William Alphonso Murrill, "The Naturalist," poses with one of his collections shortly before joining the staff of the New York Botanical Garden. *(Courtesy of the LuEsther T. Mertz Library of the New York Botanical Garden.)*

It spoke to Murrill's great personal ambition that he—long a sentimentalist about trees—would privately view the chestnut canker not as a national tragedy but as "just another timely round in the ladder of luck he was climbing to fame and influence." Sadly, his "luck" only grew, for the delicate microscopic spores spread invisibly through the skies on the winds and were carried by birds and insects, efficiently affecting all the American chestnut trees in their ever-widening path. Murrill traveled forth to assess the kill zones, an arboreal grim reaper pronouncing the death knell for American chestnuts as far south as the nation's capital and as far north as Massachusetts.

At the New York Botanical Garden only a few dozen of the garden's 1,500 American chestnuts were still standing by 1910. Giant specimens with twenty-five-foot girths that had grown long before the colonial era in these woods had fallen. As Murrill told a visiting reporter, "We have lost many beautiful trees on the lawns and along the roads, large trees which had attained their maturity." He pointed out the "trees with bare, dead branches, some with the entire top of the tree dead, others having only a few twigs growing out of the lower part of the trunks and these beginning to wither."

American chestnut trees, like redwoods, regenerate naturally from their stumps, their sprouts growing over time into sturdy trees. Murrill noted that virtually all of the dying trees were second growth, explaining, "When the giants of the forests were cut down the chestnut trees we know grew from their stumps. These trees of the second generation were much weaker and less able to withstand disease than their progenitors." He suspected that "it may be 100 years before we can have chestnut trees again."

Among those seeking Murrill's advice was ex-president Theodore Roosevelt, who had returned to New York and his estate, Sagamore Hill, in June of 1910 after a long trip to Africa. When Roosevelt had a chance in the following days to walk through the woods on his estate, he became one of many Americans who found their chestnuts looking skeletal. Alarmed, he contacted the botanical garden. Murrill came forth on the railroad to Oyster Bay, walked the grounds with Roosevelt, and confirmed the obvious: Every American chestnut on Roosevelt's estate—about a thousand trees, some of great age—was dead or would be soon. Murrill gave the president the advice he always gave now—cut down and harvest the timber, which could be sold and used in so many ways: as "railroad ties, telegraph and telephone poles, interior decorations, cabinet making, coffins, and for general construction purposes."

A few weeks later, on July 16, 1910, a group of black politicians from the South arrived by carriage at the entrance to Sagamore Hill to find Roosevelt doing just as Murrill had suggested: cutting down a huge dead chestnut tree, its trunk three feet in diameter. "The ex-President was clad in knickerbockers, a negligee shirt and an old army hat. He was wielding the axe like a pioneer and the perspiration ran off him in rivulets." The estate's superintendent stood by with a rope around the trunk to pull the tree in the right direction when it toppled. The two men were so preoccupied they did not at first notice the visitors, who were taken aback by the sight of Roosevelt, who "swung his axe, the chips flying about him in a shower. . . . [Roosevelt,] drawing his sleeve across his dripping brow, allowed that he would receive them right there."

One man asked if he might have a few of the chestnut chips as a souvenir. "Help yourself!" urged Roosevelt. And "in a minute there wasn't a chip left. Each man had an armful and their pockets bulged." Their mission completed, the delegation departed, no doubt carrying the fungus to whole new stands of American chestnuts wherever they lived. Three years later, Roosevelt would write in his autobiography, "Alas! the blight has now destroyed the chestnut trees, and robbed our woods of one of their distinctive beauties."

A year after Murrill had first identified the fungus responsible for wiping out the American chestnut, the USDA appointed Haven Metcalf as the first chief of its new Forest Pathology Lab. By 1908 Murrill was openly feuding with Metcalf, a graduate of Colby College in Maine who earned his PhD studying with a pioneer of plant pathology at the University of Nebraska. When he first took the job, Metcalf was advising that individual trees could be saved by cutting off blighted limbs and spraying what remained. He also believed the wider war could still be won if states created broad quarantine zones where every chestnut was removed, thus creating fungus-free buffers.

In May of 1908 Murrill made his opinion clear: "There are absolutely no remedies against [the chestnut canker]. In spite of what Secretary Metcalf has stated on the subject. He has advised people who have chestnut trees to spray them and otherwise seek to stay the spread of the canker. This would involve needless expense." Murrill explained that he and Merkel had already tested these solutions, and the treated trees "died exactly as other trees had been dying in the woodlands where no treatment was given them."

Still, Metcalf felt government should not simply let the blight engulf other regions, above all the Appalachians, where huge chestnuts dominated the forests and served as a linchpin of the local economies of lumbering and nut gathering. He warned, "Unless this disease is controlled by human agency or unless some natural enemy appears to check the disease—and there is no hope of this—the chestnut tree will become extinct within the next ten or fifteen years." In the fall of 1908, Metcalf and his agency established a thirty-five-mile quarantine zone around Washington, D.C., after identifying and removing 1,014 infected American chestnuts.

Even as tens of thousands of trees were dying of the virulent canker throughout New York City and out on Long Island, Metcalf would report in the fall of 1911 that his buffer zone was working: "The disease has not reappeared at any point where eliminated and the country within a radius of approximately 35 miles from Washington is apparently free from the bark disease, although new infections must be looked for as long as the disease remains elsewhere unchecked." He felt sufficiently optimistic that in December he called upon the twenty-two states where the American chestnut was a native tree to jointly create other wide buffers to halt the spread of the canker.

Murrill was not by disposition diplomatic. At the botanical garden he had become ever more powerful, assuming the director's duties when his superior was away on long research trips to the tropics. The Naturalist now routinely supervised the garden's hundred-plus employees and any new construction, reviewed money matters with accountants, and twice a day "drove or walked about the [botanical garden] to see that all was well, and he often visited it at night." When Metcalf claimed his buffer zone was holding, Murrill promptly accused him and the USDA of deceit, telling reporters that on *his* recent visit to this very zone he had seen obviously infected trees, noting, "Some of the trees affected being of large size and apparently having suffered for several years from the attacks of the fungus. The trees most conspicuously affected there have been cut and burned so that the presence of the disease is not readily apparent."

No matter how many trees Metcalf and his allies cut down, said Murrill, "it would do no good as the disease spores would already have been distributed to adjacent trees. . . . There is not an instance known where an individual tree or a grove has been saved by the methods they propose for forests. . . . The effects of the disease are even more disastrous than was at first supposed. It has now been known since 1905 and has swept like a tidal wave over the woodlands about New York City, leaving not a single healthy native tree standing."

But, of course, given Americans' pride in their optimism and can-do spirit, Metcalf's call to action stirred Pennsylvania's governor John Tener to lead the charge, convening a conference to "consider the ways and means of preventing the spread of the chestnut tree disease . . . [through] some concerted action." And so it was that William Alphonso Murrill found himself walking up the marbled steps of the Beaux Arts statehouse, with its gilded dome, in Harrisburg, Pennsylvania, on the cold, gray morning of February 20, 1912.

Murrill was in a foul humor, as all around him other scientists, foresters, local commissioners, and politicians streamed into the rococo interior to discuss how best to save the American chestnut. For seven years now he had seen firsthand the swift death the fungus unleashed, destroying one of the nation's most fruitful and noble trees in city parks, private yards, and woodlands, forever altering and diminishing those landscapes. How, the Naturalist asked himself, did mere men expect to stop such a ruthless and efficient enemy? Murrill had nothing but contempt for this "bunch of Pennsylvania politicians" trying to "persuade the chestnut states to waste a lot of money on a project that was foredoomed to failure."

Governor Tener, a successful baseball player and then banker who had won office in 1910, set the tone when he opened the meeting by declaring, "It seems unthinkable that a disease of this character should have invaded so large an area and that no means of preventing the spread is yet at hand." The governor announced that Pennsylvania would lead the fight with a $275,000 war chest (close to $6 million in today's dollars) to pay battalions of tree cutters to cut down millions of American chestnuts and create a several-hundred-mile-long buffer zone between the state's southeastern sector (including Philadelphia, where the American chestnuts were already dying by the tens of thousands) and its still-healthy western forests and such cities as Pittsburgh. New York State's former agricultural commissioner adopted similar language, bellowing to the assembled conferees from twenty-two eastern states: "It is not in the spirit of the Keystone State, nor the Empire State, nor the New England States, nor the many other great States that are represented here, to sit down and do nothing when catastrophes are upon us." To just accept the calamity, to put up no fight, would be "un-American."

When the assembled politicians, foresters, and other men of action heard several scientists, including Murrill, testify that "frankly, we know of no way to control the disease" and that buffer zones were as doomed as the beloved chestnut, they balked at such defeatist attitudes. Pennsylvania's deputy commissioner of forestry, Irvin C. Williams, could not bear the thought of doing nothing. "Let us get busy all along the line and when we have utterly tried out every method and are absolutely and abjectly defeated, then it's time to talk about impossibilities."

Williams even denounced New York State for "allowing this scourge to go on. . . . [Had they] gone out and done what they could, this thing would probably not have come upon us." As the conference concluded, every single attendee but one voted to urge the states and the federal government to adopt Pennsylvania's plan to establish quarantine lines and buffer zones. The lone dissenter was William Alphonso Murrill. And so the battle was now joined.

CHAPTER FIVE

"Washington Would One Day Be Famous for Its Flowering Cherry Trees": Eliza Scidmore and David Fairchild

Descriptive catalogue of the Yokohama Nursery Co., Ltd., for 1905. *(Courtesy of the Henry G. Gilbert Nursery and Seed Trade Catalog Collection, Special Collections, USDA National Agricultural Library.)* Inset: Eliza Scidmore. *(Courtesy of the Washingtoniana Collection, Special Collections, District of Columbia Public Library.)*

Even as William Alphonso Murrill was prophesying the grim extinction of the beloved American chestnut, in Washington, D.C., travel writer and photographer Eliza Ruhamah Scidmore was in far more cheerful spirits, expectantly awaiting the arrival of thousands of rare Japanese cherry trees. Although few Americans had ever seen a single such tree, in the land of *The Mikado* Scidmore had strolled beneath miles of them, their ethereal blossoms creating an enchanted realm of billowing clouds of pink, rose, and white, all faintly perfuming the air.

Miss Scidmore was first smitten with cherries in 1885 when she was twenty-nine and visiting her older brother, George Hawthorne Scidmore, an attorney and U.S. counsel in the port of Yokohama, Japan. Eliza declared the blossoming trees "the most beautiful thing in the world." She loved both their dreamy arboreal loveliness and the ancient Japanese celebrations their blossoms inspired. Poetry contests, moon viewings, special kimonos, dances, songs, royal garden parties, torch and lantern festivals, and sake drinking all played their part in this ancient cult of tree worship.

And now, almost a quarter century after she first began lobbying the federal superintendent of public buildings and grounds to adorn the capital, the City of Trees, with several thousand of the exotic cherry trees, they were en route on a steamer crossing the Pacific Ocean. "No other flower in all the world is so beloved, so exalted, so worshipped," Scidmore explained to her fellow Americans, "as *sakura-no-hana*, the cherry-blossom of Japan. It is not only the national flower, but the symbol of purity, the emblem of chivalry . . . [whose] vernal celebration has been observed with unflagging zeal for at least two thousand years."

Scidmore was not the most obvious candidate to be seduced by exotic flowering trees. A tall, striking woman born in Iowa, raised largely in Washington, D.C., and educated for two years at Oberlin College, she already had gained fame as an intrepid pioneer of the Alaskan wilderness, the first "white woman" to voyage by boat into remote Glacier Bay, one of a series of expeditions recounted in her first book, *Alaska: Its Southern Coast and the Sitkan Archipelago* (1885). She returned from these Alaska expeditions with unusual souvenirs—a peak and a body of water named after her: Mount Ruhamah and Scidmore Bay.

But in 1885, as Eliza Scidmore's steamer coursed up lovely Yeddo Bay to Yokohama and Japan, she had watched Mt. Fuji slowly grow "from a pin point's size to a majestic peak" and began to feel wonderfully enveloped, observing, "Japan encircles one." Not long after Scidmore debarked to meet her brother the career diplomat, the cherry-blossom festivals began. First there were the blossoming trees themselves: "In the April sunshine—better still by moonlight, and best of all by the poet's pale, pure light of dawn," she rhapsodized, "the blooming cherry-tree is the most ideally, wonderfully beautiful tree that nature has to show, and its short-lived glory makes the enjoyment the keener and more poignant. Light radiates from it. There is a soft, pink electric glare overhead, beneath, and all around when one stands under branches laden with masses of flowers."

And the celebrations and customs and fetes were as lovely as the blossoms themselves. In Tokyo, Scidmore wrote, the major festival was to be found on the Mukojima, the "world-famous avenue of cherry blossoms along the river bank." There for two miles cherry trees planted on either side mingled their branches overhead, creating an arching canopy of dazzling pale-petaled light. "Under the great grove of cherry-trees teahouse benches are set close," she wrote, "and there the people lunch and dine and sup; and though the sake flows freely, the most confirmed drinker is only a little redder, a little happier, a little more loquacious than the rest. . . . Tattered beggars gaze entranced at the fairy trees, and princes and ministers of state go to visit the famous groves. Bulletins announce, quite as a matter of course, that Prince Sanjo or Count Ito has gone to Nara or Kioto, a three days' journey to see the blossoming trees." Families in their finery gathered beneath the trees for picnics, savoring special bean-paste treats, while boats full of revelers singing cherry-blossom songs floated past upon the river. "Then," said Scidmore, "every one with any soul or a spark of poesy in him writes poems to the lovely flowers, dashing off a few characters . . . , twists the end of paper into a string, and ties it to the nearest branch, where it waves honorably with the blossoms it celebrates."

Farther south, in Kyoto, one special, enormous three-hundred-year-old cherry tree—named *yozakura*, or Cherry Blossom of the Night—was so revered that its worshipers "have built bonfires and strung lanterns around it to further adore it by night," which seemed only to heighten its loveliness. The "raining" down of the cherry blossoms offered its own poignant commentary on the fragility and impermanence of beauty and life.

When Scidmore, now a devoted votary of *sakura*, returned home later that year to Washington, D.C., she began lobbying the federal superintendent of public buildings and grounds (SPBG; always a graduate of West Point) to plant Japanese cherry trees. The capital was then in great tree-planting mode. "It was in the first year of President Cleveland's first administration," Scidmore would later recall, "and I addressed myself to Marshall A. A. Wilson with my photographs of Japanese cherry trees and the plea, that since they had to plant something in the great stretch of raw, reclaimed ground by the [Potomac] riverbank, since they had to hide those old dump heaps with something, they might as well plant that most beautiful thing in the world—the Japanese cherry tree.... He listened amiably and sent me on. In the succeeding administration of President Harrison, I took the pictures of the Japanese cherry trees to Col. Ernst, the [new] superintendent of public buildings and grounds, and made a plea for something of Spring beauty down in the waste space at the foot of Seventeenth street. He listened patiently and seriously to my fairy tales. Nothing happened."

Over the next ten years Scidmore became increasingly famous as a well-traveled correspondent for *National Geographic* and the *Chicago Daily Tribune* and the author of such books as *Jinrikisha Days in Japan* (1891) and *China, the Long-Lived Empire* (1900). She became the first woman board member of the National Geographic Society. Admired for her "superb physique... and super abundant vitality and health," she was prized as a guest for her "inexhaustible fund of anecdote and reminiscence of interesting people." In 1895 Scidmore socialized in London with the Princess of Wales, and in 1900 made the acquaintance of Helen Taft, wife of Philippines governor general William Howard Taft, in Yokohama.

Whenever Scidmore landed back in Washington, D.C., from her far-flung voyages, if a new superintendent of public buildings and grounds (SPBG) had taken charge, she would renew her crusade, armed with her photographs of Japan's cherry-blossom festivals. She would rhapsodize that "like Tokio," the American capital could welcome spring with "avenues of cherry-blossoms along the river-bank." She would urge the superintendent to envision the pleasure of festive crowds wandering amid the dream-like froth of cherry blossoms, the waters of the Potomac mirroring the poetic clouds of flowers. When Cleveland returned for his second term after President Harrison's tenure, Scidmore met with General John M. Wilson, the new SPBG. "That delightful person had ideas of his own," she wrote,

> and he swept the pretty pink pictures aside and turned his grim official countenance, his practical engineer's eye on me and said, "Yes! And when the cherries are ripe we would have to keep the park full of police day and night. The boys would climb the trees to get the cherries and break all the branches!"
>
> "But these cherry trees do not bear any cherries. Only blossoms," I ventured.
>
> "What! No cherries! Huff! What good is that sort of cherry tree?"

The SPBGs under Presidents McKinley and Roosevelt proved equally unreceptive. "So fixed was the type," wrote Scidmore, "so standardized their minds on park planting, so much they resented the unknown and the untried, that I gave up the quest, turned from the deaf ear and the hard heart, the wooden routine, of the whole lot."

By 1900 Scidmore had added to her considerable CV the role of war correspondent in China during the Boxer Rebellion. In 1904 the *Chicago Daily Tribune* engaged her as its "special correspondent in the far east. . . . [She would be] traversing Manchuria and northern China, and making her way . . . to the French colonies." If war were to break out between Japan and Russia "she will abandon her present itinerary and proceed to the firing line."

April of 1904 found Scidmore again in Tokyo, reporting on the ongoing war and "Japan's Carnival in Cherry Season" and "Games, Sports and Picnics for Old and Young While Famous Trees Are Blooming." Once again she reveled in strolling along the "cherry blossom resort of all the people . . . the Mukojima on the river bank." On the Sumida River itself, schoolboys engaged in spirited boat races. Friends and fans crowded into the riverside teahouses to yell and cheer. "Then," wrote Scidmore, "all the thousands went on with their cherry viewing," the women dressed in their cherry-blossom kimonos, young geisha in training playing their samisen. Among the men the sake flowed more freely, the singing grew more spirited. Some families and groups of friends picnicking under the clouds of blossoms were reciting famous poems and composing their own efforts on long white strips of paper to tie onto the branches. *Hanami*, or cherry-blossom viewing, was a lively affair, but *sakura* poems by ninth-century literary immortals such as Lady Ki no Tomonori reflected more somber thoughts: "Cherry blossoms / on this quiet / lambent day / of spring, / why do you scatter / with such unquiet hearts?"

In 1908, after almost a quarter century of rejections, Scidmore finally acquired a powerful ally in the form of U.S. Department of Agriculture plant explorer David Fairchild, the man who introduced American farmers to soybeans, mangoes, nectarines, bamboo, and dates (to name just a few crops). Born in 1869 in Lansing, Michigan, where his abolitionist father was a professor of literature, Fairchild moved at age ten with his family to Manhattan, Kansas. There his father became president of Kansas State College of Agriculture, which Fairchild attended before going east with his aunt and his uncle, a well-known mycologist, for graduate study in plant pathology at Rutgers University.

In July of 1889 Fairchild, just nineteen, was thrilled to join the U.S. Department of Agriculture as an assistant in the Office of Plant Pathology. There he began to make a name for himself investigating "the various rots of the sweet potato." After four years Fairchild, a studious, awkward young man of swept-back dark hair and thick glasses, resigned so he could travel to Europe to pursue further studies in plant pathology (this time seaweed) at the famous laboratory at the Naples Zoological Station.

It was during the stormy crossing to Italy on the S.S. *Fulda* that Fairchild first met Barbour Lathrop, "the star passenger of the ship and a friend of the Captain. He was very good looking, standing there in the sunlight, quite the distinguished man of the world." Lathrop, always to be found with a Bohemian Club cigarette in a Turkish Weichsel cherrywood holder in one hand and a novel in the other, was an imperious, blunt-spoken millionaire who at age forty-six had already been around the world eighteen times. A brief career as a San Francisco newspaperman had stoked his curiosity about everything. Upon inheriting the family fortune, he had the means and connections to satisfy his wanderlust.

Lathrop first noticed plants during a trip to Peru in the 1880s, when he observed that his Andean mountain guides managed their heavy loads in part by chewing a plant called coca. Lathrop collected its leaves and dispatched them to the California Academy of Sciences for analysis and study, a request that was ignored. Now, as Lathrop trekked through Japan or Siam or Australia, he began to appreciate "how valuable many fruits and vegetables and trees were to many countries, and he began to wish that they could also be introduced into the United States." Yet when he sent these new varieties back to the U.S. Department of Agriculture, they too went unnoticed.

When young David Fairchild confessed to a Harvard academic on board the *Fulda* his ambition to study termites at the Java Garden in Bogor, Indonesia, the professor insisted he meet Barbour Lathrop, who had already been twice to Java to hunt rhinoceros. Lathrop had little patience for this gangly young man, who showed no interest in important matters like botany and the fruits and vegetables of the world but instead spoke glowingly of the *Atta* ants of Brazil, which grew and ate mushrooms. Fairchild and Lathrop went their separate ways until a month later, when Lathrop materialized in the Naples laboratory where Fairchild was deeply, happily immersed in his studies of seaweed cells.

"I've come to tell you that I've decided to invest a thousand dollars in you," Lathrop announced. "I think you are a good risk. I'm not interested in you personally, understand, it's just an investment. When you are ready to start for Java, let me know and I'll send you a letter of credit for the amount. When do you think you want to go?" Another two years would pass, during which Fairchild prepared by studying botany and zoology at several German universities, before he finally debarked in Java in the spring of 1896. He soon found himself riding a small train up into the mountains to the botanical garden, where he spent the next eight months absorbed in his studies of termite fungus gardens.

In early December Lathrop arrived and was not pleased. He ordered Fairchild to wrap up his studies, as they would shortly be departing together for a trip around the Pacific. Lathrop had a mission. "As the new year of 1897 began," wrote Fairchild, "I had promised Mr. Lathrop that I would take up the study of plants useful to man and, together with him, find a way to introduce their culture into America. . . . From that time on I began to pay attention to the economic plants around me, and to work with Mr. Lathrop."

Fairchild returned that summer to Washington, D.C., a confident, magnetic cosmopolitan, a scientist with an infectious enthusiasm for the "great green world of nature," and persuaded the USDA to set up an Office of Plant Introduction. As he worked to establish it, Lathrop swooped in once again, surveyed the situation, and declared Fairchild no more fit for the task "than I am to run a chicken farm. You have no contacts with the rest of the world. You can't do that sort of thing by correspondence." He proposed to take Fairchild on a two-year tour around the world to meet top scientists, plantsmen, and agricultural officials and to visit major botanical gardens and experimental stations.

In spring of 1902 Fairchild arrived in Japan to learn he had just missed the cherry blossoms. To assuage his disappointment, Dr. Matsumura, head of the Tokyo Botanical Garden, introduced him to a Mr. Takagi, a "great authority on flowering cherries . . . [who] received me most cordially and produced his water-color sketches of the cherry blossoms . . . and explained the drawings to me one by one. I have rarely been so thrilled, for I had no idea of the wealth of beauty, form, and color of the flowering cherries." While he had seen one or two weeping Japanese cherries in Washington, trees brought home by American naval officers decades earlier, they were nothing compared with these. Fairchild ordered thirty Japanese cherry trees, which were duly shipped to the USDA's garden in Chico, California, where most of them died in the "burning summer sun," though a certain number of offspring were propagated and sent elsewhere.

By 1905 Fairchild was back in Washington, D.C., where he soon came to know Charles Sprague Sargent. Sargent was always looking to add new specimens to the Arnold Arboretum's herbarium and to identify new trees that could thrive in the Boston climate, all of which he could then dispense to his network of botanical gardens, arboretums, and commercial nurseries. While Fairchild and the USDA were focused on introducing new kinds of practical foods and crops, Sargent was looking for ornamentals and city-tolerant trees of any kind.

The years had not made Professor Sargent any less acidic. "I had learned to enjoy the Professor's sarcasm," remarked Fairchild years later, "and the pitfalls he set for my ignorance about some species with which he was familiar, even though his remarks often seemed rather severe and almost brutal. Sargent's points were usually well taken when pertaining to ornamental plants, but . . . he showed an indifference to economic species which was often annoying to me. . . . Sargent [believed] it was impossible to get people to eat new foods. In fact, he once told me that it was useless to try to introduce any new vegetable as it simply could not be done." Fairchild, a plant pathologist with the soul and tender heart of a poet, who cherished the beauty of the Japanese cherry as much as he did a useful new kind of Chinese wheat, was deeply offended by one of Sargent's sarcastic remarks, "which I never forgot, to the effect that we botanists in Washington were interested only in things to eat."

Sargent and the Arnold Arboretum, not surprisingly, were enthusiastic about Japanese ornamental cherries and had first planted seeds during the 1890s from the Imperial Botanic Garden in Tokyo and also from Dr. William S. Bigelow, a Boston physician who had spent seven years in Japan documenting its culture and amassing a vast art collection. By the time Fairchild had fallen under the spell of the cherries, the arboretum had "the oldest trees in cultivation outside Japan," including the *higan* (*Prunus subhirtella*) and one named Sargent's cherry (*Prunus sargentii*).

Fairchild, meanwhile, had fallen in love. In April of 1905 he married Marian Bell, the wealthy daughter of inventor Alexander Graham Bell. Fairchild was now no ordinary bureaucrat but a man of consequence at the USDA and one half of a Washington power couple. The Fairchilds bought thirty-four sylvan acres at the very end of Connecticut Avenue, ten miles beyond the city, where they began building a simple wooden house, to be replaced a few years later by a Japanese-influenced stucco home open to their gardens and the surrounding woods of scrub pines, tulip poplars, oaks, and dogwoods. Of this hectic time Fairchild had two strong memories: an evening with Marian when "I saw the moon through the branches of a drooping cherry tree in full bloom, as the Japanese love to do, and a tea party which [childhood friend and USDA colleague] Charles Marlatt gave, in Japanese style, for an old double-flowered cherry tree in his yard."

In 1906 Fairchild wrote to a Japanese nurseryman in Yokohama and ordered for his new grounds 125 Japanese cherry trees of twenty-five different varieties. They arrived "beautifully baled with bamboo staves," he wrote. "It was all an experiment in those days. Doubt of their hardiness had been expressed by so many horticultural experts. . . . On a perfect spring day, Marian and I motored out to open the crates of cherry trees, which had cost us, landed there, about ten cents apiece. . . . Fastened on each one was a little cloth tag, on which was painted in Japanese characters the Japanese name with its English equivalent. The planting of these trees was the real beginning of our experiences on the place which Mr. Lathrop named for us, 'In the Woods.' . . . I tried to coddle the [trees] by planting them in sheltered spots." At just this time the Fairchilds "acquired an unusual servant, a delightful Japanese peasant boy named Mori" as their gardener, and he helped plant and nurture the cherry trees while also translating into English all those little tags identifying the new cherry tree

David Fairchild and his wife, Marian, with their cherry trees in later years at their estate in Washington, D.C. *(Courtesy of the Fairchild Archives, Special Collections, Fairchild Tropical Botanic Garden.)*

varieties, including the Tiger's Tail, the Milky Way, and the Royal Carriage Turns Again to Look and See.

Once Fairchild could see that his Japanese cherries were thriving in this foreign climate, "Marian and I wanted to do something towards making them better known in Washington." Fairchild knew Eliza Scidmore and in the spring of 1908 invited her to enjoy their new *sakura-no*, or cherry field, blooming at In the Woods. Mori had cleared places for the imports here and there among the native pines and cedars, and the young trees were flowering.

Eliza Scidmore was now fifty-two years old and by then had spent twenty-three years promoting—to no avail—planting Japanese cherry trees along the Potomac. She explained to the Fairchilds that she had devised a new idea to break through the long-standing wall of official resistance. Why not raise money from other travelers who had seen the Japanese cherries, enough to plant one hundred cherry trees each year, so that after ten years "there would be a great showing in Potomac Park—a rosy tunnel of interlaced branches, a veritable Mukojima along the river's bank?" Fairchild liked her plan and then shared how he and Marian planned to make the cherry trees better known. He showed her a sizable cleared field where Mori had "heeled in" hundreds of two-foot-tall newly arrived Japanese cherry trees of the hardiest "drooping" variety. The following day, Arbor Day, he would be presenting each school in the District of Columbia with one of these trees for its schoolyard. Might she like to attend the ceremony where he would be speaking and showing slides at the Franklin School?

The next morning, Friday, March 27, Fairchild watched in delight as eighty-three Washington schoolboys came single file through his forest, led by Miss Susan B. Sipes, the school system's "Nature Study" teacher. Fairchild, who kept a supply of magnifying glasses for children so they could study nature up close, was pleased that the boys were able to be there when all the sweet hepaticas and dogtooth violets were blooming, bursts of subtle color pushing up through the dead leaves on the forest floor. The schoolboys assembled at the field, each one armed with a yard of twine and a two-foot square of burlap. "Each boy was shown how to dig and plant his tree; we gave them a little talk on tree culture, and then they went back on the special car which the Street Car Company had provided for their transportation," wrote Fairchild.

That afternoon at the Franklin School, Fairchild presented a slide show with views of "the new, unplanted Speedway, which I threw on the screen at the close of the lecture saying that the Speedway would be an ideal place for a 'Field of Cherries,' and further quoted Miss Scidmore, the great authority on Japan, who was there as the distinguished guest of the occasion." Fairchild, being a man with connections, could count on his friend the editor of the *Washington Star* to send a reporter, whose Arbor Day story quoted Fairchild predicting: "Washington would one day be famous for its flowering cherry trees."

During the following months Scidmore continued to promote the idea of raising money from the many Americans who had visited Japan to buy a gift of one hundred Japanese cherry trees. She took heart when William Howard Taft was elected president, for she had met the First Lady and knew "Mrs. Taft had once lived for several months in a pretty garden and bungalow on the bluff at Yokohama, had spent a Summer at Chiunzenji [before taking up her husband's post in Manila], and knew Japan." Encouraged by Fairchild, Scidmore wrote a letter to Helen "Nellie" Taft some weeks after the president took office in early March, asking the new First Lady's "approval and aid in getting an avenue of Japanese cherry trees planted in Potomac Park."

Mrs. Taft, a native of Cincinnati, Ohio, turned out to be a kindred soul. While she, like Fairchild, had arrived in Japan too late to experience the extravaganza of the cherry trees, she had been enchanted by its people and culture. On April 7 Mrs. Taft wrote back to Scidmore: "Thank you for your suggestion about the cherry trees. I have taken the matter up." The new First Lady accordingly informed the superintendent of public buildings and grounds, Colonel Spencer Cosby, of her desire to plant Japanese cherry trees. She further noted that the White House landscape gardener had managed to locate at Hoopes Brothers and Thomas Co. in West Chester, Pennsylvania, a grand total of ninety such trees (price $106) to line the Speedway. In keeping with tradition, Colonel Cosby responded dismissively, stating, "I have planted elm trees along that road." Ordered Nellie, "Take them up then."

David Fairchild had also sprung into action, offering fifty additional flowering cherries to Colonel Cosby, writing in an April 4 letter (which included several hand-colored photos of Japanese cherry blossom scenes), "You know Mrs. Fairchild and I are quite crazy about Japanese cherry trees and have made something of a study of them. . . . I don't know whether you have been in Japan or not but in any case the beautiful photographs which I sent you will show you, or remind you, of how beautiful the roadways lined with cherry trees are in that country of floral appreciation."

On April 12 Mrs. Taft met with Scidmore to further discuss plans. By coincidence, Dr. Jokichi Takamine, the chemist who had made a $30 million fortune licensing his discovery of purified adrenaline to American drug companies, was visiting from New York City along with Japan's New York consul general, Kokichi Midzuno. Dr. Takamine lived in Manhattan with his American wife, had studied in Glasgow, and now said to Scidmore in his Dutch-accented English, "Will you find out if Mrs. Taft will accept 1,000 cherry trees for her Mukojima. In fact, I had better give 2,000 trees. She will need them to make any show." Mr. Midzuno diplomatically suggested the trees come not from this wealthy private expatriate citizen of Japan but from the City of Tokyo, whose mayor, Ozaki Yukio, was happy to assist. By December Colonel Cosby was writing to formally accept the "magnificent gift" of two thousand Japanese cherry trees from the "Capital of the wonderful Japanese Empire to the Capital of this Republic."

Scidmore's plan inevitably became entangled with politics. U.S. relations with Japan had become strained in recent years as labor leaders on the West Coast complained about Japanese immigrants displacing white Americans from their jobs. In 1907 riots had broken out in San Francisco. Then President Theodore Roosevelt had persuaded the Japanese to accept the so-called Gentlemen's Agreement, whereby the Japanese would restrict emigration, while the United States would not enact exclusionary laws.

President Taft took pleasure at the offer of the two thousand cherry trees, argues historian Philip J. Pauly, because "a field of Japanese flowering cherries in Washington was preferable to a settlement of Japanese working families in California." The Japanese, in turn, were happy to be planting their national symbol in the American capital. As one letter writer to the *New York Times* explained, the cherry blossom "is a symbol of the very soul of the manhood of Japan. . . . That is why people flock to the cherry gardens in crowds, in order that, while gazing upon the outward beauty, their souls may be baptized afresh with a baptism of the real Japanese spirit. There is nothing in American life to illustrate just what the cherry means to the Japanese people. But suppose we had some symbol of nature that would embody all that Plymouth Rock, all that the Declaration of Independence, all that the Emancipation Proclamation means, of liberty, patriotism, union."

The United States officially accepted the gift, and Eliza Scidmore traveled to Japan, where her brother was now U.S. consul in Nagasaki, to personally supervise the selection and shipping of the trees. David Fairchild was concerned that Scidmore "wanted bigger trees than Marian and I had imported to make a show as soon as possible, but I cautioned her not to order large trees, because of the difficulties in making them live." As for Fairchild, he did his part by offering "the services of [the USDA's] importing agent" in Seattle and arranged for his Office of Plant Introduction to handle the shipment across the continent into Washington. On December 10, 1909, the two thousand flowering cherries arrived in Seattle, where they were given a preliminary inspection. On Christmas Eve they were loaded onto three refrigerated railcars for their two-week journey east.

Charles L. Marlatt, Fairchild's boyhood friend from Kansas and acting chief of the USDA's Bureau of Entomology, had been trying with no success for almost a decade to get Congress to enact a bill entitled "Inspection of Nursery Stock at Ports of Entry of the United States." Fairchild and Marlatt had grown up together on the prairie, and like Fairchild, Marlatt had arrived at the USDA in 1889, taking a post as an assistant in the Bureau of Entomology. But during the years when Fairchild was studying abroad or voyaging to exotic climes, Marlatt was out in the field in California and Texas, investigating the foreign pests that were attacking American cotton farms and fruit orchards. Most recently he had seen the devastation wrought in New England by the gypsy moth, an insect that had been imported with dreams of establishing a silk industry. While he became progressively more alarmed at new alien threats to American crops and forests, the nursery industry repeatedly thwarted his efforts to introduce quarantine legislation. Fairchild himself was not particularly sympathetic to these efforts.

On January 6, 1910, the two thousand ornamental Japanese cherries from the City of Tokyo arrived in Washington, D.C., and were immediately transported via horse-drawn wagons to the USDA's garden storehouses on the grounds of the Washington Monument for unpacking and storage. Marlatt had previously notified Colonel Cosby that the trees would have to be thoroughly inspected upon their arrival "to ascertain whether they are free from insect pests new to this country." As Fairchild and four USDA inspectors eagerly unpacked the wooden boxes, ten trees per bamboo-lined packing case, they were dismayed to find how severely the root systems had been pruned prior to shipping.

For the following week Marlatt's three inspectors methodically subjected each of the young cherry trees to the kind of careful scrutiny that Marlatt believed necessary for every foreign plant entering the United States. The fourth inspector, mycologist Flora Patterson, searched for any signs of worrisome fungi via microscopy. On January 19, 1910, Marlatt wrote to the secretary of agriculture: "I regret to report that the condition of the trees included in this importation is the worst, from the standpoint of insect pests and root gall, of any imported stock I have ever examined. The difficulty is largely due to the fact that very old stock has been sent with the object being to give large, showy trees." His inspectors found the trees to be badly infested with two kinds of scale, a wood-boring moth that posed a threat to the peach industry, root and crown gall, and a number of lesser insects. Moreover, most of the trees were in poor health. The entire shipment should be destroyed by burning as soon as possible. . . . "The destruction of the entire shipment," advised Marlatt, was "imperative."

"I had been worried about the trees," an appalled Fairchild would write later, "fearing they might prove too large [to grow and thrive], but I had not dreamed of any difficulty with the Quarantine authorities; this was in the early days of the existence of the Quarantine and it had not yet assumed the role which it now plays. The crates arrived . . . and immediately came under the inspectors' eyes, with the result that almost every sort of pest imaginable was discovered, and I found myself in a hornets' nest of protesting pathologists and entomologists, who were all demanding the destruction of the entire shipment." The matter went all the way to President Taft, who had little choice but to accede.

Fairchild endured numerous sleepless nights in the wake of this calamitous news. "Ghastly, as it seems," he lamented, "the trees were all [to be] burned." As if this were not bad enough, Commander Richmond Hobson chose this moment to "speak on the floor of Congress in a derogatory way of the Japanese. This added to the disgust of Miss Scidmore, the annoyance of Major Cosby . . . and the astonishment of my Japanese friends. . . . My only comfort was the knowledge that the trees had been so large, and their roots had been so cut, that I felt sure the greater number of them would have perished in the raw soil of the Speedway."

On May 17, 1909, at age forty-seven, First Lady Helen Taft had suffered a serious stroke that left her unable to speak properly for almost a year. The White House concealed the severity of her condition, and she saw no one but close family and staff as she slowly recovered. By Christmas of 1909, not long before the Japanese cherry trees arrived, she gamely began making public appearances at White House events, though not in a role that required her to say anything. Her condition was a severe blow to her intentions to be an activist First Lady. The cherry trees were part of a larger ambition to create in Potomac Park a beautiful space "where all Washington could meet, either on foot or in vehicles, at five o'clock on certain evenings, to listen to band concerts and enjoy such recreation as no other spot in Washington could possibly afford."

By January of 1910 Mrs. Taft had been well enough to write to a friend, "I am delighted to think that there is a chance of the cherry trees arriving soon. I am anxious to have them set out as soon as possible so as to secure their successful growth next spring." She had no comment when the long-awaited gift was condemned.

Charles Marlatt made a highly public example of the infested Japanese cherry trees. On January 28, 1910, Charles Henlock, the chief White House gardener, rode over to the Washington Monument grounds and there, next to the USDA greenhouses, directed the building of many large bonfires to consume the cherry trees, along with all their bamboo wrappings and crates. All day the huge fires roared. The *New York Times* observed, "We have been importing ornamental plants from Japan for years, and by the shipload, and it is remarkable that this particular invoice should have contained any new infections." But Marlatt had made his point, though it would take him another two years to get his bill passed. A week later a *Washington Post* article decried the "weeping and wailing and gnashing of teeth among the esthetically inclined when the order went out that the

2,000 cherry trees" had to be destroyed. But those trees harbored "a menace, which in a few years, might have swept over the entire country."

Oddly, the USDA did not in fact destroy every tree. Some twenty ornamental cherries were saved and sent to grow in one of the agency's nearby experimental plots. The *Washington Star* wondered if these trees would be watched over by "an expert entomologist with a dark lantern, a butterfly net, cyanide bottle and other lethal weapons placed on guard, to see what sort of bugs develop."

As for Scidmore, her disgust was only compounded when her lead story in the March issue of the prestigious *Century Magazine*, "The Cherry Blossoms of Japan," intended as a triumphant tale of a horticultural coup, became instead an obituary of sorts. The editors appended a small footnote on the front page of her article, noting, "As this paper goes to press it is announced that the experts in Washington have found it necessary to destroy the cherry-trees which have arrived, as a protection against certain kinds of infection attaching to them. This disappointment, which will be national in extent, does not impair the graciousness of the gift or the interest attaching to this article."

Scidmore consoled herself with the fact that the ninety Japanese cherry trees from the New Jersey nurseries, planted the previous spring at Mrs. Taft's behest, bloomed profusely in their very first spring in Washington, serving as floral diplomats in the whole debacle. When Mrs. Taft saw the trees blossoming for the first time during a drive with the president, she "actually clapped her hands in delight," reported a White House aide.

Tokyo mayor Ozaki Yukio made light of the public auto-da-fé, telling the U.S. ambassador, "To be honest about it, it has been an American tradition to destroy cherry trees ever since your first president, George Washington! So there's nothing to worry about. In fact, you should be feeling proud." Determined to present the gift again, Yukio ordered that "cherry tree seeds should be planted in sterilized beds under the jurisdiction of the Ministry of Agriculture and Commerce. In this way we could nurse a new batch of cherry trees. After three years the saplings were ready." These, in turn, were grafted onto special understock. And just to show there were no hard feelings, for this shipment Dr. Takamine and the Japanese had prepared six thousand trees, half from the City of Tokyo for

Washington, D.C., and half as a gift from the Japanese Society of Tokyo for New York City.

In late December of 1911 this second shipment of cherry trees was sent via the Japanese steamship *Awa-Maru* to Seattle. The trees arrived in early March and were repacked into refrigerated railcars, unloaded March 26 in the capital, and driven through a spring snowfall to the USDA's Propagation Gardens at the Washington Monument grounds. Once again Marlatt's inspectors descended upon the cherry trees, but this time Secretary of Agriculture Wilson was able to inform Colonel Cosby on March 28, 1912, that "so far as the inspection has gone, the trees seem singularly free from injurious insects or plant diseases." The White House, presumably not wishing to allow Marlatt further opportunity to identify some dread pest, immediately scheduled a ceremonial tree planting.

The very next afternoon, on Wednesday March 28, 1912, Mrs. Taft, largely recovered from her stroke, stepped into the chauffeured presidential automobile and motored to the Speedway, where she stepped forth at the bottom of Seventeenth Street, not far from the Tidal Basin, and walked toward a grassy area west of the soon-to-be-dedicated John Paul Jones Memorial. The snow and bitter weather had melted away, and the day was breezy, crisp, and sunny. Eliza Scidmore and Colonel Cosby were fittingly present as the First Lady greeted the recently arrived Japanese ambassador, Count Sutemi Chinda and his wife, the Viscountess Iwa Chinda.

Two ten-foot-tall cherry trees were awaiting them. Mrs. Taft "planted the first tree with her own hands," reported Cosby in a letter to Tokyo mayor Ozaki. "At the invitation of Mrs. Taft, the Viscountess Chinda planted the second tree. I take pleasure in sending you by this same mail, in separate package, a blueprint on which I have indicated by small red crosses the locations in which most of the trees have been planted in Potomac Park. . . . Twenty Gyoiko [cherry trees] will be planted in a few days in the grounds of the White House." These latter trees were prized for their unusually large fifteen-petal flowers, which opened a yellowish green with darker green stripes and flushed with pink as they came into full bloom. (*Gyoika* was the ancient name of Japanese emperors' yellowish green costume.)

Finally, after twenty-seven years, Eliza Scidmore had her wish—the capital of the United States would soon have a waterfront roadway surrounded on both sides by groves of Japanese cherry trees that would in time create an enchanted realm of billowing clouds of pink, rose, and white, all for the pleasure of ordinary citizens. And yet . . . Colonel Cosby had, in her opinion, thoroughly botched the plantings. For one thing, he had planted only half the trees, relegating the remaining 1,500 to a nursery "in order to replace any others which may not thrive." Those he had planted had been crammed in so close together that any artful effect was spoiled. By mid-June Scidmore was setting sail from San Francisco to Korea, where she would rendezvous with her brother, now posted there as consul general.

Again she waited, and when Woodrow Wilson won the presidency, the following spring, in 1913 Scidmore made the now-familiar journey to the office of the superintendent of public buildings and grounds, this time to see the new occupant, Colonel Harts. "I went to his office on almost the first day," she would later write, "to beg him to thin out the crowded trees around the basin and to rescue the pretty fledglings relegated to the unseen nursery at the corner of Seventeenth Street and B street. The crowded hedges around the basin were thinned out one-half at once, and during that and the following spring the 1,500 leftovers were set out along the road beyond the railroad bridge down to the point and back along the inner basin facing the wharves. The twelve varieties were carefully chosen for successive blooming, so that cherry blossoms can be seen from early in April far into May."

As Fairchild had predicted, Washington soon became famous for its flowering cherries. Each year, as the trees grew larger and the clouds of blossoms more lovely, the crowds swelled, with Scidmore noting that "the elect" opted to view them "by moonlight and by dawn." It not only was a lovely arboreal fairy tale with the happiest of endings but also marked the beginning of the American love affair with the ornamental cherry tree, which would become an urban staple, gracing city streets, parks, and yards all across the country.

CHAPTER SIX

"I Knew That There Were No Roads in China": Plant Explorers Frank Meyer and E. H. Wilson

USDA plant explorer Frank Meyer in Wutai Shan, Shanxi, China, February 26, 1908. *(Courtesy of the Isabel Shipley Cunningham Collection on Frank Nicholas Meyer, Special Collections, USDA National Agricultural Library.)*

In mid-July of 1911 William Alphonso Murrill and a visiting reporter had walked up through the forest of the New York Botanical Garden to where the Bronx River flowed through a steep gorge. Before the arrival of the American chestnut canker, the surrounding woods would have been afroth in the summer warmth with creamy chestnut flowers, teeming with insect life. This, however, was a more sober occasion, for the Naturalist wanted to bear witness to the garden's two surviving American chestnut trees, all that remained of the 1,500 magnificent specimens that had so recently towered above the hemlocks and everywhere defined the garden's landscape.

The reporter duly memorialized the two gigantic survivors, writing, "They are fine old trees, the last of the garden's beautiful chestnut groves, seven feet in diameter and over twenty-two in circumference. A large proportion of the garden [chestnut] trees were large, more than six feet in diameter. The trees were in beautiful clumps on ridges in the garden, and are an irreparable loss." Murrill hastened to remind him that this pair of chestnuts was also infected and condemned to be chopped down that winter. They would bear nuts one last time in the coming fall. Murrill still had no doubt that the blight would in time wipe out all the nation's American chestnuts, save a few isolated cases. "It may be 100 years before we can have chestnut trees again," he told the journalist.

The origins of American chestnut blight were still a mystery, and the fungus continued to spread as virulently as ever. At the end of 1912 a New York City forestry crew clearing dead trees in Fort Washington Park in northern Manhattan by the Hudson River took down a stricken chestnut. A passerby who stopped to marvel at the fallen giant learned from "the overseer, a most intelligent man, one of the city's expert arboriculturists," that the specimen that lay before them had been, "in all probability, the largest tree on Manhattan Island." They estimated the old tree's circumference as seventeen feet when measured three feet up the trunk, where its massive limbs began to fork. The overseer regretted that no reporters were present to document this historic loss, and so the passerby paid tribute in a letter to the editor of the *New York Times*.

Even as the Pennsylvania Chestnut Tree Blight Commission had unleashed its martial game plan of destroying all American chestnuts in a wide swath to save those outside its quarantine zone, USDA plant pathologists C. L. Shear and Haven Metcalf (who had already tangled with Murrill over his quarantine zone around the District of Columbia) had come to suspect that the deadly fungus had slipped into America on nursery stock imported from Asia. At their behest David Fairchild, whose title was now Agricultural Explorer in Charge, Foreign Explorations of the USDA Plant Introduction Unit, dispatched a letter in early May of 1913 to Frank Meyer, his top plant explorer in China, with an unusual request: Would Meyer look for incidents of the disease there?

Meyer knew where to find Chinese chestnut trees (a species distinct from the American), for some years earlier he had sent back a quantity of its nuts for planting. Fairchild reported that the Chinese chestnut "had grown rapidly and trees of it had been distributed. I had it growing [in 1913] at 'In the Woods,' and expected it to bear in a year or so." Although he acknowledged that Meyer was not a trained pathologist, he felt "confident that he would find the disease if it were common on Chinese trees. Doctor Shear sent Meyer a fragment of bark with a description of the disease and I instructed him to send back samples of Chinese bark in case he discovered any which seemed infected with the blight."

Meyer was himself one of a rare Western species: professional plant hunters active in Asia. He followed in the footsteps of Dutch East India Company physician-botanist Engelbert Kaempfer, who in 1690 had been among the first Europeans to collect botanical treasures from Asia, including the first ginkgo seeds, and to publish in the West descriptions of ginkgo and many other unknown Japanese plants. A half century after Kaempfer, the French Jesuit botanist priest d'Incarville, living in Peking, had sent home the seeds of the ailanthus and of numerous other trees.

In the ensuing centuries only a handful of other botanists and intrepid plant explorers had dared (or managed) to follow in d'Incarville's footsteps, for travel inside China had long been forbidden to foreigners and, when it was finally permitted, remained a parlous affair in a nation lacking real roads and beset by dangerous plagues, bandits, and political unrest. In 1848 the British East India Company had dispatched botanist Robert Fortune, who had already made his way into the Celestial Kingdom once, on the first of several secret missions to steal Chinese tea plants. Thanks to Fortune, the company wrested control of this valuable commodity from the Chinese throne, using the purloined plants as the foundation for an empire of tea in colonial India. After Fortune, China opened its borders further, but it still took a rare sort to brave its many perils, identify new plants, return when their seeds could be harvested, and manage to send them to the West in a viable state.

Back in 1905, when David Fairchild had first been seeking a China plant explorer to track down "economically useful plants," he had heard reports of a young Dutchman named Frank Meyer. Not only had Meyer spent eight years training under the botanist Hugo de Vries in the Amsterdam Botanical Garden, but he seemed to be a restless soul who loved nothing better than walking hundreds of miles through the countryside looking at plants. "I was much impressed with the fact that Meyer was a great walker," recalls Fairchild, "for I knew that there were no roads in China, and a man must either be carried in a sedan-chair or walk if he is to get anywhere throughout the interior."

By the time Meyer accepted Fairchild's invitation to meet in Washington, he was thirty years old and had just spent four years in California, Mexico, and Cuba, working in nurseries and taking long rambles. "He cared nothing about his dress," wrote Fairchild. "Somewhere he had picked up a striped shirt, and when he came to see me it was wringing wet and the stripes had run. But he sat on the edge of the chair with an eagerness and quick intelligence that won me in an instant. His lack of pose, his willingness to work for any reasonable sum, and his evident passion for plants, all were evident in that first interview."

Meyer told Fairchild he had been in California when some bamboos that Fairchild sent had arrived at a USDA plant station. Meyer described how a "stubborn plant pathologist who did not know enough to mulch them would not let Meyer do it either. They had died in consequence, and," related Fairchild, "as Meyer told me about it, his eyes filled with tears. From that moment Meyer and I were friends." Fairchild had at that point been married for only a few months to Marian, but he wanted to bring this fascinating new hire home, as he thought his new in-laws, the Alexander Graham Bells, would enjoy him. Meyer "was so unconventionally dressed that I tried to spruce him up a bit. . . . I even presented him with a tuxedo thinking he would need it in the Orient, but he brought it back three years afterwards . . . green with mold." During his stay in Washington in that summer of 1912, Meyer loved to wander the grounds of In the Woods with Fairchild, checking on the progress of the Chinese fringe tree, the Nikko Japanese maples, or the rare Chinese dove tree that Fairchild had obtained. Meyer would approach a new plant "'with the eager interest of a child,' looking, feeling, smelling, and tasting in his attempt to identify it. Once acquainted with a new plant, he never forgot it."

During the thirteen years he roamed China on foot for the USDA, with a few side trips to Japan and Korea, Frank Meyer would become one of the world's great plant collectors. In the spring of 1913 Meyer was back in eastern China, where, armed with Fairchild's instructions and the diseased bark from the USDA pathologists, he set out on a six-day journey by mule to the mountains to locate again the trees from which he had already collected Chinese chestnut fruits. On June 13, 1913, Meyer reported in by cable to say he had found infected chestnut trees. In a follow-up letter that arrived three weeks later, he told Fairchild, "Here I am sitting in a Chinese inn in an old dilapidated town to the northeast of Peking. . . . [I] have been busy for several days collecting specimens of this bad chestnut bark disease and taking photos of same. It seems that the Chinese fungus is apparently the same as the one that kills off the chestnut trees in north eastern America. . . . This blight does not by far do as much damage to Chinese trees as to the American ones. Not a single tree could be found which had been killed entirely by this disease, although there might have been such trees which had been removed by the ever active and economic Chinese farmers." He tucked a two-inch square of blighted bark into his report, as well as more nuts.

"I looked at it doubtfully," Fairchild later wrote of his examining the Chinese bark, "not being familiar with the character of the disease, and wondered why Meyer had cabled with such confidence. When I took it to Metcalf's office, he lifted the bit of bark in his slow, deliberate way, and studied it carefully with his hand lens. 'It looks like it,' he said. 'The fungus strands are quite characteristic. However, cultures will soon show. We'll let you know as soon as they mature.'" The following week the two pathologists came into Fairchild's office holding six test tubes, three inoculated with Meyer's Chinese fungus, three with canker from a diseased American tree. Fairchild peered into the slender glass tubes. "The cultures all looked alike to me. Apparently Meyer had found the chestnut blight endemic in China. . . . Ten days later, Shear found the characteristic ascospores of the fungus on Meyer's cultures."

Later in the summer Fairchild introduced the pathologists' cultures into the Chinese chestnut tree he had been growing at In the Woods and saw that while the fungus did "a great deal of damage," it did not "kill the tree outright as it had the American chestnuts. In less than three months, the fact was established beyond a doubt that chestnut blight was an introduction from the Orient, and had probably arrived on nursery trees some time in the late eighties or nineties."

Once Meyer's work was known, says science journalist Susan Freinkel, the "plant explorer's discovery helped explain the astonishing rapidity with which the pandemic had engulfed the eastern seaboard. It was not that the blight had emanated from a single outbreak in New York." Asian chestnut trees from Japan had first been offered through nurseries as far back as 1876, and wherever someone had planted one the chestnut blight might have gained a foothold. "New York City," Freinkel wrote, "just happened to be the first place anyone had noticed the stealthy organism at work. By 1911, when Pennsylvania set out to contain the unruly invader, it was already too late."

Tracking down the chestnut blight had been an atypical task for Frank Meyer, for his usual mission was searching out new or better kinds of food crops or trees—economic plants. In his thirteen-year stint as an Asian plant hunter, Meyer would gather 2,500 new kinds of plants, everything from Chinese persimmons spotted at the Ming Tombs and the Chinese pistache tree "collected as bud sticks in the Western Mountains near Peking" to one hundred pounds of wild Callery pear seeds from Jingmen. With those pear seeds, gathered during his final collecting foray in 1917, Meyer was hoping to aid the American pear industry, for the blight-resistant Callery pear served as sturdy rootstock. Today his name is familiar to us, thanks to the (now trendy) Meyer lemon, descendant of a diminutive lemon tree he discovered at Fentai near Peking on March 31, 1908.

When the USDA's David Fairchild first sent Frank Meyer off to China in 1905, Charles Sprague Sargent—back from a round-the-world trip with his oldest son and John Muir—was working on finally hiring Ernest H. Wilson, a veteran of several highly successful China forays for a British nursery. Wilson always remembered his "first meeting with Sargent. . . . [It] took place in the shadow of the large Pig-nut Hickory on Bussey Road in the Arboretum. . . . After formal greetings he pulled out his watch and said, 'I am busy now, but at 10 o'clock next Thursday I will be glad to see you. Good morning.' I voted him autocrat of the autocrats, but when our next interview took place I found him the kindliest of the autocrats."

In late 1906 Fairchild and Sargent agreed to join forces, and Fairchild cabled Meyer to be in Shanghai by mid-February to meet Wilson, a cheerful, robust Englishman passionate about plants. Sargent was eager to have Meyer explore the Wutai Mountains southwest of Peking for certain trees and ornamentals, with the understanding, wrote Fairchild, that "In exchange . . . [Wilson] will get for us small quantities of grains, forage plants, fodder crops and other cultivated plants and scions or cuttings of fruit varieties which [he might find] . . . in explorations in the upper reaches of the Yang'tsi Valley."

In Shanghai, Wilson, by now an old China hand, advised Meyer on the arboretum's need for herbarium specimens, field notes, and photographs from Wutai. Wilson wrote back to Sargent that, while Meyer was hardworking and enthusiastic, the USDA's plant explorer had no sense of humor and "very little sympathy for the yellow race, & ignores their manners & customs. . . . [But] while he glories in roughing it & the superiority of the white race, I fancy he will modify his views and methods. . . . I seriously hope so . . . [for] his present methods are not calculated to get the best from his men."

Wilson spent most of 1907 through 1911 in the wilds of China on two expeditions for the arboretum, amassing 65,000 botanical specimens and documenting his work and travels with thousands of photographs. He was notably lighthearted about his job's many perils. "The roar of the rapid was deafening," he wrote as his collecting junk–houseboat, christened *The Harvard*, prepared to brave major white water on a Chinese river. The chaotic scene was, he reported, "heightened by the yellings of hundreds of trackers as they strained and tugged in their endeavor to haul the boats over. Of the ten junks in front of me three were wrecked." His boat came through safely, whereupon he walked ashore to note "a few plants of passing interest."

On Wilson's first exploration of Mount Wawu, he took shelter for the night in an old wooden temple, where the "rooms, though dingy and damp, were alive with fleas. . . . I had my bed arranged in a large hall where three huge images of Buddha looked down benignly upon me." They brought good karma, for among the botanical booty Wilson collected during that trip was the Chinese or Kousa dogwood, which in time would prove to be one of America's most ubiquitous and beloved small urban trees. Its showy white blooms (quartets of creamy petals that each end in an elegant point) flower at the end of spring, a final floral tribute before the summer descends.

During his lengthy wanderings through China and later Japan, Wilson became enamored of the ginkgoes, amazed at their size and antiquity—hundred-foot-tall trees with fifty-foot girths that were reportedly a thousand, even two thousand years old, invariably growing around temples and monasteries.

Until 1896 botanists had no inkling how truly unique and ancient the *Ginkgo biloba* species might be. That year a Japanese laboratory technician named Sakugoro Hirase peered through a microscope at the inside of a female ginkgo fruit's ovule. The previous spring in Tokyo, a mature male ginkgo tree (twenty years old or more) had released its pollen into the wind. That pollen had wafted toward the old female ginkgo at the Koishikawa Botanical Garden at Tokyo Imperial University and then onto a pair of the female tree's round, dangling drupes, where secreted drops of sticky fluid captured and ingested the pollen into an interior pollen chamber. That particular tree happened to be near the laboratory of Dr. Matsumura, the garden's director.

Hirase, who had mastered the microscope, had asked his superior, Dr. Matsumura, for a scientific problem on which to work. Matsumura suggested that he study the ginkgo fruit's process of fertilization. All that summer Hirase pursued detailed observations of the seeds from that nearby female tree. On September 9 he placed another ginkgo fruit slide under his microscope. "What Hirase saw was a botanical sensation," writes Peter Crane, dean of Yale's School of Forestry and Environmental Studies and author of *Ginkgo*. The pollen grains had, in the course of the summer, developed into a "branching tube," and at its bulging base were a pair of sperm cells.

Hirase watched as "these two sperm were released into a cavity in the top of the ovule," and then, he was astounded to see, the multiflagellated ginkgo sperm (three times larger than human sperm) began *swimming* to fertilize a waiting egg cell. Only ferns and cycads—which predate the dinosaurs—were known to reproduce with living sperm. Average plants have an evolutionary run of a few million years. Scientists now had proof that the modern *Ginkgo biloba* was an evolutionary superstar that had somehow survived virtually unchanged across epochs.

"The order to which the tree belongs, the *Ginkcoales*," writes Peter Del Tredici of the Arnold Arboretum, "can be traced back to the Permian era, almost 250 million years ago," thanks to the study of many ginkgo fossils found in the Northern Hemisphere. "The genus *Ginkgo* made its first appearance in the lower Jurassic period, 180 million years ago. . . . At least four different species of Ginkgo coexisted with the dinosaurs during the Lower Cretaceous." One of these four species, *G. adiantoides*, possessed leaves and female ovules, which are considered identical to those of *G. biloba*, the species that exists today. In short, the ginkgo has probably existed on earth longer than any other tree that is now living.

The USDA's Fairchild, familiar as any botanist with Hirase's historic coup, sought him out during a 1902 trip to Japan. "Hirase was [by then] a teacher in a secondary school at Hikon," wrote Fairchild. "We had luncheon together in a tea house and he gave me, as a souvenir of that afternoon, a microscopic slide on which was mounted a section of the ginkgo flower, one of those which first revealed to him the motile character of the male nucleus."

More than a decade later, Marie Stopes, a young British botanist who worked for a time at Tokyo Imperial University, wrote in her journal on September 9, 1908, of the still-fresh thrill of these rare sightings under the microscope: "There is grand excitement over Ginkgo; the sperms are just swimming out, and they only do it for a day or two each year. It is no such easy business to catch them, in 100 seeds you can only get five with sperms at the best of times, and may get one and be thankful. I spent pretty much well the whole day over them, and got three. . . . It is entertaining to watch them swimming, their spiral of cilla wave energetically." (One is not so surprised that Stopes became a pioneer of family planning, a scientist whose work and book *Married Love* are "credited," says Crane, "with almost single-handedly leading women in Britain out of the repression of the Victorian Age into a more enlightened age of sexual awareness.") Today patient souls still thrill to this sight through the microscope, while the more harried or lazy among us can resort to watching on YouTube.

A few years before Meyer and Wilson began their Asian explorations, Sargent wrote that the ginkgo "is now found in most temple gardens, both in China and Japan; and it is possible that this once widely distributed type has only been preserved by [human] cultivation, for if we are not mistaken the Ginkgo is nowhere to be found in a wild state." "Chinese" Wilson (as the arboretum's seasoned veteran came to be nicknamed) never found a wild-growing ginkgo tree despite logging thousands of miles roaming China's most remote valleys and mountains.

In May 1915 Frank Meyer became the first westerner to see "the ginkgo growing spontaneously in rich valleys over some ten square miles near Changhua Hsien, about seventy miles west of Hangchou, in the Chekiang Province. There were many seedlings and the trees here were so common that they were cut for firewood, something which had never been seen before in China." Sargent noted this eureka moment in an article, cautioning, "it is by no means certain that this is the original home of the Ginkgo as these trees may all have descended from a planted tree." Still, he conceded, it was "exceedingly interesting . . . to find that there is at least one place in China where the Ginkgo grows in the woods and reproduces itself spontaneously." Oddly, Meyer himself made no mention of his discovery beyond his report.

Wilson, who never returned to China as an explorer, clearly discounted Meyer's report, for he would later write of the ginkgo: "It no longer exists in a wild state, and there is no authentic record of its ever having been seen growing spontaneously. Travelers of repute of many nationalities have searched for it far and wide in the Orient but none has succeeded in solving the secret of its home. . . . In Japan, Korea, southern Manchuria, and in China proper it is known as a planted tree only, and usually in association with religious buildings, palaces, tombs, and old historic or geomantic sites. . . . What caused its disappearance [in the wild] we shall never know."

Thanks to Eliza Scidmore's persistence, the whole of the Speedway along the Tidal Basin in Washington, D.C., was finally abloom in the spring of 1914 with the billowy white and pink clouds of all several thousand cherry trees. Out at In the Woods, Fairchild was as enthralled as ever with his own Japanese cherries. Now seven to eight years old, they were coming into their own. "I used to roam among our trees at dawn," he wrote, "and gaze at the individual flowers through the darkness of my enlarging camera before the dewdrops had vanished from their petals. . . . I must be alone or with someone who cares for them as I do. . . . The cherries are delicate and one must stand beneath their branches and see the dainty blossoms against the blue or gray of the sky to fully appreciate them."

On April 25, 1907, Fairchild had started a small black leather-bound "Record Book" in which he intended to record the progress of the new shrubs, trees, and flowers planted at In the Woods. It became largely a record of the progress of his adored cherries. In a typical entry he noted that in the fall of 1906, he had received from the Yokohama Nursery a Japanese cherry known as *Ojochin*, or "big lantern." By April 9, 1910, he wrote that the tree was thriving: "Full bloom, fragrant semi-double white with pink tinge." On April 30, 1914, he noted that the tree was in full bloom but had become diseased. Fairchild kept track of two dozen different kinds of Japanese cherries growing on his land, noting which did well surviving frosts and the color and size of their blossoms, and rendering judgment on which were suited to life in the United States.

Fairchild's chief mission remained, as ever, to send forth the Good Things his agency found in foreign plant kingdoms to enrich America. While these were most often edible or economic plants, he now intended to exercise his considerable influence on behalf of Japanese cherry trees, whose poesy he was certain every soul and city needed. As he prepared to launch this campaign, he knew Sargent could be an invaluable ally. But although Fairchild had been collaborating with Sargent and the arboretum for years, he had mixed feelings about the great dendrologist.

Plant explorer E. H. Wilson *(left)* poses with Charles Sprague Sargent in front of a Higan cherry tree (*Prunus subhirtella*) at the Arnold Arboretum in 1915. *(Photograph by Oakes Ames. Courtesy of the Arnold Arboretum Archives.)*

On one hand, Fairchild admired Sargent's passion and encyclopedic knowledge. When he visited the Arnold Arboretum in the company of the famous ophthalmologist Dr. Wilmer, the two men "trailed Professor Sargent as he stalked from tree to tree pounding his cane on the ground as he went. His marvelous memory for names and his interest in every characteristic of his plants made the experience memorable. . . . I filled two notebooks with the scientific names of the plants we saw that day." Fairchild also appreciated Sargent's generosity, for in the wake of such visits, "I always carried away something interesting and valuable, not merely in the way of information, but also seeds and plants which he gave freely to our organization and which we propagated and distributed throughout the country." In the fall of 1913 the professor had given Fairchild a Japanese cherry, duly planted at In the Woods.

But Fairchild acknowledged that he could never really forgive Sargent for calling him and his colleagues plant philistines—USDA bureaucrats interested only in plants that could be eaten. Fairchild further resented Sargent's inevitable scoldings, complaining that the director "expected every botanist who visited the Arboretum to know all the plants by name, and by his manner made them feel ashamed to pass under any tree without knowing what it was. As I never had a particularly good memory for names, I received many harsh words from the Professor." Sargent also felt free to chide Fairchild in detailed letters about the USDA's sloppy identification of new plant materials. In a typical missive he wrote: "I can't help feeling regret that the Department is distributing unnamed or wrongly named material, for this means errors will be perpetuated indefinitely by nurserymen who are already responsible for too many errors."

Fortuitously, E. H. Wilson was just then plant exploring in Japan, an assignment he considered so easy that he had taken his wife and daughter along. For the prior two years, Wilson had been hors de combat, having narrowly survived a landslide in Szechuan Province in September 1910 with multiple breaks in one leg. After a slow recovery, Wilson returned to Boston to endure further surgery. He was also subject to recurrent bouts of malaria contracted during earlier China sojourns. By the end of 1913, though, Wilson was well recovered and en route to Japan with his family.

Sargent proved amenable to Wilson's collecting Japanese cherry seeds for Fairchild, which delighted Wilson, for he was as captivated by the trees as Fairchild, writing, "No language can exaggerate the beauty of the Japanese cherries," ornamental trees he was certain all "must have yearned deeply to possess." Wilson relished his assignment, traveling throughout Japan with his family during the festivals when "old and young, rich and poor put on their best raiment, visit and entertain their relatives and friends. There is something peculiarly gay and cheery about the white and pink Cherry blossoms and the prodigality of flowers and joyousness of color is ravishing." During 1914 Wilson carefully located, photographed, and, in time, gathered seed of scores of different varieties of flowering cherries, determining their proper names and writing a detailed monograph on them.

That September Fairchild and his agency "seriously took up the propagation and distribution of the beautiful Japanese flowering cherry trees. We had brought in No. 32860 from the island of Oshima, and we used it as stock for the best of the many varieties in our collection. It was this distribution to nursery firms and private individuals that started the universal planting of Japanese cherry trees throughout the United States; a movement of which the Arnold Arboretum played an important part by establishing E. H. Wilson's collection of named sorts in Boston."

When some of these trees did not flourish, in February of 1916 Frank Meyer brokered a second shipment with an illustrious provenance for Fairchild, who described this windfall of prize flowering cherries: "54 varieties from the municipal collection of Tokyo near Arakawa, which represent the loveliest of the hundreds of varieties known to the Japanese . . . and these will be propagated and distributed under the same varietal names as they bear in the Arakawan collection."

Thanks to Fairchild's efforts, the Japanese cherry tree would, in coming decades, emerge as one of America's most ubiquitous and beloved city trees, a harbinger of spring commonly planted as a street tree, set near water in parks, and cherished as a beloved object of fleeting beauty in many a front and backyard. This new arboreal immigrant provided some pretty recompense for the continuing loss of the American chestnut.

On the afternoon of June 4, 1918, the phone rang at Fairchild's office at the USDA. A State Department official was calling to read aloud a cablegram received from the U.S. consul at Nanking: "Frank Meyer, Department Agriculture, disappeared from a steamer in this consular district en route Hankow to Shanghai June 2." Five days later another cablegram arrived from the consul saying he was "proceeding up river with Chinese to search for Meyer." That same day Fairchild wrote Sargent a letter about other matters, appending a handwritten postscript about Meyer's disappearance. "No details are given and I hardly know what to guess about it. It's a serious blow to my work here. I hope he will turn up alive somewhere."

Two days later Fairchild learned the bad news in another cablegram: "Found Meyer's body thirty miles above Wuhu." Shortly thereafter Meyer's final letter arrived at the USDA, describing traveling through "villages that had been looted and burned, and food was hard to obtain, but to an old hand like myself, these things have so often been encountered that one is used to them. . . . The whole country is so fearfully upset that travel has become a gamble. . . . I am awfully glad I got away from Ichang; the situation began to depress me." Fairchild was distraught, unsure if Meyer's death was murder or suicide, for Meyer's recent letters revealed his low spirits regarding the war grinding on in Europe and the chaos and violence enveloping China. He lamented his own uncertain future and the way "uncontrollable forces seem to be at work among humanity."

Some weeks later the U.S. consul wrote a letter describing how he had located Meyer's remains. At Wuhu, the crew of a Standard Oil Company launch reported that the body of a bearded foreigner had been fished out of the river farther north. The next morning the consul set out with that crew on their boat, locating the mound where the body had been buried. After removing the earth piled upon it, he could see "the body had been placed on two planks, and one plank covered it. . . . The head was already badly decomposed, but a beard was seen, and the body was attired in a white undershirt and a pair of grey trousers." Shoes and suspenders were produced by local officials and appeared to be American, and so the consul, who had met Meyer and knew he had a beard, concluded this must be Meyer and made arrangements to put the body in a coffin and have it shipped to Shanghai, where he was buried. Later investigation did not resolve how Meyer died.

Oddly, that same summer William Alphonso Murrill also disappeared. When the Naturalist, who had so gloried in his ascendancy as a power in the world of plants and science, failed to return from a trip to Europe, where the Great War was still raging, the botanical garden contacted colleagues he had visited there, but none knew of his whereabouts. When Murrill reappeared months later in New York, he explained that he had almost died from a kidney condition in a hospital in a small town in France. The garden's director, Nathaniel Lord Britton, was skeptical of this account and demoted Murrill to a low-paid position. The Naturalist was so distressed that he soon quit, retreating to a cabin in the Virginia woods that he built himself, disappearing altogether from his professional world.

The chestnut blight, meanwhile, swept on, killing every American chestnut in its path, removing from the city parks, homes, and woodlands of New York, Boston, and Washington, D.C., what had long been one of the nation's most fruitful, noble trees. With time it would appear that Murrill might have actually been too optimistic when he predicted it might take a century before the American chestnut would once again flourish in its native land.

CHAPTER SEVEN

"A Poem Lovely as a Tree": Cherishing Memorial and Historic Trees

A tree planting in honor of dead marines. In memory of the gallant marines who had "gone west," mothers of the marines marked Mother's Day in New York City by planting memorial trees in honor of the sons who gave their lives to their country in World War I. The trees were placed on the mall in Central Park with impressive ceremonies. *(With permission from American Forests.)*

When the brutal trench warfare of World War I came to an official end on Armistice Day, November 11, 1918, the grieving nation sought suitable ways to honor its almost shocking number of dead. All told, 117,000 young Americans had died in the Great War. As a way to pay homage to these lost lives, the American Forestry Association proposed planting trees to create a new kind of living memorial. As *American Forestry: An Illustrated Magazine About Forestry and Kindred Subjects,* argued: "The trees will be, in their very greenness and robust strength, reminders of the youths who gave their vigor to win the big war. There will be no gloom about them."

American Forestry board member David Houston, Wilson's secretary of agriculture, wrote in March of 1919 to the governors of every state: "We shall seek many ways to perpetuate the memory of those who made the great sacrifice. It has been happily suggested that we do this by adorning with young trees, each named for a fallen soldier, our waysides, our yards and our pleasure places. This can be done on Arbor Day.... Such an observation of the day will give it a meaning more profound, a purpose more exalted, than it ever had before."

Officials at American Forestry were stunned by the fervent grassroots response to their idea of memorial trees: "Never before in [our] history," they confessed, had they received "so great a number of inquiries in regard to tree care and tree planting." By the spring of 1920 Arbor Day memorial plantings were sweeping the nation, as families, friends, and officials sought to assuage their sorrow and honor the dead with community tree planting on an unprecedented scale. In Middleton, Delaware, a huge throng, including uniformed soldiers, turned out for a ceremony as the high school students followed the prescribed American Forestry program: The students began by singing "The Planting Song" to the tune of "America," which began: "Joy for the sturdy trees / Fanned by each fragrant breeze, / Lovely they stand." School superintendent Wilbur H. Jump followed with a brief tribute.

Then a student stepped forward to recite "Trees," a poem written by Joyce Kilmer, who had died in France. Next fourteen children each declaimed two lines of Helen O. Hoyt's "What the Trees Teach," the first beginning with "I am taught by the Oak to be rugged and strong / In defense of the right, in defiance of wrong" and the last concluding with "The firm-rooted Cedars, like sentries of old, / show that virtues deep-rooted may also be gold."

The high school students then helped plant three young trees. The Reverend F. H. Moore dedicated "the linden, to J. J. Hoffecker, Jr., of Company B, 9th Infantry, who was killed in battle near Soissons; the maple, to Rupert M. Burstan, of the marines, who died of pneumonia six weeks after reaching France; the catalpa, to David Manlove, who fought in several battles, went over the top safely—then, after the armistice was signed, was killed by an exploding shell while engaged in reconstruction work." Then Dr. Moore and "a number of ladies went to the negro school where a maple was planted, dedicated to the memory of Jeremiah Jackson, the only negro boy from Middleton who died in the service."

Each month *American Forestry* featured page after page of black-and-white photos showing crowds of children, uniformed soldiers, top-hatted dignitaries, and local citizens in suits and straw hats watching as a small child or a mayor or a governor's daughter shoveled dirt onto a memorial tree, while all around American flags marked these solemn observances of lost lives. "Trees are being planted everywhere," exulted the magazine, "in honor of the men of the war. Those men of war carried the message of freedom and now the trees will carry the message of the men through the coming generations. . . . [And] the trees will mark the remaking of the cities just as those men marked the remaking of the world."

The American Forestry Association's helpful Arbor Day program booklet launched Joyce Kilmer's "Trees" to new status as one of America's best-loved poems. When Kilmer had enlisted in 1917 in New York's Sixty-ninth Infantry Regiment, he was a staffer at the *New York Times Magazine* as well as a sought-after lecturer and poetry editor at *Current Literature* and the *Literary Digest*. After Kilmer was killed by sniper fire on the western front on July 30, 1918, at the age of thirty-one, Charles Willis Thompson, a *New*

York Times colleague, wrote, "The German bullet that slew Joyce Kilmer at the Ourcq slew a brilliant promise."

Kilmer, a 1908 graduate of Columbia University, first published "Trees" in the August 1913 issue of *Poetry* magazine. Listed in *Who's Who* by the age of twenty-five, he was viewed by many as the leading Roman Catholic poet of his generation. He was married to Aline Murray, another well-regarded poet, and together they had five young children. While Kilmer had published four books of poems, his work—including "Trees"—though critically acclaimed, was not well known to the broader public until the *New York Times* first published "Trees" on December 26, 1918, in an article titled "Urge Memorial Trees" that mentioned Sergeant Kilmer's tenure on the paper's staff and *American Forestry*'s use of his "little poem" in its memorial trees literature.

In coming years, "Trees" would become not only the most famous such arboreal ode in the English language but also one of the most cherished and recognized of all American poems. Throughout America, citizens of every age were soon reciting, hearing, and reading its twelve lines of iambic pentameter:

> I think that I shall never see
> A poem lovely as a tree.
> A tree whose hungry mouth is prest
> Against the world's sweet flowing breast.
> A tree that looks at God all day
> And lifts her leafy arms to pray:
> A tree that may in Summer wear
> A nest of robins in her hair;
> Upon whose bosom snow has lain;
> Who intimately lives with rain.
> Poems are made by fools like me,
> But only God can make a tree.

Its first two lines were especially indelible, and in the ensuing decades American schoolchildren by the millions learned the verses by heart. In 1926 the New York City Board of Aldermen voted to name what had been Concourse Plaza Park, at 161st Street and the Grand Concourse in the Bronx, Joyce Kilmer Park.

In 1920 American Forestry Association president Charles Lathrop Pack, heir to a lumber fortune, proposed a far more ambitious vision of memorial tree plantings: "Roads of Remembrance," mile after mile of street trees that would shade city avenues, the nation's many new motorways, or the historic Lincoln and Jefferson Highways. In Washington, D.C., the American Legion whirled into action, planting 507 young Norway maples to honor the local war dead. As the *Washington Herald* reported on May 31, the Legion trees ran for "almost three miles, on both sides of Sixteenth street, from the north line to Webster street to Alaska avenue, the long line of sturdy saplings stand as an army in double file, looking to the north where stands their leader, the beautiful maple dedicated to Edward D. Adams, the first of the comrades to fall." At the foot of each young tree, the Legion embedded a bronze plaque.

On April 30, 1921, new First Lady Florence Harding, as ever fashionable in bobbed hair, brimmed dark hat with white plumed feathers, and patterned long coat, arrived at American Forestry headquarters at 1214 Sixteenth Street to lend presidential glamour to a memorial tree ceremony. She planted a tiny tree from the Hardings' home state of Ohio in a lilliputian Road of Remembrance fronting American Forestry's yellow brick offices. After wielding a sturdy trowel to ceremonially dig a few clods of dirt, Mrs. Harding presented the tool to her hosts. Thus the First Lady Trowel began its long career (which continues to this day), lending presidential luster to tree plantings far and wide.

Whether Mrs. Harding loved trees or was just being politically astute has been lost to history, but several weeks after her appearance on Sixteenth Street she joined American Forestry's board. President Warren G. Harding duly endorsed the Roads of Remembrance, telling the *Chicago Tribune*, "It would be not only the testimony of our sentiments, but a means to beautify the country which these heroes have so well served," noting that tree-lined byways were "one of the useful and beautiful ideas which our soldiers brought back from France." Not long afterward the First Couple attended a ceremony in New York's Central Park, where the president was photographed shoveling dirt onto a large-caliper tree honoring the fallen soldiers. That year he also planted a southern magnolia at the east entrance to the White House, memorializing the tens of thousands of American horses who perished in the Great War.

On the fiftieth anniversary of Arbor Day in 1922, President Harding designated April 22 as the official golden anniversary. Tree lovers had come a long way since 1882 in Cincinnati, when, *American Forestry* acknowledged, "to be a 'tree enthusiast' was to be a 'crank' . . . in the same category with those persons who worked for women's suffrage, prohibition, [and] believed in flying machines." Now, however, with the entire nation united around these living memorials, "We are just awakening to the possibilities of tree planting. The trees are monuments with a meaning, for they live gloriously just as did those for whom they are planted. The glory is the thing to tell the world."

In the fall of 1922 the *Bulletin of the Arnold Arboretum* weighed in on the issue of using trees as memorials, which it recognized had become a popular movement, "judging by the number of letters which come to the Arboretum on the subject. . . . Clearly the essential thing in a memorial tree is its ability to live long." The arboretum (presumably reflecting Sargent's opinion) had its own vision of such monuments: "If memorials are to be erected for soldiers and other men in the form of trees, the Redwood-forest offers the best opportunity in the beauty and permanency which can be found anywhere in the world." As ever, the Arboretum's *Bulletin* complained about the lack of real knowledge about trees and the consequent bungled plantings. "There is nothing more laudable than to plant a tree," averred Wilson, ". . . provided the right kind of tree is planted."

With Arbor Day so firmly established and Roads of Remembrance planted or planned in almost every state, the tree lovers at the American Forestry Association added yet another dimension to Arbor Day—the celebration of individual historic trees. The American Forestry Association so liked this idea that it established a nationwide arboreal Hall of Fame. "Zest is given to Arbor Day tree planting," declared the *New York Times* as the holiday neared in the spring of 1926, "by the fact that famous trees of long ago still flourish and engage popular interest, especially in Washington, where there are more historic trees than in any other city in the world. Visitors may still behold Washington's elm, Lincoln's European hornbeam, the tree spared by Senator Simon Cameron of Pennsylvania, [and] the oak from the tomb of Confucius."

When it came to designating trees with an illustrious enough pedigree to qualify for *American Forestry*'s Hall of Fame, those with a George

Washington provenance trumped all others. The Washington Elm at the U.S. Capitol rose in "majestic symmetry, the greatest in dimension of all the historic elms" on the Capitol grounds, even though it was uncertain whether the father of our country had actually planted the elm in front of the east entrance to the Senate Wing or just used it as his outdoor office when supervising the Capitol's construction. Erle Kauffman, author of *Trees of Washington: The Man, the City* (1932), opted for the latter: "The story goes that the noon repast was often laid beneath the branches of this elm and that the First President would sit in their shade and talk with the builders." An opposing view was taken in 1902 by the *Washington Post*, which reported that "the elm that once stood on Capitol Hill and which George Washington is said to have planted was cut down under order of the landscape architect Frederick Law Olmsted, in 1878, but Superintendent [William R.] Smith secured the roots and has since raised eight trees from them."

Kauffman had made it his mission to track down every famous elm, oak, horse chestnut, and willow where Washington had sheltered, tied his horse, held meetings, or eaten breakfast, from Valley Forge to Charleston to Cambridge. Of all these none was more famous than the Cambridge Elm, "a mighty symbol of the dawn of the Republic" that Kauffman believed to be "undoubtedly dearer to the hearts of Americans than any other historic tree."

It was under this spreading elm on July 3, 1775, that General George Washington had reportedly assumed command of the Continental Army. "Artists have painted it," wrote Kauffman, "poets have sung its praise, and historians have recorded its association with the great Continental soldier and patriot." And yet when roads needed to be widened near Harvard Square at Garden Street and Mason, local leaders thought nothing of confining this living monument to an isolated island of soil, fenced in against the traffic swirling all around, the earth above its far-flung roots paved over, depriving them of water and air.

Predictably, the Cambridge Elm began to die, and city authorities sent forth crews to minister to the revered tree. "More and more dead branches were cut off," wrote one observer, "the wounds smeared with tar, the hollows filled with cement, the remaining limbs braced with iron bands and rods, until it became a truly pitiable object. Finally, on October 26, 1923, the whole wretched ruin was accidently pulled over by workmen trying to remove another dead branch, and it crashed against the iron railing surrounding it. Examination showed that the trunk was hopelessly rotted below the ground, a mere mass of punk: the wonder was that it had stood so long." The carcass was cut into a thousand pieces and sent forth like so many holy relics to all the states, legislatures, and fraternal organizations. The following year, on Washington's birthday, an offspring of the tree was planted, but it did not survive.

Today, almost a century later, the Cambridge Elm is still remembered in its former habitat. But only the cognoscenti will know to peer under the wheels of the passing vehicles on Garden Street to catch a glimpse of the large manhole cover–like plaque embedded in the road, all that remains to mark the fabled tree.

CHAPTER EIGHT

"The Two Great Essentials for an Arboretum, Soil and Money": Chicago, D.C., and Boston

Joy Morton plants a tree honoring his father at the Dawes Arboretum in Ohio in 1927. *(Courtesy of the Sterling Morton Library at the Morton Arboretum.)*

In early June of 1921, Charles Sprague Sargent was delighted to welcome to the Arnold Arboretum Joy Morton, sixty-four, founder and president of the Morton Salt Company of Chicago. Sargent, still formal in manner and appearance at age eighty, now walked with a limp due to gout, but his mind and temper were sharp as ever, and his passion for trees undiminished. About to publish the second edition of his *Manual of Trees of North America*, updated from its 1905 version, he wrote: "The new edition contains the results of forty-four years of my continuous study of the trees of North America carried on in every part of the United States and in many foreign countries. If these studies in any way serve to increase the knowledge and love of trees I shall feel that these years have not been misspent."

Sargent's wife, Mary, had died two years earlier, and he dearly missed her "vitality, sympathy, and cheer." As he wrote a friend, "The glory and true meaning of Holm [their house] have gone forever, but I have the Arboretum to work for, and enough work laid out to occupy every day of my remaining years." A week after his wife's death, he was back in the office "dictating instructions for packaging seeds to Joseph Hers in China and requesting seeds of *Cupressus guadalupensis* from Alice Eastwood in San Francisco." He had little enthusiasm for anything modern, though he had given up his horse and carriage, and now his chauffeur drove him in a Rolls-Royce with polished brass headlights and a large rubber bulb horn.

Sargent was thrilled at Joy Morton's decision to create a midwestern living tree museum. In 1873 Morton, the oldest of four sons of J. Sterling Morton, founder of Arbor Day, had visited the Arnold Arboretum with his father (dead now twenty years) when it was first incorporated and still a worn-out farm. The Morton family motto had long been "Plant Trees," emblazoned on its heirloom blue-and-white gilt-edged china. The stout, balding Joy Morton looked very much what he was: a wealthy, important civic leader. In 1917 Joy had married Margaret Gray, the longtime caregiver for his invalid first wife, Carrie, a woman afflicted with mental illness and fragile health until her death in 1915. Margaret had then stayed on to manage Thornhill, Morton's 1,250-acre property twenty-five miles west of Chicago. Joy had purchased the estate near Lisle, Illinois, attracted by the land's beauty and trees, and with the hope that it might revive his first wife's spirits. Thornhill was situated in "a wide valley with the east fork of the DuPage River running through it, bordered on either side by rolling hills of considerable elevation largely covered with a splendid growth of native forest. In this forest are found oaks, ash, maples, elms, lindens, apples, viburnums, dogwoods and other trees and shrubs. There are large specimens of many of the trees named, some of which are of great age."

When Joy Morton came to the Arnold Arboretum seeking Sargent's counsel on how best to transform half of Thornhill into an arboretum, he and Margaret had just returned from a trip on which they had visited botanical gardens in Britain and Europe. From Biarritz Joy wrote to his son, Sterling, "I have seen so many fountains in these European gardens I am stuck on them. A little flowing water makes all the difference in the world in gardens, and I mean to have plenty of it." He also coveted conifers, insisting that "*all* the attractive places" had them. On his honeymoon in China, Japan, and Korea, he had been much impressed by Japanese planting techniques.

As for Thornhill's distance from Chicago, that was a deliberate choice. When Morton inspected the Missouri Botanical Garden in St. Louis, he confided to Sargent, it "was a disappointment to me. The smoke of bituminous coal, used almost exclusively in St. Louis, has ruined most of the trees: there are hardly any conifers left; nothing in the Garden looks good to me except the plants under glass and they are very well kept and look fine. . . . [It] fully convinces me that the Morton Arboretum is not a bit too far from Chicago and its coal smoke." Sargent agreed: "The distance . . . has the advantage of giving you air free from smoke and protection from the severe winds from the lake." Sargent predicted that in time the Morton Arboretum would "be the wonder of the world . . . for it has the two great essentials for an Arboretum, soil and money."

In October of 1921 Joy and Margaret Morton again visited the Arnold Arboretum, and Sargent returned to Lisle with them to see how the proposed arboretum was shaping up. In the late spring Morton had engaged the services of Chicago architect and engineer Ossian Cole Simonds, who had designed Graceland Cemetery and consulted on the Lincoln Park neighborhood. Morton had been "quite pleased" when in June Simonds had submitted an arboretum design that honored the local landscape; favored native, naturalistic plantings; and also created water features dear to Morton's aesthetic sensibility. Morton looked forward to working closely with Simonds, telling his son, Sterling: "I am going to have a lot of fun doing it." Sargent was very enthused by all he saw in Lisle, and soon after his return to Boston, Morton further endeared himself by again donating $2,500 to the Arnold Arboretum.

For the next six years Sargent advised and encouraged Joy Morton's project, providing lists of books essential for a library, as well as seeds and plants for an herbarium. Sargent recommended and even trained some early personnel: a young Austrian propagator named John van Gemert and a German botanist, Henry Teuscher. He reviewed Morton's incorporation papers, dispatched large numbers of botanical materials, and made himself useful any way he could.

In 1922 the Arnold Arboretum celebrated its fiftieth anniversary. Under Sargent it had doubled in size to almost 250 acres, and over the decades its staff had planted almost six thousand trees and shrubs (324 genera in eighty-seven families), all tagged and labeled and cataloged.

During this anniversary year Sargent compiled a nineteen-page list containing the 1,932 tree (and some shrub) species and varieties that the Arnold Arboretum had introduced into cultivation in the United States, with 778 being introductions from the wild. "Chinese" Wilson, now the arboretum's assistant director, had personally gathered many of them in China, Japan, and Korea. As part of this ambitious botanical enterprise, Sargent had overseen the creation of the herbarium, with 200,000 sheets of dried samples and seeds of trees and shrubs, located on the second floor of the Bussey administration building.

Sargent had also designed on that same floor the capacious wood-paneled library. He had furnished the first 5,000 books, including some rare items from his personal collection. A half century later the library boasted some 35,500 volumes; some 900 periodicals or serial publications devoted to botany, horticulture, and forestry; 8,000 such pamphlets; as well as 10,000 mounted photographs (2,800 taken by Wilson on his multiple collecting forays), all cataloged.

In his 1922 edition of *Manual of the Trees of North America,* Sargent noted that the book included eighty-nine new species, while two others, *Amelanchier obovalis* and *Cercocarpus parvifolius,* had been demoted from trees to shrubs and thus removed. Sargent used the book's preface to air an ongoing grievance with his federal counterparts, which had not adopted the new rules of plant nomenclature established in 1905 and 1910 by the International Congress of Botanists. "It is unfortunate," Sargent scolded, "that the confusion in the names of American trees must continue as long as the Department of Agriculture, including the Forest Service of the United States, uses another and now generally unrecognized system."

In 1924, even as Morton's arboretum was taking shape outside Chicago, the long-discussed National Arboretum was finally gaining legislative momentum in Washington. On December 4 U.S. Senator George W. Pepper of Pennsylvania introduced a bill authorizing the secretary of agriculture to establish such an institution. For the USDA's David Fairchild, this was a long-awaited day. "Since my arrival in Washington in 1889," he wrote, "we had all talked a great deal about an arboretum, but nothing had materialized. There was, it is true, a so-called National Botanical Garden established in 1853 on ten acres near the base of Capitol Hill when that region was virtually a swamp. This garden provided the trees when 'Boss' Shepard was planting the streets of Washington."

Why a National Arboretum? First, as Fairchild had long argued, the United States was "one of the few countries in the world which has no national botanic garden worthy of the name." Second, D.C.'s many government scientists would benefit from and could participate in the work of such a conveniently located institution. Third, the capital was situated at the juncture of north and south and thus could "permit cultivation of a wide range of specimens."

This lack of a federal arboretum had long been a sore subject between Fairchild and Charles Sprague Sargent: "Professor Sargent occasionally came to see me in Washington and generally started the conversation with the facetious remark that he had come to Washington to see what we had in the way of plants. Then I would tell him that he ought to know better, for we had no plants to show him but only people; people, however, who knew a great deal about plants."

By 1914, as Fairchild noted, "the professional botanists in the [USDA] resented having the title 'National Botanical Garden' applied to a ten-acre park filled with greenhouses not associated in any way with the Department of Agriculture or the Smithsonian Institution. Furthermore, they were annoyed that it was under the management of the Committee of the Congressional Library and got its appropriation through that channel."

When the longtime head of that botanical garden had died in 1914, the Department of Agriculture had seized the moment to create a "National Arboretum Committee." Fairchild was appointed, along with USDA scientists Frederick Coville, botanist and breeder of blueberries; Carl Scoville, head of irrigation agriculture; and Walter T. Swingle, a plant explorer specializing in citrus and dates. Soon thereafter Fairchild was contacted by Mrs. Frank Brett Noyes, a Washington power player described by one contemporary as a "very remarkable woman, a born leader and a lobbyist . . . [motivated by] her pride in her Federal City, her long range point of view, her indomitable spirit *and* [having] the influential *Washington Star* in the hollow of her hand." Janet Noyes, twenty-eight, was married to Frank B. Noyes, publisher of the *Evening Star*, then the capital's most important newspaper, and she advanced all her causes in its columns.

Fairchild recalls her asking in 1914 "if I would address a group of women at her house on plant introduction and our ideas of what an arboretum would be. This seemed a good opportunity to further our cause, and I gladly accepted." Despite years in the public spotlight, Fairchild had always felt anxious about any kind of public speaking and compensated by overpreparing his "lantern" slide talks. Having journeyed all the way from Florida, where he was busy with the USDA's tropical Experimental Garden, he was disappointed to walk into Mrs. Noyes's elegant drawing room at 1239 Vermont Avenue and find as his audience a mere twelve "sleepy" ladies.

But Mrs. Frank B. Noyes turned out to be a formidable ally. "From the day of that lecture," Fairchild later recalled, "she threw all her energies into the fight for an arboretum and even besieged the White House during two administrations until her husband told her that she would never be invited socially there again!" An activist member of the new Garden Club of

America and chairwoman of its Capitol Committee, Janet Noyes was an ardent advocate for gardens, trees, and floral good taste. During World War I, the crusade to found a national arboretum had been suspended, but before the war ended, the USDA had identified, mapped, and soil-tested the proposed site: a 367-acre lot overlooking the Anacostia River, just two miles northeast of Capitol Hill.

The proposed arboretum's site's "high point, called Mt. Hamilton, on the west side was heavily wooded, principally with oaks. This species extended down the valleys south of the hill. In this area also were other deciduous species, including tulip poplar, elm, beech, gum, maple and dogwood, with an undergrowth of Kalmia. On the east side was Hickey Ridge with valleys and cuts extending down to the Anacostia River. Between these two high points was a broad central valley traversed by a stream called Hickey Creek. There were a number of springs. . . . The whole aspect encouraged imaginative planning of plant collections." A further 433 acres of adjacent land could be added. To an outsider the parcel appeared to be all dusty country roads, working farm fields, swamps, woodlands, and sharply rising hills.

By 1924, a decade after David Fairchild first spoke to the twelve "sleepy" ladies, Janet Noyes had rallied to the arboretum cause the American Association of Nurserymen, the American Horticultural Society, the Washington Academy of Sciences, the American Forestry Association, and the Wild Flower Preservation Society, among a long list.

Not long after Senator Pepper of Philadelphia (brother-in-law to one of the Garden Club's founders) introduced the National Arboretum bill that December, Katherine Sloan, president of the Garden Club of America, wrote to her second vice president that Mrs. Noyes had hosted a large luncheon where "I pleaded to the wives of the Congressmen and Senators to help pass the bill. . . . Mrs. Noyes has seen President Coolidge and I think every possible pressure is being used to at least buy the Mt. Hamilton site." Added Mrs. Sloan, "I found all the Senators' and Congressmen's wives most agreeable and willing to cooperate in every way." By mid-February of 1925 Janet Noyes was reporting in a letter to President Coolidge that "you advised me to go to Congress to try to get their approval. . . . I have succeeded, after a great deal of personal effort, in winning the approval of both houses of Congress, I think, for the Bill."

A year later, on January 19, President Coolidge indicated his approval of a version of a National Arboretum bill "drawn up in Mrs. Noyes' living room." Frederick V. Coville, chief botanist of the USDA, whose work on blueberries was creating a new commercial industry, wrote to Mrs. Noyes the following day giving her full credit for the success of the plan: "Without your support . . . the whole enterprise probably would have been postponed for a generation." The letter was premature, however, for a year later certain House members were still holding the bill up. At the end of 1926 the *Washington Star* ran four editorials championing the National Arboretum. In 1927, with the reintroduced bill again before the House and Senate, the *Star* ran "a new article in support of the Arboretum at approximately three-day intervals for over a month and a half."

By now Mrs. Noyes had also rallied to her ranks the U.S. Chamber of Commerce. Finally, on March 4, 1927, President Coolidge signed a bill into law establishing a National Arboretum, but it would take another two years for the USDA to complete the purchase of the first 250 acres of the Mt. Hamilton land.

Late that winter Charles Sprague Sargent, age eighty-five, had come down with the flu. When he died on Tuesday evening, March 22, he had lived just long enough to know that the nation now would have not just two arboretums but three. One of his daughters described Sargent's final moments: Raising himself up in his bed, he pointed to something he saw in the distance. She put her arm around him and asked what it was. "He answered faintly but surely, 'locust and honey,'" she wrote. "He was already botanizing, it seems, in the fields of Paradise."

As Sargent's biographer wrote, "Nothing demonstrates the absolutism of Sargent's office more dramatically that the panic that seized the Arboretum's staff after his death." They feared, wrongly, that Harvard would abandon the arboretum and its work, allowing it all to lapse back into a mere city park. Instead, "Chinese" Wilson was given the new title of "Keeper of the Arboretum" and continued in charge of day-to-day matters. Harvard then appointed Professor Oakes Ames as new director. A longtime botanist and member of the Harvard faculty, Ames was an old friend of Sargent's and was himself a Boston Brahman. As instructed by President

A. Lawrence Lowell, Ames gently but firmly brought the Arnold Arboretum into the university's orbit.

The generation of original tree evangelists was passing. John Davey, the Tree Doctor, had preceded Sargent in death by four years, stricken at age seventy-seven by a sudden heart attack. As he was widely beloved, Davey's death on November 8, 1923, in Kent, Ohio, had provoked an outpouring of heartfelt tributes, news stories, and obituaries. Mourners marveled at and applauded his passionate devotion to his cause, many citing—as did Pittsburgh's *Gazette-Times*—his visible legacy in their city of "thousands of vigorous trees where otherwise there would have been declining or dead ones."

"Hundreds of great men and women have publicly paid tribute to Father John as a genius," wrote Davey employee David Quincy Grove on behalf of the firm's employees, expressing their awe that this one humble man had persuaded so many to see the natural world with new eyes and "to look upon trees as living things that deserved to be protected and preserved. It was he who originated the science of tree surgery which has made it possible to prolong trees whose beauty we admire and shade we enjoy."

Two weeks before his death, Davey had sent a letter to be read aloud to "My Dear Boys" at a company banquet, alerting them to his next crusade: "I am now completing a book 'How To Plant And Care For Trees.' I have been in actual work, shaking hands with old Mother Earth, for more than 67 years (for I have worked incessantly since I was 10 years of age) and now I am . . . putting it into a book of 192 pages . . . [with] 150 high-grade photographs. The sale of *one million* copies must be our first goal. Does one of you say, 'Can't be done'? Oh, *Get out!* Throw out 10,000,000 numskulls of our population, we have one hundred million left, and don't you think *one* in a hundred of our people ought to know HOW to plant and handle trees? Well, what do you say? Let 'Here goes!' be the answer."

Father John, as he often styled himself, could concentrate on this summing up of his life's tree knowledge because the Davey Tree Expert Company was well run by his son, Martin L. Davey. The young Davey was a natural leader who also loved politics and in 1913, at age twenty-nine, had been elected the "Boy Mayor" of Kent while continuing to run the family business. When Father John died in 1923, Davey was running the firm *and* serving as the Democratic congressman from Ohio's Fourteenth District.

The Davey Tree Expert Company would eventually become a nationwide company, and Bartlett Tree Company entered the field in 1907. By 1923 Bartlett had opened its own Tree Surgery School, as well as what would prove to be its strength, a research laboratory. The three Asplundh brothers opened Asplundh Tree Expert Company in 1928, devoted largely to clearing trees around utility lines. In 1924 this new world of tree specialists, working mainly in cities and suburbs, convened at the first Shade Tree Conference, launching the professionalization of arboriculture. John Davey could depart this life knowing that his passion for arboriculture had won over millions.

On April 28, 1927, a month after Charles Sprague Sargent's death, Massachusetts celebrated Arbor Day. "Chinese" Wilson, by now a notable adopted Bostonian, stood on the statehouse grounds just above the Boston Common, watching, appalled, as Governor Alvan T. Fuller, a Republican, honored Sargent by planting a white spruce tree. As Wilson would later write, "The site chosen was in the heart of a city, and the tree, a native White Spruce, lover of pure air and of cool forest soils. I was invited to the ceremony but was tongue-tied." Wilson conceded that the tree was beautiful and had been planted properly. However, that memorial "White Spruce from the moment it was placed in the earth was doomed to a lingering death by suffocation and slow poison." Why, Wilson wondered, had they not planted an English elm? Or a European beech, or a ginkgo? "One and all of these are well suited to city conditions and could be looked upon as promising to thrive for a hundred years or more."

While Wilson may have been incredulous at this misguided memorial, Sargent would have found it par for the course. Back in 1889, the April 17 issue of Sargent's periodical *Garden and Forest* had been positively dismissive on the subject of Arbor Day: "The reports of Arbor Day observances which come to us from some places indicate that the talk and ceremonies sometimes border closely on the farcical and ridiculous. . . . We observe that in some towns which last year celebrated Arbor Day with much sentimental oratory many fine trees have since been unintelligently and barbarously slaughtered." Nor had Sargent been any less churlish toward this ever-more-popular school holiday the following year, when he had instructed: "Arbor Day will prove most beneficial in those places where the trees are not forgotten as soon as the songs have been sung and the poetry recited." One aspect of Arbor Day Sargent did approve of was the planting of memorial trees: "It is a beautiful custom—this planting of memorial trees—but in order to make it impressive," he counseled, "the trees must live to vigorous and venerable old age."

The year 1922 had marked the fiftieth anniversary not only of the Arnold Arboretum but also of Arbor Day. Despite Sargent's frequent curmudgeonly critiques, year by year it had become ever more popular and elaborate, inspiring an entire oeuvre of sentimental poems, declamations, and songs. Not only was Arbor Day now celebrated in every state, but it had been declared a legal holiday in Colorado, Florida, Idaho, Minnesota, Montana, North Dakota, and Wyoming. Orchestrating ornate Arbor Day festivities became something of a competitive sport, with cities and states vying to outdo one another. In 1911 Washington, D.C., had boasted that on Friday, April 17, "School Children Are to Plant 60,000 Catalpa Trees," financed by the Woodward & Lothrop department stores. It was bested by Pittsburgh, whose pupils that year would "Plant 100,000 Trees for Arbor Day." These too were catalpas, paid for by mercantile Kaufmann Brothers.

"Chinese" Wilson himself did not long outlive Sargent. The intrepid plant explorer who had joyfully braved bandits, famines, rapids, and plagues in China died in an automobile accident outside Worcester, Massachusetts, on October 15, 1930, at the age of fifty-four. His car was said to have skidded off a road slick with autumn leaves.

Like Sargent, Wilson left his own particular legacy—in his case the much-loved Kousa dogwood and the wide selection of Japanese ornamental cherries. A great favorite first planted in 1889 was *Prunus sargentii*, of which horticulturist Michael A. Dirr would later write: "Many gardeners consider this the crème de la crème of the flowering cherries."

While Eliza Scidmore had introduced the cherry trees to the capital, Wilson took special pride, as had Fairchild, in having expanded the choices and made them available to the commercial nursery trade. By the mid-1920s Americans were expressing their delight in these lovely ornamentals by planting them by the millions. "I never dared to imagine," Fairchild wrote, "the popular enthusiasm which the Washington [cherry] trees have caused throughout the country."

As the thousands of Japanese cherry trees along the Tidal Basin grew tall enough for their collective blooms to create engulfing clouds of white, pink, and rose blossoms, a spontaneous Washington tradition of cherry tree "worship" arose. When the days warmed and the delicate buds began to unfurl, crowds assembled to enjoy their beauty, just as Scidmore and Fairchild had envisioned. Fairchild and Marian had by now begun to spend their winters in Florida and in 1926 built a tropical house, "The Kampong," in Coconut Grove, near the USDA's Chapman Field Experiment Garden. "Consequently," he would later write, "it seemed necessary to sell our flowering cherry trees and 'In the Woods.' To exclude from our lives such glorious living things as those cherry trees caused a painful wrench of spirit, but the lure of the many tropical flowering and fruiting trees on our Florida place made us more content to entrust the cherries to Doctor E. A. Merritt of Washington."

Like Fairchild, Scidmore had also left the capital, in her case resettling in Geneva, Switzerland, in 1923. There she hosted a pro–League of Nations salon in her apartment at 31 Quai du Mont-Blanc, a popular gathering place for "distinguished Americans." In 1927, when the first Cherry Blossom Festival was held in Washington, a modest homage to Japanese-American friendship, neither Scidmore nor Fairchild was present to see the sweet tableau of a girl in a pale kimono patterned with five-petaled *sakura* and a large flowered headdress—"Spirit of the Cherry Flower"—link hands with "Spirit of America" in her white-sashed gown and broad-brimmed hat decked with flowers. On November 3, 1928, Scidmore died at the age of seventy-two, following an emergency appendectomy. "It is probable," noted one U.S. newspaper, ". . . that no American woman had a more cosmopolitan assembly of friends or more varied interests of work than Miss Scidmore enjoyed."

The American love affair with flowering Japanese cherry trees offered some recompense in the final gathering twilight of the American chestnut. In 1923 Charles F. Thurston had lamented in *American Forestry,* "Good-bye, chestnuts! The trees are rapidly dying everywhere. What was formerly a majestic, soul-inspiring landmark is now but a rotting stump. No more are they seen on Main Street; no longer do they stand in battalions in the forests. They are as few as the veterans of the Civil War and just as decrepit." That year the blight struck the hills of Appalachia, a region whose forests and woods were defined by chestnut trees. A pathologist told Thurston, "Nothing can be done for them. The United States government has spent thousands—no, hundreds of thousands—of dollars to check the blight, but its efforts are futile; it can do nothing to save the most beloved of all our trees."

In American cities, with the chestnut often only a memory, maturing American elms now dominated, their soaring canopies arched above avenues, yards, parks, and campuses, the stately architecture of these incomparable trees making them the city tree of choice. John Davey had written:

> I love all trees, but there are three that seem to surpass all others: the maple for its richness of foliage, the oak for its strength, and the elm for its arching habits and peerless drapery.
>
> Who—that loves a tree—has not stood spell-bound and compared the giant elm to the human frame? From the majestic trunk extend the powerful arms. There are the "joints" and, finally, a division into the fingers on the tops of which are borne the modest flowers and those marvelous structures we call "leaves." Oh, the beauty of a leaf! How charming its veining! How divine its mission of preparing oxygen for the animal kingdom!

While the American elms now reigned over boulevards, parks, and civic spaces, the once-admired tree of heaven had fallen completely from grace. An 1885 Washington, D.C., law banning *Ailanthus altissima* was still on the books, threatening citizens who failed to "abate the same" with fines up to ten dollars. Sargent was one of its few defenders, but he swayed no one by arguing that "ailanthus is one of the most commonly planted, and most highly esteemed trees in Paris and other European cities." Americans viewed this tree as an unwanted but tenacious interloper. Yet the ailanthus did bring joy and shade to those who did not know to despise it: most famously Francie Nolan, the young heroine of *A Tree Grows in Brooklyn*, who tells of her coming of age during World War I. She loved to spend Saturdays on her third-floor fire escape engulfed by an ailanthus. There in nice weather she read her library book and "imagined that she was living in a tree."

As the book ends, Francie describes how the landlord's men chopped down and then burned her tree when housewives complained it got entangled in their laundry lines. "But the tree hadn't died. . . . It hadn't died," she says, "A new tree had grown from the stump and its trunk had grown above the ground until it reached a place where there were no wash lines above it. Then it had started to grow towards the sky again. . . . It lived! Nothing could destroy it." For urban tree snobs, however, these fast-reproducing trees were little more than renegade, out-of-control weeds that replicated uncontrollably, occupied any unsavory, undefended urban terrain, and could not be killed. Few city dwellers grieving the swift destruction of their American chestnuts would ever view the ever-proliferating tree of heaven as any solace. Nor could they dream that they would soon face far worse losses.

CHAPTER NINE

"Imagine the Wiping Out of the Beautiful Avenues of Elms": Battling to Save an American Icon

American elm-lined walk in Lincoln Park, Chicago, in 1900. *(Courtesy of the Prints & Photographs Division, Library of Congress.)*

On Monday, August 7, 1933, L. M. Scott, chief inspector of the USDA's Bureau of Entomology and Plant Quarantine at the Port of Baltimore, opened a crate containing ten massive Carpathian elm burl logs destined for a furniture veneer manufacturer in Kentucky. Stamped on their sawed ends was their provenance: PRODUCT OF FRANCE. As the inspector, noticing discolored streaks in the wood, began examining the logs, a shiny brown red beetle the size of a grain of rice, an insect new to him, wandered forth. Under closer scrutiny, Scott saw that the elm logs were alive with the tiny creatures. He carefully removed a few, sealed and embargoed the crate, and requested entomological backup. By Tuesday morning Scott had determined that the beetles were the European species *Scolytus scolytus* and telephoned this discovery to headquarters in Washington.

On Tuesday evening the USDA's plant pathologist, Curtis May, arrived at the Port of Baltimore with another colleague, and all the men began dissecting the embargoed logs. What they found was not reassuring: "Most were scored with insect tunnels, and sample sections cut from these displayed the rusty-brown striations that are often characteristic of the Dutch elm disease. Several of the samples were then subjected to microscopic examination." May headed north with samples of the wood and beetles to his new laboratory in Morristown, New Jersey, an abandoned speakeasy whose bar served as his lab bench, and soon confirmed what they all suspected: The logs were infested with *Graphium ulmi*, a mortal fungal affliction known as Dutch elm disease (DED), spread by these bark beetles. For plant pathologists this was exceedingly bad news.

In Europe, when "elm death" first struck war-wracked Rotterdam in 1918, some theorized that the old trees were dying because "the movement of heavy trucks along the roads during the Great War had jarred the root systems loose; others felt it was the result of atmospheric contamination from the use of war gases." Four years later graduate student Maria Beatrice Schwarz of the Scholten Phytopathologisch Laboratorium in Baarn, Netherlands, identified a new fungus as the culprit, but peers were skeptical. In 1927 another graduate student at Baarn, Christine Buisman, confirmed Schwarz's findings and further determined that the fungus could thrive in both live and dead elms. Entomology student J. J. Franzen showed that two species of European elm bark beetle were the "vectors."

Unlike the chestnut blight, whose spores drifted by the billions on the wind, the DED fungus spores hitched a ride by sticking to the emerging adult beetles, who preferred to dine on elm twig crotches and small stems. As the insects masticated their woody meals, the fungus spores rubbed off in the newly exposed wounded elm tissue, germinated, colonized, and clogged up the tree's water-carrying vessels. The elms, starved of water and nutrients by the fast-proliferating fungus, slowly died. The beetles would then fly off to other elm trees, laying eggs, eating, and breeding, a traveling cloud of arboreal extermination. In the course of the coming decades, most of Europe's millions of elms, and then those of the British Isles, would be infected and killed.

While no known treatment or cure existed—skepticism of "female science" had hindered early responses—"sanitation," or the removal and destruction of sick and dead trees, did slow DED's spread. In 1919 Charles Marlatt and the USDA had imposed a strict quarantine on importation of live elm nursery stock. But DED had breached American defenses; too late they learned they had not been strict enough. Marlatt, almost seventy, was retiring as chief of the USDA's Bureau of Entomology and Plant Quarantine. His final years had been a triumph, for he had personally directed the eradication of the Mediterranean fruit fly, which had threatened Florida citrus groves in 1929. Using $6 million in emergency funds, Marlatt had overseen all planting, harvesting, and massive spraying, and a year later the fruit fly had been defeated. But now, just as he was departing, the USDA was facing the worst threat to a native tree species since the American chestnut blight.

The Baltimore findings were not, as it happened, the first appearance of Dutch elm disease in the United States. But the beetles in the Carpathian elm burl logs did solve the four-year-old mystery of the dead Ohio elms. In

June of 1929 Ohio plant pathologist Paul E. Tilford had been contacted about a group of dying elms along the railroad tracks in Creston, Ohio, near Cleveland, where steam trains took on coal and water. When Tilford stood by the rail line and looked up into the noticeably thinning elm canopy, he had the awful fear that he was witnessing the effects of Dutch elm disease. He cut out a few dying twigs and leaves and sent the samples to Curtis May, who was then an associate plant pathologist at Ohio's Agricultural Experiment Station at Wooster. When May put the wood under the microscope, he confirmed Tilford's suspicions. It so happened that Dr. Christine Buisman, the Dutch scientist who had now become the world's leading DED expert, was serving a one-year fellowship at Radcliffe College as well as the Arnold Arboretum in Boston, and she was able to corroborate May's findings.

The only positive (if puzzling) news was that, despite determined scouting, pathologists had located only four more elms in Cleveland and one in Cincinnati stricken with Dutch elm disease. Crews were swiftly dispatched to cut down the stricken trees. Every part was piled and burned to ashes in roaring fires. May wrote in the spring of 1931, "None of the infected trees came from nurseries and there is at present no satisfactory explanation for the presence of the disease in these two localities." No one then knew the origins of the "causal fungus" either. Noted May, "It is possible that the fungus may be a native of Asia, or even of North America. However, this latter source seems highly improbable."

Not long after this initial discovery in Ohio, F. A. Bartlett wrote in 1930 in *American Landscape Architect* about this unsettling but limited discovery: "It may be a flash in the pan, but inasmuch as the American elm represents as high as 75 percent of the tree growth in many eastern cities and holds an important feature position in landscape design, the question of probable disease attack is one of greatest scientific, economic, and aesthetic importance."

By the time the sixth annual National Shade Tree Conference was held in Cleveland in August 1930, Curtis May and his colleagues were still baffled by this small outbreak of DED. "We have no way at present of knowing where the disease came from," he announced in a talk. "There has been no common source of infection that we can find. We can't trace it to a nursery. We can't trace it to imported stock. . . . Our plan is to destroy diseased trees as soon as we find them. . . . It may be possible that we can eradicate the trouble here if it is confined to only those local centers of infection. I am not yet ready to hang crepe on the American elm."

In 1931 another four stricken trees in Cleveland were found and destroyed, but beyond that no further infestations were identified. Wary officials were relieved but remained perplexed: How had the disease entered the United States? Then, in June of 1933, a park foreman in Maplewood, New Jersey, noticed that, despite all the heavy spring rain that year, an elm in Memorial Park still had no leaves, while other branches were turning yellow and wilting. He had personally seen the ravages of DED in Germany and was sufficiently suspicious that he sent samples to May's laboratory in Ohio. By the time the disease was confirmed for the first time on the East Coast, the New Jersey tree was dead. On June 26, when it was cut down, it was "found to be heavily infested with the European bark beetle."

As Curtis May would write to his colleagues, "The seriousness of the situation can hardly be exaggerated." The Ohio lab immediately sent east to New Jersey Dr. O. N. Liming, who for three years had been the chief DED scout in Ohio. Throughout New Jersey officials, arborists, and nurserymen sounded the alarm, and within weeks crews had located sixty-nine infected American elms in fourteen New Jersey communities within twenty miles of New York City. "The actual work of removing trees," wrote May, "is greatly increased by the fact that those so far found infected have been in cities, frequently in close proximity to telephone and other wires. The trees must thus be taken down a limb at a time rather than felled at once."

On July 10 local New Jersey shade tree commissions, fortified with hundreds of workers from the federal Works Progress Administration's Civilian Conservation Corps, launched an all-out search-and-destroy campaign. With each passing week more afflicted and dying elms were found across New Jersey—in Paterson, Morristown, Newark, the Oranges, and Jersey City—and then in New York—Staten Island, Brooklyn, and southern Westchester County. Any hope that this was a minor outbreak gave way to the crushing fact that an epidemic was under way in the New York–New Jersey region.

Dr. May, tapped to lead the USDA's anti-DED pathology services, came east from Ohio to open the Morristown laboratory and several field offices. It was while he was in Washington conferring with his new superiors that Inspector Scott raised the alarm at the Port of Baltimore, solving the mystery of how DED had gained entry to the country. While the United States had moved swiftly to ban live elm nursery stock, it had failed to recognize the perils of dead elm wood. On high alert, federal inspectors examined the next such shipment from France when it arrived on August 15 at the Port of Norfolk. These bark-covered logs also contained the fungus, as well as seven adult beetles that carried the fungus and fifty-two larvae. A third shipment a week later was similarly infested.

Now agency "pathologists-turned-detectives combed through the maze of records, checking every shipment that might have included elm," tracking down when and where such shipments (destined to be decorative wood veneer for furniture) had been received. Week after week USDA officials scouted ports of entry, rail distribution yards, and veneer plants, looking for clues and further information. They determined that as far back as 1926 infected logs had entered through four ports, been loaded onto sixteen different railroads, and then traveled thirteen thousand miles into twenty-one states. As logs sat in rail yards or outside furniture factories, some beetles had emerged and flown off to feed and breed on nearby elms. No region, scientists learned as they searched the surrounding landscapes, was more infested than that around the Port of New York.

It would be hard to overstate the potential looming calamity. The American elm was not only ubiquitous across the whole nation; it was *the* beloved city and suburban shade tree. A. J. Downing had viewed it as the ultimate in arboreal elegance and grace, the definitive "expression of classical beauty." Charles Sprague Sargent had found it "indescribably beautiful in winter . . . a fitting ornament to stand by the stateliest mansion or the humblest farmhouse." Its historical associations ranged from General George Washington exhorting the revolutionary troops beneath the famous elm in Cambridge to an American elm far out in Atchison, Kansas, where Abraham Lincoln was said to have delivered his first campaign speech in 1860. Just the previous year, on November 11, 1932, the Disabled American Veterans had planted an American elm at the Tomb of the Unknown Soldier in Arlington Cemetery.

The American elm's large stature and architectural glory enabled it to more than hold its own in the modern city. "No other tree looks like the

American elm against the sky," wrote tree expert William M. Harlow. "Its branch configuration—serpentine, Medusa-like—is unique." Equally important, it possessed practical qualities for thriving in tough urban environments. "Its crown rides high above the tallest traffic," explained *New Yorker* writer Berton Roueche, "and its forking trunk provides a natural passage for power and telephone lines. Moreover, its roots are so shallow that it can flourish in the meager earth of parks and malls built over subways, tunnels, and underground garages. It is, in addition, as rugged as a weed. It can live in almost any filth of smoke and soot and noxious fumes that man himself can tolerate."

In addition to these characteristics, Roueche declared: "As a shade tree, the American elm has no equal. The quality of its shade is unique. Most comparably robust trees are densely foliaged and branch close to the ground, and the shade they give is dark and heavy, and often all but breathless. The arching elm reaches high and wide before it breaks into leaf, and its leaves, though numerous (a big tree will put out a million or more), are widely spaced and ranged on a single plane, forming an open, latticelike pattern. Elm shade is thus an exquisite dapple of airy light and shifting shadow. Grass will grow to perfection beneath an elm."

On September 8, 1933, Dr. Curtis May stood before the scientists, arborists, and nurserymen attending the ninth National Shade Tree Conference at the New York Botanical Garden in the Bronx to announce that the Ohio elm mystery had been solved. Surely many present noted the sad irony of learning of the blighted elm burl logs at the New York Botanical Garden, original epicenter (with the adjacent Bronx Zoo) of the American chestnut blight. Twenty-seven years earlier, on these very grounds, William Alphonso Murrill had cracked the riddle of that other tree-killing fungus. Now all who remembered the devastating loss of the chestnuts were facing another ecodisaster, but one that would be calamitous for beautifully canopied city streets and parks.

Murrill's own life since then had taken some odd turns. After Nathaniel Lord Britton, the garden's director, had demoted Murrill for disappearing for months, the Naturalist had quit, retreating to a cabin in the Virginia woods. No more was heard of him until 1926, when another mycologist was amazed to recognize "a frail and haggard-looking Murrill playing a piano concert . . . in the 'Tin Can Tourist Camp' in Gainesville, Florida, a dingy trailer park for tourists. . . . Murrill had been drawn to Florida by the rich variety of mushrooms there, but he had no real means of supporting himself." The colleague found Murrill a house and a little space off a stairwell at the local university. There Murrill would devote the rest of his days to identifying and classifying Florida mushrooms.

It was another New Yorker, the idealistic curator for public instruction at the Brooklyn Botanic Garden, a Yale-trained botanist named Arthur Graves, who in 1931 had appointed himself to pursue the seemingly quixotic quest of breeding a blight-resistant American chestnut. At his summer home in Hamden, Connecticut, on the south side of a woody slope, Graves had started with an American chestnut from Washington, D.C., as the "father" tree and a Japanese chestnut from Long Island as the "mother." In coming years he would plant more and more varieties of chestnut trees and patiently cross them, seeking a hybrid that looked like the American tree but fought off the blight like an Asian tree.

The USDA had responded to the tragedy of the American chestnut by dispatching plant explorer R. Kent Beattie in 1927 to gather as many possible varieties of viable nuts from chestnuts in Japan, Korea, and China. His shipment of 250 bushels had been grown into seedlings. "Even as the USDA was promoting Chinese chestnuts as a replacement for the native trees, its scientists were trying to create new Asian-American trees, often in collaboration with Graves," wrote Susan Freinkel in her history of the American chestnut. For those working to create a viable American chestnut, these were still very early days.

As for the American elm, on October 21, 1933, the "barn door" closed, as the USDA issued Quarantine No. 70, forbidding the importation of elm wood. Naturally, in the wake of the American chestnut debacle, some argued that it made no sense to waste scarce Depression-era dollars to save the American elm. But the optimists prevailed. Unlike the chestnut blight, argued Richard E. White of the New Jersey Agricultural Experiment Station, the "seeds or spores of the [DED] fungus are not spread by the wind or picked up by birds or squirrels. . . . It is evidently also in the nature of Dutch elm disease to skip about, attacking a tree here or one there, leaving many trees in the vicinity untouched. The chestnut blight, on the contrary, took all the chestnuts as it advanced, making a clean sweep." Still, the sheer scale of the task, given the number of elms involved, would be daunting.

In the course of the year after May's talk at the New York Botanical Garden, the USDA, working with a coalescing DED army of state and local tree specialists and Civilian Conservation Corps workers, had identified 6,500 infected trees in the five-thousand-square-mile epidemic zone radiating out from the Port of New York. Homeowners and city arborists watched in horror as towering elms, revered ancients that had shaded yards and streets for more than a century, suddenly showed the telltale yellowing, wilted leaves high in their visibly thinning crowns. Because of the lack of funds in the hard times of the Depression, though, the DED troops had cut down and burned only 2,500 of the known affected elms. "The salvation of the elm in America," one alarmed USDA official told the American Forestry Association, the nation's oldest conservation organization and most vocal champion for saving the elms, required swift and aggressive removal to ensure "the complete eradication of these insects." Every day that an infected elm continued to live, it served as an arboreal Typhoid Mary, its bark a breeding ground for beetles and fungus spores.

The very same Hermann W. Merkel who had alerted America to the chestnut blight thirty years earlier while serving as forester at the Bronx Zoo was now general superintendent of the Westchester County Park Commission, overseeing a locale rich in elms. On July 1, 1934, he launched an organized hunt for stricken trees. Within a month his DED troops had located 275 sick specimens, and they feared that by season's end the toll would reach more than a thousand. "The task of control," Merkel worried, "is getting to be enormous, as the spread of the disease is greater than had been suspected in the beginning."

The afflicted New York counties pressed Governor Herbert Lehman for more funds to halt the contagion. "The value of our elms cannot be estimated in terms of money," urged one Westchester official. "Their death would destroy the charm and beauty of thousands of highways and village greens, country clubs and private estates, and would leave even greater scars upon this countryside than did the loss of our chestnut trees." Moreover, what of the moral obligation to stop DED before it imperiled the famous elms of New England?

Even in such hard times, financial support was able to be pooled from federal, state, and local coffers. By the following spring the USDA's

Lee A. Strong, who had taken over from Marlatt as chief of the Bureau of Entomology and Plant, was declaring a tentative victory over DED, announcing on April 21, 1935, "the destruction of the last of the 7,786 trees known to have been infected. . . . Some diseased elms were undoubtedly missed in the area of infection, 5,000 square miles around New York City. . . . For this reason a vigilant watch will be maintained." But as the region's millions of elms leafed out that spring, it did not require much vigilance to perceive that the beetles were again on the wing and wreaking havoc.

In the second week of July, Mrs. F. W. Stevens of 65 Beach Avenue in Larchmont, a leafy Westchester County village, hosted a series of farewell parties for her ancient, doomed elm. Long a local landmark, this gigantic tree had flourished for centuries, fed by a natural spring. Tradition held that "Indians often met at the scene when they were being driven northward from New York City." Her elm, its gnarled, massive trunk fifteen feet in circumference, its canopy a vast umbrella shading her stately home and greensward, was believed to be the largest in the county.

What made the tree truly noteworthy, Mrs. Stevens said, was that it had been often admired by the writer Joyce Kilmer, who had briefly lived nearby with his wife and children. Even the *Washington Post* took note of this elm's impending demise: "Since news of its removal has become known, scores of persons are visiting the spot from miles around to see the venerable elm for the last time. Obviously the tree has become famous, not for its own beauty, but because of the beautiful lines written about it."

On the afternoon of Saturday, July 13, 1935, when tree man Frank Patterson arrived with his crew to execute her beloved elm, Mrs. Stevens found he was equally distraught. As he told a *New York Times* reporter, "My main interest in life has been trees every since I was a small child. To put a saw into this tree is the hardest job I have ever had. I delayed it as long as I could, having first destroyed fifteen other elms that are in our contract. Now I am obliged to begin on this one." When Mrs. Stevens asked permission to have a table cut from the trunk, "she was told that it was in such a segment of a tree that the disease was first brought to this country."

For several hours that day Mrs. Stevens, neighbors, and sympathizers from five nearby states watched as Patterson's men scaled the tree, swooping about in roped harnesses, sawing off its hefty but graceful branches and dropping them with earth-shaking thuds atop a growing pile on the ground. It was terrible to look up and see the destruction of such aged grace. The men departed at sunset, leaving the partially disfigured tree. They returned Monday morning while mourners came by to bear witness and the postman delivered letters of condolence from strangers. But such was the monumentality of the elm that by evening the job was still not done.

Tuesday morning the tree men returned to complete their $147 contract. By late that afternoon, Mrs. Stevens could look up into an open sky and watch the last branches as they were sawed off and dropped to her lawn. Finally only a stark skeleton of the massive trunk was silhouetted against the sky. The crew then sawed through much of the trunk, rigged it with ropes, and carefully maneuvered to pull what was left of the giant to earth. A few minutes before seven o'clock, Mrs. Stevens "watched mournfully as workers gave the last tug which pulled the tree to the ground" with a thundering crash. Where once there had been a rustling canopy, a green world of birdsong, was now open summer sky.

When the tree's raw stump was exposed, Mrs. Stevens and others could count the tree rings and learned that the elm was more than four hundred years old. It had been growing in this spot since Giovanni da Verrazzano sailed into New York harbor. Every part of the cut-up elm would be carted away, saturated with oil, and burned to ash. The heartsickness that Mrs. Stevens and Mr. Patterson experienced at the loss of that one elm was shared by millions as the world of familiar, cherished American elms slowly but steadily disappeared.

Mrs. Stevens's historic elm was just one of the 616,933 dead or diseased elms felled and burned in those first two years as the USDA and its DED army—now three thousand men strong—struggled to locate any American elm tree in the quarantine zone that was dead, diseased, or a possible haven for bark beetles and the fungus. On November 2, 1935, the USDA's Lee Strong announced that in the coming winter another 382,031 elms were slated for destruction in New York, New Jersey, and Connecticut.

By spring of 1936, the USDA's resolve began to waver, its experts suggesting at congressional hearings that stopping DED was not a realistic goal. Conservation groups immediately protested. "From a spiritual, esthetic and historical standpoint, the American elm is unquestionably the nation's greatest tree heritage," wrote Ovid Butler, executive secretary of the American Forestry Association, in a cri de coeur. "Its close and long association with all classes, rich and poor, young and old, has given it a place in the hearts of our people that no other tree can ever fill. Indigenous mainly to the eastern United States, it spread westward with the pioneers until today it is in point of range and numbers the most national shade tree in the entire country.... The expenditure of a few million dollars now, when it is possible to control and eradicate the disease, the Association maintains, is wise public policy in that it will save a national asset dear to all the people and valued at almost a billion dollars."

The Arnold Arboretum brought its Harvard prestige to the fight that spring, urging the USDA to press on with an "intelligent, persistent, aggressive... policy of eradication" to preserve the country's wealth of "American elms of all species." Joseph H. Faull, professor of forest pathology, had joined the arboretum in 1928 to open a laboratory dedicated to better understanding and combating tree diseases. While his specialty was pine rusts, he joined the DED fight, arguing, "By dogged persistence the number of cases can surely be reduced to zero, just as has been true of an eradication campaign against citrus canker in Florida." Moreover, two years into the anti-DED campaign, he pointed out, "the number of trees showing symptoms of the disease appears to be somewhat fewer in 1935 than 1934.... What a contrast with the doleful efforts to get rid of the chestnut blight!" The elms were a more hopeful story. "Our confidence is strong because we believe the United States has begun its campaign in time."

In October 1930, after her year at Harvard, Professor Christine Buisman had returned to Holland, where she began pursuing a completely different goal: finding and breeding DED-tolerant elms. Unlike the American tree experts, she had no hope of saving the continent's existing elms through any kind of sanitation or quarantine—by then DED had spread everywhere, and the elms were dying by the millions. As a leader of the Committee for the Study and Control of the Elm Disease, she assessed thousands of elm seedlings at her laboratory in Baarn, inoculating them with high doses of the fungus. By 1935 she had found two promising DED-tolerant candidates, one from France and one from Spain, and she and colleagues began hybridizing and testing. Progress was promising, especially for clone no. 24.

On March 27, however, Buisman died of an infection following gynecological surgery, a few days after her thirty-sixth birthday. To honor this brilliant scientist, her colleagues felt a fitting memorial would be to name clone no. 24 "Christine Buisman" and release it that autumn to the commercial nurseries, even if it had not been thoroughly tested. Such was the clamor for an elm that could replace all those still-mourned trees that Dutch nurseries were soon selling tens of thousands of "Christine Buisman" elms.

American scientists took note, and in 1937 Curtis May had 35,000 young American elms planted next to the USDA's Morristown headquarters. Emulating Buisman, May and his colleagues began exposing those trees to intense doses of the DED fungus, hoping to find one that could stand up to the disease.

While the U.S. government continued its anti-DED campaign, its haphazard nature infuriated the civic groups who had organized themselves into the National Conference on Dutch Elm Disease Eradication. The USDA, railed the chairman of that gathering, did not know "from year to year what funds will be available or when they will be allotted. . . . For example, in the latter part of June, 1938, all the WPA workers had to be dropped. . . . Even when funds were made available July 1, it was not known until August just exactly how they could be expended. . . . Federal and state officials have done an excellent job under the most discouraging conditions. There is ample evidence in the New York area that the disease can be conquered wherever work is adequately financed."

Nonetheless, when Professor Faull came down from the Arnold Arboretum with Yale forestry professor J. S. Boyce on September 15, 1938, to spend several days touring the quarantine zone, he came away feeling reasonably optimistic. The two men examined field maps and records, interviewed the top campaign generals, and sallied forth to see the battlefront for themselves. Faull had high praise for Herman Merkel's bailiwick—Westchester County—and areas south. In greater New York City, he enthused, sanitation had so slowed DED that only 55 sick elms had been found in 1938, versus 1,264 in 1933–34. In some New York counties it was clear that "the work is dragging," which both professors found "disturbing, because on the outcome of that part of the undertaking depends the future of America's elms."

Faull did write bluntly of another concern—namely, the men supplied by the WPA: "The quality of the field forces, welfare recipients, has been distinctly inferior for scouting purposes. . . . Most of the men now employed are city men. They generally dislike the work and many of them are afraid and helpless in wooded tracts." But given the steady progress in New Jersey, Connecticut, and most of New York, Faull declared, "I still think we can preserve our elms."

On September 18, as Faull was finishing his inspection of the DED front lines and heading home to Boston, days of torrential rains began. On September 21 at 3:30 p.m., the most destructive hurricane in New England history smashed ashore on Long Island, roaring across Connecticut and into Massachusetts, its winds reaching 180 miles an hour. The storm's eye was forty miles wide, and when the worst passed and stunned residents emerged to survey the damage, they could not fathom the scale of ruination. Buildings, cars, docks, boats, and trees were wrecked or upended, and almost seven hundred people were dead.

That period of lashing rains left the trees waterlogged and vulnerable. "Their leafy crowns caught the full force of the coming storm," writes historian Thomas J. Campanella. "Century-old trees were tossed about like matchsticks. The elm-bowered towns of western Connecticut and the Connecticut River valley were the hardest hit." New Haven alone lost 13,500 elms in the ferocious winds, while 7,000 were severely damaged. One resident stared out at the wreckage and lamented: "Here, where the lordliest of trees stood, all is waste and desolation. . . . The glory of the years is gone, the beauty so long in the building is vanished in the twinkling of an eye." The second-largest city in Massachusetts, Worcester, was particularly hard stricken, the storm having uprooted 8,000 mature street trees and mortally damaged another 15,000 public trees along avenues and in parks.

Faull had returned home to Boston in time to witness both the hurricane and its aftermath: 8,000 dead or damaged street trees, and at the arboretum another 1,500 trees ripped from the earth or shattered into pieces. For the arboretum the storm ultimately proved a blessing in disguise, as it cleared out masses of decrepit trees and inspired a rejuvenation and rethinking of the grounds. This was not the case elsewhere. "Many of the historic trees on Boston Common and in the Public Gardens at Boston were reduced to a mass of splinters," reported E. P. Felt in *American Forests,* noting one superstar survivor: "The Rugg elm, in Framingham, Massachusetts—the largest elm in the state—came through the storm practically without damage." Planted in 1774, the Rugg Elm was admired for its size and style: It had two massive trunks (one seventeen feet in circumference, the other fourteen feet) and a remarkable but elegant crown measuring 145 feet across.

The famous Wethersfield Elm of Connecticut—the nation's largest American elm tree—also survived, though this sturdy ancient of deep-buttressed roots and a dozen giant trunks had lost two of those arching trunks to the ferocious storm, marring its classic shape. The Wethersfield's history dated back to 1758, when twelve-year-old John Smith was driving family cattle home from nearby Hangdog Hill and pulled an elm sapling from the ground to use as a switch. "That small twig was planted on the east side of the Broad street green" in front of the Smith home, wrote the *Hartford Times* in 1925, ". . . until now, after two centuries, it stands in its grandeur, far exceeding in size any tree of its kind in New England or even in the United States. For miles around the old Wethersfield elm is known and is a landmark dear to the heart."

Fittingly, Wethersfield's great elm had not one but two origin stories. The first told of the young boy bringing back a sapling to his home. The second was far more colorful: "When Sarah Saltonstall came from New London to be the bride of David Buck of Wethersfield, she intended to bring, after an old Connecticut custom, a bridal tree to plant, but ice on the river prevented the transportation of any gift except herself," relates Donald Culross Peattie in his classic *A Natural History of Trees of Eastern and Central North America.* "Next spring she encountered an Indian bearing an Elm sapling in his hand, and after a pow-wow in sign language, secured it in exchange for a quart of rum. Perhaps more impressive than the fame and stature of Sarah's Elm is the mystery of how a church-going lady would have a quart of rum about her!" Legend had it that this was yet another of the elms under whose capacious limbs General George Washington had once stood.

In 1908 the Wethersfield Elm had gained more modern renown as an old tree nursed back to good health by the new art and science introduced by John Davey, tree surgery. The townspeople, alarmed at the tree's visible decline, had voted to adopt this new scientific approach to address it, hiring for $150 one H. L. Mead, a forester and tree surgeon from nearby Hartford, "to preserve [the elm] for future generations." Town clerk S. F. Willard described the work involved: "All broken or decayed limbs have been cut away to the extent of nearly six cords of wood. After cleaning out and cutting away all decayed substances from knots and holes a coating of thick paint was applied and the apertures were filled with stone and covered with cement shaded with lampblack, to resemble as closely as possible the bark.... One large limb has been weakened and split so this has been chained to another large limb, the chain being fastened by bolts passing through the limbs, the holes being bored by a specially made augur.... The lives of trees may thus be prolonged."

And indeed, the Wethersfield Elm flourished anew until in 1925 it was "threatened with death by poison due to a leaky gas main." The gas and light company paid a $1,200 fine for killing other public trees, and the town used this money to again employ tree surgeons to shore up the gargantuan elm. "More than a thousand feet of cable will be used to brace it," reported the *Hartford Times*. "For protection against windstorms, dead branches will be removed and the old cement will be replaced by reinforced cork after the cavities have been scientifically treated, and the tree will be sprayed and fertilized. The tree will be under the care of the tree company for two years."

Such was the national celebrity of the Wethersfield elm that film crews from Pathé, Famous Players, Fox News, and International News arrived that summer and set up on the town green with their bulky movie cameras to capture for newsreels the tree surgeons climbing high up among the twelve gigantic trunks, cleaning, pruning, and cabling. By 1930, when the rejuvenated tree's circumference was 36.5 feet, its crown spanned 165 feet across, and the U.S. Geodetic Survey confirmed its preeminence as the nation's largest American elm.

While the Wetherfield Elm and others survived the fierce storm, it had long-ranging consequences throughout the Northeast. "Dutch elm disease and the Great Hurricane," wrote Thomas Campanella in *Republic of Shade,* "converged with remarkable precision, as if an unseen hand had guided each to a fateful rendezvous. In the summer of 1938, Dutch elm disease was known to have reached the Bridgeport area, and an infected tree had been recently discovered northeast of New Haven. . . . The hurricane met the plague to roll out a mighty red carpet of dead wood from Connecticut to Vermont." The storm's gale-force winds had transported beetles far beyond the existing quarantine zones, while mountainous piles of newly dead elms now sprawled across New England, awaiting the beetles and the DED fungus. Faull acknowledged the storm but still wrote optimistically: "In formulating the [USDA] program for 1939, the damage done to the elms within the infected areas by the hurricane of September 21 will have to be taken into consideration."

With Dutch elm disease imperiling millions of American elms along streets and in yards and parks, arborists found themselves almost overwhelmed by all the stricken elms that needed to be attended to. It had now been almost thirty years since John Davey published his seminal *Tree Doctor,* and in addition to the large firms of Davey, Bartlett, and Asplundh, every city and state had many independents with a truck or two. This essential DED work became the catalyst, wrote Richard J. Campana, for a "significant expansion of arborist services to cities, town, institutions and homes."

What little enthusiasm the USDA retained for the anti-DED campaign only continued to wane. By 1940 the quarantine zone had expanded from the original 2,464 square miles in 1934 to 10,900 square miles. To anyone but the most determined at the American Forestry Association, the prospects were grim. In January of 1941 Ovid Butler still insisted, "The fight to save [the elm] is on the way to being won." He conceded that with war looming, "many federal activities can and should be set aside," but those did not include fighting DED, for "abandon the battle for even one season and most certainly all will be lost." But as one USDA pathologist at that time conceded, "Amidst war preparations in 1941, however, Congress virtually scuttled the effort by failing to appropriate further funds for eradication."

CHAPTER TEN

"A Forest Giant Just on the Edge of Extinction!": Discovering the Dawn Redwood

Professor Ralph Chaney of the University of California, Berkeley, in China on his 1948 expedition to see the dawn redwood. *(Used with permission of the University of California Museum of Paleontology [UCMP], Berkeley.)*

As America emerged victorious from World War II, citizens like Elmer D. Merrill, seventy-one, the third director of the Arnold Arboretum, could again worry about the fate of incredibly rare Chinese trees rather than that of entire nations. On January 27, 1947, Merrill, recently retired as professor emeritus, wrote to his Long Island nurseryman friend Henry Hicks: "I'm personally tremendously intrigued with a recent discovery in Szechuan. A giant tree like Sequoia and Taiwania, representing a new genus to be described as Metasequoia or some such name." Merrill was sharing an exciting discovery: "Only three living trees in the stand! Later another grove with perhaps 20 trees located. We got a botanical specimen recently & Dr. Hu has promised to send seeds as soon as he gets them. Here is a forest giant just on the edge of extinction!"

A passionate botanist of Asian flora, and possessed of an unrivaled knowledge of the subject after half a century devoted to hands-on collecting and study, Merrill was thrilled by Dr. Hu's news. "If C. S. S[argent] were alive," he wrote Hicks, "and learned of such an extraordinary thing he would probably send out a special expedition to bring home the bacon. I can do it at practically no cost." Here was the first report of an entirely new tree species identified in remote rural China, but one that was facing imminent extinction at the hands of local peasants needing wood. In a country falling into civil war, could this tree be saved? So began one of the more surreal episodes in modern urban dendrology.

Merrill was born, raised, and educated in Maine, where he and his twin brother roamed the countryside but also helped on their grandfather's farm. And as they explored the woods and fields, young Merrill collected birds' eggs, rocks, minerals, Indian relics, and shelf fungi. Within three years of graduating from the University of Maine at Orono in 1899, he settled in Manila to work as the USDA's chief botanist for the bureaus of both agriculture and forestry in the Philippines. Though slight in stature, the bespectacled Merrill had the necessary iron constitution to thrive in the tropics and reveled in travels to the wilds of the islands, as well as to China, Borneo, and Japan. In 1907 he married and soon had four children, though after his second son died, his wife and family moved back to live in the United States in 1915.

During his two decades in Asia, Merrill was an enthusiastic botanizer, and through his own ventures, constant collaborations, and networking established an herbarium of 275,000 mounted specimens. When Chinese botany students first began to return home from studies in the United States in 1918, Merrill used his vacation time, he told Hicks, to travel to Canton and Nanking to "help train Chinese botanists in field methods.... The result: there is now in China a body of trained men who know their way around field work appertaining to both horticulture and botany."

In 1923 Merrill returned to the United States as dean of the School of Agriculture at the University of California at Berkeley, where he was much admired as "a scholar, a scientist, and above all a lovable human being." He proved a talented administrator who reorganized a faculty of 350, overhauled the curricula, added buildings, and emphasized fundamental research. Given his impressive track record, the New York Botanical Garden recruited Merrill to take over when Dr. Nathaniel Lord Britton, head since its 1895 inception, retired in 1929.

Arriving in New York at the onset of the Depression, Merrill wondered "why I ever left such a dynamic progressive expanding institution to cast my lot with an institution that was practically static, very badly underfinanced and more or less under fire for its past policies and accomplishments." He immediately engaged the Works Progress Administration and secured one hundred laborers to revive the botanical garden's grounds with new walkways, rock gardens, and floral displays. He welcomed 120 WPA women, who were assigned to the library and herbarium working as artists, clerks, secretaries, and technicians. Their sheer numbers enabled Merrill to eliminate the herbarium's large backlog of unmounted plant specimens, and also to dramatically reorganize and upgrade the entire herbarium system and vastly enlarge its specimens.

In 1935 Harvard lured away Merrill to be the Arnold Arboretum's third director. It was his task to foster greater integration and coordination among eight separate botanical fiefdoms, further advancing what Oakes

Ames had achieved. Merrill proposed that Harvard, with three of the largest botanical libraries in America and its three great herbaria, start by combining in a new building all three libraries with those of the Arnold Arboretum, while maintaining the three herbaria as separate but all housed in the new building. The botanists were amenable, and Merrill launched planning. Unlike Sargent and "Chinese" Wilson, Merrill was not particularly interested in the arboretum's tree museum, so he appointed a horticulturalist, Donald Wyman, to manage it and engage with the public. Merrill was, however, delighted to underwrite new botanical expeditions. Drawing on his many years in Asia, he cultivated a valuable corps of Chinese botanist-collaborators. A mere $50 to $250 could fund a single plant expedition, and Merrill's institution received half the botanical booty.

During World War II the U.S. secretary of war regularly summoned Merrill, with his familiarity with the Pacific islands, to share his now-invaluable botanical knowledge. Merrill prepared a handbook, *Emergency Food Plants and Poisonous Plants of the Islands of the Pacific*, which was reproduced in many survival manuals issued to the troops. He even encouraged American servicemen whiling away dull hours on atolls to take up botanizing on behalf of the Arnold Arboretum, a pastime "commanding officers and postal censors considered . . . [a] violation of security regulations and many shipments were confiscated."

When the war ended, Merrill was delighted to hear again from Dr. Hu in China, and eager to relaunch his Asian plant explorations. Merrill dispatched $250 to Hu, who had studied at the arboretum and received a Harvard PhD in 1925, to fund further pursuit of the rare giant tree in the remote valleys of Szechuan Province.

In 1937, after Japan invaded China, the Chinese government had retreated to the interior and, anticipating a long struggle, began dispatching surveyors to remote regions to scope out natural resources. It was in 1941 that a young forester on just such a mission noticed a hundred-foot-tall tree that looked like a redwood with a distinctive buttressed trunk flare. The natives called this towering tree a *Shui-sa* or "water fir." With war raging, it would be several years before another Chinese botanist managed to return to the village of Maodaoqi to secure a branchlet of the tree, along with two of its cones. He concluded that the specimen was just a common water pine.

To resolve the issue, one Hsueh Chi-ju, a new forestry graduate from the National Central University in Nanking, set forth by steamboat in 1946 from Chungking on the Yangtze River, debarking two days later at Wanxian. "After crossing over, I had to walk 72 miles to my destination," wrote Hsueh. "There was no highway. . . . My trip was very difficult, and the trails threading through the mountains being less than one foot wide." Several peddlers joined him the second day but turned back after the local innkeeper warned of bandits and gave "a vivid and horrible description of a murder."

"Finally, at dusk on the third day," Hsueh continued,

> I reached my destination safely. I set out immediately to search for that colossal tree despite hunger, thirst, and fatigue, and without considering where I would take my lodging. It was February 19th and cold. The tree was located at the edge of the southern end of a small street. In the twilight nothing was discernible except the withered and yellowed appearance of the whole tree. My excitement cooled.
>
> "Am I to bring back some just dried branches?" I asked myself.
>
> The tree was gigantic. No one could have climbed it. As I had no specific tools, I could only throw stones at it. When the branches fell from the tree, I found, to my great surprise, that there were many yellow male cones and some female cones on the leafless branches. I jumped with joy and excitement.

In May the dedicated young forester again flew to Szechuan and retraced his arduous journey to "collect cone-bearing specimens in addition to ascertaining the [tree's] natural distribution." He measured the "colossal tree," which soared 122 feet, while its huge buttressed trunk turned out to be twenty-three feet in girth. The villagers, who revered this tree as quasi-divine, had built a shrine next to it. They viewed the tree's fruit bearing as a foretelling of coming crop yields, and withering twigs or branches as "a forecast of someone's death." When passing missionaries had once asked to cut down the tree and buy its wood, said Hsueh, the villagers had refused: "Feudalistic superstition . . . [saved] the tree." Hsueh estimated its age as four hundred years. However, villagers thought nothing of chopping down the other local water firs for firewood and building, and Hsueh located only twenty more such trees. He did hear rumors of other stands many miles deeper into the mountains.

By that fall Dr. Hu at the Fan Memorial Institute of Biology in Peking had received the collected twig and cone specimens from Professor Cheng, Hsueh's teacher. Cheng wondered if Hu agreed that this "colossal tree" was closely related to the American redwoods. Hu, in fact, had just been reading newly available articles by Japanese paleobotanists describing fossils they had found near Tokyo and Manchuria of an ancient but now extinct redwood that they named *Metasequoia*. These relics differed from the still-living American sequoia by virtue of "an opposite arrangement of the leafy twigs and cone scales . . . and the cones were attached to tree branches by long stems called peduncles." In a rare eureka moment, Hu examined photos of the then extinct fossil, matched them against the herbarium specimens from Cheng, and concluded that the "colossal tree" in Maodaoqi village was actually the supposedly extinct *Metasequoia*.

While it was not as ancient as the ginkgo biloba, *Metasequoia glyptostroboides*, as Hu named it, also boasted an unbroken lineage dating back to the dinosaurs. And like the ginkgo, this rediscovered "living fossil" had in long-ago epochs populated temperate-zone forests throughout the world, including America.

Professor Hu promptly published "Notes on a Paleogene Species of *Metasequoia* in China" in the *Bulletin of the Geological Society of China* and sent the exciting news and some herbarium specimens to Elmer Merrill in Boston to get his opinion and ask for funding for a further expedition, which Merrill happily provided. Financed by the arboretum's $250, yet another forester, Ching Tsan Hwa, spent three months scouring ever-more-remote valleys of Szechuan, finding one hundred large *Metasequoia*. Over the Hupeh Province border, in a temperate region of considerable rainfall with occasional ice and snow in the winter months, Hwa came across almost a thousand of the trees growing on forested slopes at altitudes of about three thousand feet. Many of the trees were small and scattered among other species. By early December Hwa was back in Nanking with seeds, and the warning that local farmers were rapidly cutting down the remaining water firs for fuel and lumber.

On January 5, 1948, Professor Merrill received the first seeds from Dr. Hu of the rare Chinese *Metasequoia*. As was the tradition, he reserved a certain number for his arboretum, where they were quickly planted in the heated propagating house. He then dispatched the remaining seeds to almost fifty arboretums, botanical gardens, and nurseries around the Western world—institutions as disparate as the Morton Arboretum; Edinburgh's Royal Botanical Garden; the Batumi Botanical Garden in Batumi, USSR; the Botanischer Garten Rombergpark in Dortmund, Germany; and the University of California at Berkeley. The goal, said Merrill, was "to establish this ancient but now nearly extinct type in various parts of the United States and elsewhere."

Out on the West Coast, Ralph Chaney, professor of paleobotany at the University of California at Berkeley, had also been hearing from his old friend Dr. Hu, who had earned his undergraduate degree at Berkeley, about the discovery of *Metasequoia*, but he had been skeptical. Sitting in his office on the top floor of the Hearst Mining Building, Chaney might have looked like the classic mild-mannered academic with his thinning hair, pouchy eyes, tweed jacket, and tie, but he had traveled widely in South America and Asia hunting fossils and in 1925 had been among those on a major Mongolian expedition, crossing the Gobi Desert from Peking with 125 camels.

"I really didn't believe any of this," Chaney said in an interview with his old friend, science editor Milton Silverman of the *San Francisco Chronicle*. "Surviving Metasequoias? It was completely preposterous." Even receiving seeds from Professor Merrill had not convinced Professor Chaney. But now, to his wonder, he had received a new envelope from Dr. Hu and held in his palm a few greenish twigs and needles, as well as more rare *Metasequoia* seeds—which looked like small, rolled, dried corn kernels. Examining these actual specimens from the tree he had heard so much about, the professor looked up at Silverman and said, "If this is confirmed, it will rank as the greatest botanical discovery of the century." While holding the botanical evidence in his hand, Chaney decided he would "visit [the tree] in its native home. I wanted to see how it lived."

Chaney told Silverman to come along to the main library, where they spent time poring over maps of China. Having located Wanxian and seen that it had an airfield, Chaney, undaunted by remote locales, told the journalist, "There'd be absolutely nothing to it. We could fly to Shanghai. Then we could get a pilot. . . . You can always find some joker who'll fly us to a place like Wanxian. We'll land there, maybe stay overnight, then the next day pack a couple of picnic lunches and hire a few coolies . . . walk over to where the trees are, collect cuttings and seeds, and maybe some samples of the wood, take a lot of photographs, eat our sandwiches, and then get back to Wanxian. Next day return to Shanghai and catch the first plane back to San Francisco." Silverman's editors were game to send him on the quick in-and-out expedition to see these ancient trees, and the two were soon arranging their trip.

Merrill, who knew Chaney well, was not pleased to hear that he was off to see the trees in their native habitat. By now the seedlings had sprouted in Boston, Berkeley, and many other locales, and Merrill thought these tiny treelings should suffice for research. If Chaney did insist on traveling to China, why not go in the summer, when he could collect more seeds? But Chaney, who had spent his professional career having to imagine what ancient trees looked like, was determined to see the living version as soon as possible, especially because China, engulfed in a civil war, was facing a potential imminent political collapse.

In late January Merrill and Harvard issued a press release about the Arnold Arboretum's funding of the Chinese foresters and their discovery of "Dinosaur-Age Trees" believed to be extinct. While the Boston-based *Christian Science Monitor* respectfully reported the story, few other newspapers picked it up or grasped what a botanical sensation the finding was.

On Friday, February 13, Professor Chaney and Silverman set off for China in pursuit of what a *San Francisco Chronicle* editor had now christened "the Dawn Redwood," in homage to the tree's antiquity. In Shanghai Chaney's well-laid plans began to fall apart when local pilots said they knew of no airfield near Wanxian, a fact confirmed in Nanking by the American ambassador. Chaney and Silverman resigned themselves to reaching Wanxian by plane, steamer, and foot. Ching Tsan Hwa, the forestry student who had collected all the *Metasequoia* seeds the previous fall, joined them as their guide, along with a translator and eleven "coolies."

This large entourage debarked in the rain and set out on narrow paths on a journey described to them as "not too far." Wrote Silverman, "We sloshed through mud, slid down precipitous slopes, and crossed one

The dawn redwoods found in China (note bare winter branches) by the 1948 Chaney expedition (left), and a descendant of those trees on the Berkeley campus (right). *(Left photograph used with permission of the University of California Museum of Paleontology [UCMP], Berkeley. Right photograph courtesy of Rowan Rowntree.)*

rain-drenched, mist-covered ridge after another. The farms had become fewer, farther apart, and poorer." As they rose the next day to again follow the tortuous path, the Americans finally understood they were on a major pilgrimage, ascending hour after hour through rain and thick mist to cross peaks and passes that Chaney's altimeter showed to be up to 5,500 feet. Chaney's asthma kicked in, but there was no turning back. After two more days on the steep, narrow paths, passing through remote farms and occasional villages, they arrived in Maodaoqi, where it was market day. The streets teemed with men and women in dark, padded, ankle-length gowns, while pigs and chickens wandered the dirt streets.

"We, however," wrote Silverman, filthy and exhausted, "were paying attention to something else. At the edge of town, adjacent to a rice paddy, stood a towering tree, perhaps a hundred feet high and about ten feet in diameter at the base. It was practically bare of needles, and its long, graceful branches were outlined sharply against the sky."

"This is the big *Metasequoia*," Mr. Hwa announced. Chaney stared at the tree, incredulous.

"My God!" he said. "It's dropped all of its needles. It's the only redwood tree in the world that's deciduous, that loses its needles in winter!"

The villagers were happy to speak about the powers of their sacred tree, with its small shrine. One told how his wife of many years had been barren, but after making sacrifices to the tree, he had soon become father to a "fine son." Another man described first praying to the tree as his daughter was dying, and then brewing a tea for her from the tree's bark, a preparation that saved her life.

Chaney and Silverman were shown the two smaller dawn redwoods of thirty feet in height, and while it was still light, Silverman took rolls of photographs while their assistants climbed up into the sacred tree to collect cones and such twigs as there were. When night enveloped the village, Chaney returned quietly with his "increment borer" to extract a thin core. Back in their sleeping quarters in the village school, he and Silverman counted tree rings as best they could, concluding the tree was between five hundred and six hundred years old, "born" in the time of Marco Polo.

Chaney had decided the expedition would press on to see the forest with many *Metasequoias* that Hwa had discovered in the Valley of the Tiger, a dangerous journey of several days. He was thrilled to report that in just a short time he had seen almost one hundred big dawn redwoods there, and equally exciting was the fact that they were part of a mixed deciduous forest of birch, oak, sassafras, sweet gum, and chestnut, "the very trees whose fossils are found associated with *Metasequoia* fossils. They were growing together all those millions of years ago. They're growing together now. It's like a botanical alumni reunion. This is what much of the world looked like a million centuries ago. I wouldn't be surprised to see a tyrannosaur or a brontosaur coming out of this canyon."

On March 25, 1948, the *San Francisco Chronicle* featured a huge eight-column headline across the top of its front page: SCIENCE MAKES A SPECTACULAR DISCOVERY and 100,000,000-YEAR-OLD RACE OF REDWOODS. The first story, datelined Mo-Tao-Chi, described their amazing botanical adventure and quoted Chaney saying, "To me, finding a living dawn redwood is at least as remarkable as discovering a living dinosaur." The syndicated story was picked up by United Press, Reuters, and news editors not just in America but all over the world. NBC Radio featured it coast to coast. Subsequent articles, which detailed the expedition onward to the Valley of the Tigers, got equally high-profile play.

Chaney returned to the United States a week after Silverman, entering the United States at Honolulu Airport on April 2. At the Plant and Economic Quarantine section, the USDA's inspector demanded that he hand over for incineration the four precious tiny dawn redwood seedlings he had brought from China. Increasingly distraught, Chaney began protesting that these plants were priceless and, as the customs man became more insistent, the professor began shouting repeatedly: "Millions of years! Tens of millions of years! A hundred million years!" The four seedlings finally were permitted through as antiques.

On his return to America Chaney discovered that Silverman's front-page scoops had catapulted *him*, the paleobotanist, along with the *Metasequoia* tree, into celebrity. "It's wild," he told Silverman in amazement. "I've got interview requests from all over the country, and from London, and Paris, and Edinburgh, and Geneva. I have enough invitations to give lectures—from universities, from garden clubs, from technical societies—to last me for years."

Elmer Merrill, meanwhile, was irate at these developments, convinced that Ralph Chaney was deliberately grabbing headlines and taking credit for what the world was finally acclaiming as the botanical find of the century. As Merrill would repeatedly point out, the Arnold Arboretum had provided *further* funding for Professor Hu, an Arnold alumnus, to mount two additional expeditions to locate more stands of dawn redwoods. Moreover, Merrill had received large numbers of the tree's seeds as a result of those explorations in early 1947, a year before Professor Chaney set off for China. And it was the Arnold Arboretum—per its policy set under Charles Sprague Sargent—that had dispersed those seeds to other arboretums and plant growers throughout the world. And so began an on-again, off-again scientific feud.

By 1948 Merrill, seventy-two, was professor emeritus, retired, and no longer director of the arboretum. He had proven to be as productive as ever, expanding the arboretum's herbarium by 220,000 new specimens and keeping Asia as its focus. The war delayed the new Harvard botanical building, and by the time it was under way, Oakes Ames and Merrill both had had second thoughts and now opposed moving the arboretum's library and herbarium. They found an ally in the arboretum's new director, Karl Sax, not to mention the "many friends of the Arboretum" who viewed this as a betrayal of Sargent's life's work and a violation of the original agreement. Attorneys were engaged, and "the Controversy," as it was genteelly dubbed, went to law and the courts. Harvard prevailed, and in 1954 the major parts of the arboretum's herbarium, library, and wood collection were relocated to a consolidated Harvard University Herbaria in Cambridge.

With postwar communist China closed to Western plant hunters, the much-acclaimed dawn redwood would be the Arnold Arboretum's last major tree introduction. It was the end of an era, and the arboretum's three subsequent directors grappled with defining its modern mission and with how to make relevant this apparent (and now rebellious) stepchild of Harvard University. The fear always hovered that the arboretum would be reduced to a "mere park."

The year Harvard bested the arboretum in the courts, David Fairchild, who had devoted the final two decades of his life to tropical plants, died at age eighty-five in Miami, Florida. His August 7 *New York Times* obituary noted his role in introducing Russian wheat, Japanese rice, and other edible plants for "the American table," and also many "flowering plants and ornamental shrubs to beautify the American garden." Today Fairchild is better known for bringing us mangoes, dates, nectarines, bamboo, and, of course, his own favorites, Japanese ornamental cherries. He enjoyed a delightful retirement, botanizing and yachting with friends and family in such exotic climes as the Celebes, Bali, and the Moluccas and writing charming memoirs.

In a major second act, Fairchild shaped the collections of the lush eighty-three-acre Fairchild Tropical Botanic Garden in Coral Gables. Established in 1936, the garden was named in Fairchild's honor by his friend Robert H. Montgomery, an attorney, businessman, and avid plant collector. Today the Fairchild Tropical Botanic Garden is an important horticultural treasure and research center with major collections of palms and other rare tropical plants.

As for the dawn redwood, it was about to embark on to a most unlikely but glorious career. With World War II over, and the dirge for the American elm once again heard across the land, the naturally optimistic American public loved the story about the ancient tree saved from the brink of extinction in such a dramatic fashion. By 1951 the dawn redwood entered the nursery trade, and all the publicity surrounding Chaney's expedition guaranteed that every American landscaper, arboretum, and city arborist clamored for it. (In truth, the dawn redwood looked much like the American bald cypress [*Taxodium disticum*], and some could distinguish them only by the characteristic placement of their needles: Those of the dawn redwood were opposite each other, while those of the bald cypress were alternating.) In time the dawn redwood would become a striking but commonplace ornamental urban tree in American city parks and backyards, though homeowners often failed to appreciate what a fast-growing giant they had just planted. While the dawn redwood could never replace the American chestnut or *Ulmus americana*, this charismatic immigrant turned out to revel in the hurly-burly of city living, making itself at home in all sorts of urban settings.

CHAPTER ELEVEN

"There Was No Question That People Wanted to Save This Tree": Crusading for a New American Elm

The famous Wethersfield Elm in Connecticut in about 1860 *(above)*, and a closer shot in about 1900 *(below)*. *(Courtesy of the Wethersfield Historical Society.)*

By the summer of 1948 the park commissioners of Wethersfield, Connecticut, were waging a rear-guard battle to save the historic Wethersfield Elm, still the glory of the town (otherwise best known for its red onion fields and grim state penitentiary)—and now seriously afflicted with Dutch elm disease. As feared, during the war years the U.S. Department of Agriculture had all but abandoned the fight to save American elms, leaving the bark beetles to swarm unimpeded through all of Connecticut and Massachusetts, into Maine, and onward to the midwestern states.

Two years earlier the commissioners had confirmed infection of their own great elm, and tree surgeons had amputated the famous tree's individual afflicted limb—itself as gigantic as a century-old tree—in an attempt to staunch the disease. The following spring, though, anyone scanning the canopy could see the sickly yellow of the leaves and their meager numbers. By July 28, 1948, the tree surgeons were again aloft in the old elm, removing two more of its largest remaining branches. "The sad truth," wrote a reporter for the *Hartford Times,* was that "it looks as if the fight has been lost. . . . [The tree] is on its last limbs." The photo accompanying the article showed a disfigured, much-diminished elm, with huge gaps where one giant trunk branch after another had been amputated. Whenever the tree suffered another surgical event, the tree warden said, he was "deluged with requests for information, for pictures of the original tree and of the tree in its present state. The requests have come from tree wardens and horticultural societies in 30 states who have read of the tree's plight."

By March 1, 1953, the six final limbs had been sawed away, and Wethersfield's residents gazed in wonder upon a tree stump unlike any other, big and sturdy as a small one-story car garage and weighing thirty-five tons. Tree surgeon William George and town workers spent the next two months digging a sixty-foot trench around the perimeter of the monumental stump, methodically chopping and sawing several feet down to cut through the outer roots in order to dislodge the remains. Time and again equipment broke down as the tree refused to yield its anchorage in the earth.

In late April George brought in more powerful equipment, and the crew was finally able to slowly rip the stump from the ground. Over the next three days they carefully maneuvered the tree stump onto its side, its sawed-off root flare and truncated root ball now exposed, raw and bare in the spring sunshine. The space where the elm had long stood, in front of the Smith house overlooking the town green, was now but a wide, gaping hole. The sky felt large and empty.

The people of Wethersfield swarmed about, boys in plaid shirts and jeans clambering upon the prone stump, teenagers leaning in to run their hands over the weathered bark. The visitors came to commune with these arboreal remains, witness to almost two centuries of the everyday life of a small town. Those present knew they would not see the likes of this tree again in their lifetimes. Nor would their children, nor their grandchildren.

By Friday morning, May 29, the tree crew was ready to complete the job. They backed up a flatbed truck and with a powerful winch dragged all that was left of the Wethersfield Elm onto the platform, securing it with thick wires to keep it upright. Across Broad Street the first- and second-grade children of Welles Elementary School lined up solemnly, paying homage to the sheer scale of the tree's remains and their own loss. The end was

Death of the famous Wethersfield Elm in May 1953. *(Courtesy of Wethersfield Historical Society.)*

ignominious. The great Wethersfield Elm was "taken from its Broad St. site to the town dump, its final resting place. The trip took about an hour," reported the *Hartford Times.* Several massive branches were distributed to town groups to be made into souvenirs.

Civic leaders up and down the East Coast and in the Midwest now faced the same problem: how to rescue the American elms not yet infected? Mayor Donald A. Quarles of the New York City suburb of Englewood, New Jersey, proposed to save his town's 3,500 remaining elms by engaging the very latest in science and technology. The new pesticide DDT (dichlorodiphenyltrichloroethane) had gained fame during the war for eradicating malarial mosquitoes and typhus-spreading lice. Why not deploy it to defeat the elm beetles? On Wednesday, October 23, 1946, Quarles, other local city leaders, and officials from the state's Department of Agriculture gathered to watch a demonstration by the Accurate Tool Company of Newark, which was rolling out its "new truck-mounted mobile [extermination] unit employing an aircraft engine."

As the men watched, the truck operator aimed a nozzle at a stand of tall elms and released a billowing cloud of fine DDT mist that enveloped the treetops and then gently rained down, coating the autumn leaves and branches and then the ground below. Impressed, the town signed on for a $4,500 contract. The next spring the trucks rumbled through the town's narrow lanes and bucolic byways, spraying towering clouds of DDT on all the public and private elms. Once again the battle to save the elm was being joined.

In the spring of 1954, at Michigan State University in East Lansing, Professor George Wallace welcomed a new doctoral candidate in ornithology. John Mehner began his research by scouting the 185-acre campus with its lovely old elms, identifying 370 adult robins to observe and follow. While Mehner peered into thickets and trained binoculars up toward branches, looking for nests, the school's groundskeepers were spraying elms for the first time with DDT, typically treating each fifty-foot tree with two to five pounds of the pesticide. Each subsequent year university crews repeated the process. By 1958, four years after spraying had begun, Wallace observed: "At no time during the spring or summer did I see a fledgling robin anywhere on the main campus, and so far I have failed to find anyone else who has seen one there."

As DDT became the main weapon against the Dutch elm beetles, bird lovers were horrified to learn from scientists that once the pesticide entered the food chain, it took only eleven earthworms that had fed on treated elm leaves and ingested sufficient toxin to poison a robin. A student on an upstate New York campus recalled seeing a robin soon after a DDT application: The bird "exhibited strange symptoms: staggering, weakness, tremors as if it had a chill. Hardly had it expired in our hands than another was found—and another until we had nine of them."

From Wisconsin a birder reported, "I have had five or six pair of cardinals in the past, none now. Wrens, robins, catbirds, and screech owls have nested each year in our garden. There are none now. Summer mornings are without bird song. . . . It is tragic and I can't bear it." In 1962 biologist Rachel Carson argued in *Silent Spring*: "It would be tragic to lose the elms, but it would be doubly tragic if, in vain efforts to save them, we plunge vast segments of our bird population into the night of extinction. Yet this is precisely what is threatened."

DDT, she stressed, had no track record of actually saving elms. In Urbana, Illinois, after six years of regular DDT spraying, the "university campus had lost 86 percent of its elms, half of them victims of Dutch elm disease." The cities that had best preserved their elms, she noted, engaged in old-fashioned "sanitation"—the rigorous removal of dead branches and trees. Plant pathologists had learned from experience that the beetles thrived best in deadwood.

Among the millions of Americans grappling with the loss of the elms was John Hansel, a Princeton graduate, former Marine Corps officer, and New York City businessman. He lived with his family in a house in Riverside, Connecticut, which he had purchased in 1962 after having fallen in love with its four huge, spreading elms. "My boyhood home had tremendous elms," he said. "Those trees were symbols of my past. You remember the cathedral of shade you grew up under." At a time when state arborists estimated the nation was losing another million elms each year, Hansel's luck with his own trees had held.

But then, in 1964, like so many others, he noticed on one summer day a few leaves on high-up elm twigs yellowing, and then whole sections of its canopy looking sickly. Hansel recalls contacting the Bartlett Tree Company in Stamford: "What is being done to find a cure? They gave me a brief history of the disease, but were ashamed to report that little was being done to find a cure."

When the second of Hansel's trees began to succumb, Hansel was not the sort to impassively await the death of his remaining pair. "I found it hard to believe," he would later say, "that millions of us were concerned over the fate of the whooping crane or a stand of redwoods—which many of us will never see in our lives—while practically nothing new was being done about the fatal disease in the trees over our heads."

That year Hansel, the president of Filtrine, his family's national water-filtration firm, launched Elms Unlimited, writing thousands of letters to newspapers and magazines, urging others to join his crusade and to express their hope by replacing the dying elms with new ones. He found many allies, others yearning to do *something.* "The responses were poetic," he recalled, "they were passionate; there was no question that people wanted to save this tree."

Millions of Americans had grown up in an elm-shaded world, and as the poetry of those trees disappeared, citizens were distraught. "When you came into any town," recalled one, "... the landscape changed. You entered this kind of forest with 100-foot arches. The shadows changed. Everything seemed very reverent, there was a certain serenity, a certain calmness." As the elms came down, "you started to notice the severity of things—the wires and utility poles, the cracks in the hot pavement, which no longer was bathed in shadows." The summer temperatures were ten or twenty degrees hotter on streets without their lofty elms, and birds far, far fewer. An ethereal beauty was giving way to a graceless, even ugly, reality.

Hansel offered the prospect of somehow saving or eventually re-creating these disappearing vistas. He would often say in those early days of his campaign, "I am *not* a tree expert, an entomologist, a pathologist, a biologist, or even a horticulturist. I am purely and simply an optimist." He blamed "pessimists" for wasting precious years in "cut, burn and spray techniques, without the necessary research to find a long-range solution. As a result, the tremendous losses have convinced them that we should not waste time trying to save the remaining elms, but let them go the way of the [American] chestnut."

As Hansel liked to tell the tale, though, "One rainy morning in April [1967] the elm got a new lease on life. A meeting was held in Greenwich, Connecticut. . . . [People] came from as far away as Iowa, Wisconsin and Canada, all with a common purpose: to unite in a national campaign to develop the research needed to find a control for Dutch elm disease. . . . [Those present] founded the Elm Research Institute."

Hansel proved to be a dynamo of a leader, much admired by his fellow elm crusaders for his winning personality, organizational skills, and willingness to use the considerable resources of his family-owned firm to advance the cause. In December of 1969 he and the Elm Research Institute (ERI) appeared in the pages of *Time* magazine, where he appealed to the American belief in science, exhorting: "The fight against Dutch elm disease will be won in the laboratory." The photo that ran with the story showed Hansel standing defiantly on a giant elm stump, the sky behind him graced with the very tree that inspired his passion, an elegant winter-bare elm. *Time* reported that thirty thousand dollars in ERI funding for University of Wisconsin entomologist Dale Norris promised to yield an elm beetle repellent that might be injected or sprayed on the bark.

Time also mentioned that the U.S. Department of Agriculture was underwriting ongoing breeding programs that crossed "the disease-prone American elm with the hardy Siberian variety." Hansel, however, was dismissive of such efforts, citing the hybrid's ungainly shape: "Who will want a tree that looks more like a maple than an elm?"

Hansel soon recruited a number of high-powered allies—the Audubon Society, the American Conservation Association, the Mellon family, Laurance Rockefeller, the Davey Tree Expert Company—and was on his way to raising $500,000 for grants. "In this time of crisis," notes historian Thomas Campanella, "no approach was too absurd." Pheromones, a European predator wasp, systemic fungicides, soil treatments, "secrets" of resistance, determining how the fungus traveled all might yield possible solutions. Although DED had now killed almost 75 million elms since 1933, plant pathologists had learned only in 1960 that one elm could infect another through their intertwined roots, thus explaining why even good sanitation had its limits in cities where American elms were planted close to one another, lining street after street in a vulnerable monoculture.

Once DDT proved not to be the hoped-for magic bullet that would save the American elm, the U.S. government showed even less interest in what seemed a lost cause. By 1969 USDA DED expenditures had dwindled to not quite $167,000. As author Ronald Rood scoffed in *Audubon* magazine: "So there it is: the entire, official Forest Service DED allotment would barely take down 850 mature elms already dead at the going price of $200 per tree. There are that many waiting to be cut in Burlington, Vermont." Praising Hansel's optimism and organizational zeal, Rood declared that he too had become "convinced of a truth that's incredibly simple: the American elm is *not* doomed. . . . With half a chance [American elms] could still be there several generations hence. The Elm Research Institute hopes to help provide that half a chance. So does the USDA Shade Tree project, operating on its small budget."

Even the USDA's chief forest pathologist, George H. Hepting, wondered, "How much more knowledgeable would we now be in our methods of combating diseases and pests of forest and shade trees if that $27 million (which failed to stop the disease), or even 10 percent of that money [spent on cutting down elms before World War II], had been spent on fungicides, insecticides, and disease-resistant varieties?"

In July 1970, six months after the *Time* article, Hansel issued the first of many optimistic press releases touting imminent solutions to "save our elms from extinction." The institute announced "the first practical control measure to emerge from [ERI's] crash research program initiated in 1967 . . . simple, inexpensive plastic pots to hang in the branches of elms. . . . The pots will contain a chemical formulation designed to repel the beetle . . . the culmination of years of tireless research conducted by Dr. Dale Norris at the University of Wisconsin."

On June 22 and 23, 1971, Hansel and the ERI hosted the first-ever gathering of scientists awarded ERI grants for elm research. Even as these academics from Syracuse, Michigan State, Cornell, Iowa State, and the universities of Maine and Wisconsin assembled at Syracuse's forestry lab, the city's elms were coming down around them. "From the penthouse," said Hansel, "the chain saws were so loud we had to close the windows." Hansel was thrilled by the event, reporting to his board, "We talked elms from breakfast to bedtime, and each came away with the feeling that the outlook for a control has never been brighter." To his funders Hansel promised, "By next spring we expect to have available for large-scale testing a beetle trap, an effective fungicide, and a resistant tree . . . the three important areas in which we are working."

Hansel was especially hopeful about Lignasan (benomyl lignasan phosphate), a promising systemic fungicide owned by DuPont but unavailable in the United States. The goal was to save the millions of mature surviving elms by fortifying them against the beetles. Hansel galvanized thousands upon thousands of elm lovers to bombard DuPont with "telegrams, letters, and telephone calls to convince DuPont that saving Elms was more important than defending a patent." DuPont was leery of angering tree lovers if Lignasan proved to be a reprise of the fungicide Bidrin.

In 1965 Shell Chemical Company had introduced Bidrin, described by the *New York Times* as "one of the most lethal chemicals known." Bidrin had "been developed in its initial form in Germany during World War II as a chemical warfare agent. . . . A few drops on the hand are said to be enough to cause serious illness." Shell insisted it would be safe and save elms if properly administered. Soon trained technicians, suited up in protective clothing and face masks, began gingerly injecting Bidrin into the trunks of elms. Too often, though, distraught homeowners watched as their American elms succumbed to a slow death from this lethal cocktail.

Hansel was convinced that Lignasan would have a different outcome, as it already had a far better track record in Canada. His campaign soon persuaded DuPont to sidestep its PR concerns by granting ERI exclusive licensing rights to the fungicide. How best, Hansel then wondered, to get the chemical into the elm's vascular system? One early device, writes historian Campanella, "applied so much pressure that it blew off the bark of a test tree six feet up the trunk. Another required fifty drill points, used specialized injectors, and took hours to set up."

By 1971 John Hansel had moved his company and family to Harrisville, New Hampshire. One wintry afternoon he was in the woods riding his horse, pondering the Lignasan delivery problem, when he noticed maple syrup buckets hanging from small spouts. "Bingo!" he said. "I cantered home knowing, there's the answer. We'll tap the trees and get our stuff in just by transpiration. It was a Mr. and Mrs. Elm Owner homemade Do It

Yourself solution." A Canadian Forest Service researcher named Edward Kondo devised a simple gravity-based injection spout inserted at the base of the tree and demonstrated its efficacy by "injecting a vegetable dye into the base of an elm, turning all the leaves bright red within hours." In 1974 Hansel jubilantly unleashed Lignasan and the simple spouts to be field-tested on elms for at least three successive years.

Kondo significantly advanced the field with his new system of root flare injections, which enabled soluble fungicides to travel all the way up to the furthest reaches of the elm's crown, where beetles were likely to be feeding. But he was only one of numerous scientists who had been pursuing an effective systemic fungicide, and in the next few years a better option—thiabendazole hypophosphite—emerged. Eventually known as Arbotect 20-S, this soluble fungicide from Novartis "moved more slowly in the tree, did not accumulate in the leaves, was more chemically stable, and progressively increased in the stems for at least 12 months."

"But protecting elms is not enough," declared Hansel in his standard stump speech. "We must replant this most beautiful of all shade trees. To ensure the beauty and natural resistant tendency of future plantings we are propagating from selected two-year-old specimens. . . . Eventually, we hope to see the elm, in stock, at every nursery as it once was. The future of the species will then be secure."

While Hansel urged Americans to keep planting native elms, actually buying one was not an easy task. "No sane nursery would sell American elms," recalls Keith Warren, an Oregon nurseryman at J. Frank Schmidt & Son, starting his career at that time. "The American elm was a dead issue. Old-timers would talk about what a wonderful tree it had been, how it grew so fast and could tolerate all the insults of city life. But nobody was replanting because the fate of the American elm was sealed." The vast majority of American nursery trees begin their young lives on "farms" like Schmidt's on the West Coast. After two or three years the young trees would be loaded onto flatbed trucks and shipped east, where local tree nurseries would grow and groom them for several more years. Finally, when five or six years old, they would be sturdy enough to embark on new lives as city dwellers. Very rarely was an American elm among these young trees for sale.

The elm bark beetles, meanwhile, were completing their cross-country journey, having arrived in Sonoma County, California, in 1974, forty years after first debarking at the East Coast ports of Norfolk, Baltimore, and New York. Despite the frequent comparison between the demise of the American chestnut and that of the American elm, however, Dutch elm disease proved to be far less virulent. When the chestnut blight swept through a community, it swiftly wiped out every American chestnut in its path.

Such was not true for DED and the American elm. Almost forty years after DED began killing trees in New York City, the New York Botanical Garden still had hundreds of large elms. In 1967 a Wisconsin arborist had pointed out, "We still have most of our elms left after eleven years [of DED] and I know of very few municipalities that have really had a control program." The fact that certain American elms *were* surviving gave many hope. The new fungicides bought time for existing elms, but as Hansel himself acknowledged, "The end of the rainbow for us really was a tree that was resistant."

By now plant scientists had had several decades of experience searching for a DED-resistant American elm, though it had always been an underfunded afterthought of save-the-elm efforts. In 1933 scientists at the Boyce Thompson Institute in Yonkers had planted 21,000 trees in their American elm nursery and exposed them to DED. In 1947, after the end of World War II, those scientists had assessed the status of those original seedlings; a total of 168 seemed worthy DED survivors. Those hardy Yonkers trees had then been relocated to Ithaca, New York, where Cornell scientists would for the next twenty years subject them to inoculations with DED more than seven times and in at least five different years.

In 1967 only seventeen of those trees were still standing. "Remission of foliar symptoms had occurred in most of the 17 survivors," reported scientists, "although six developed bark beetle infestation." By the 1970s, three decades after those Yonkers elm seedlings were planted, there was only one survivor, labeled R18-2. This determined twenty-seven-year-old offspring was a "well formed, vase-shaped tree," and scientists had high hopes for its progeny.

In their own search for a disease-resistant elm it was inevitable that John Hansel and the ERI would find their way to Dr. Eugene Smalley, a plant pathologist in the Department of Forest Ecology and Management at the University of Wisconsin at Madison. Smalley had joined the faculty in 1957 with the nearly impossible mandate from the Wisconsin State Legislature of solving the problem of Dutch elm disease. He had since then become *the* American plant pathologist exclusively studying and breeding elms, gathering a wide variety at the university's fifty-acre Arlington Experimental Farm north of Madison, as well as on his own twenty-acre property, nicknamed Smalley's Mountain Research Sanctuary. Smalley was a lively, convivial sort who loved to treat his many guests to tours of the magnificent collection of rare orchids he raised in his home greenhouse.

When Hansel first visited him in 1968, he recalls, "Smalley had a plantation of 17,000 elms that he had collected from all over the world—Japan, Russia, China, Europe." As Smalley explained his methods, he began with "seed importations of elms from various parts of the world—from various government or provincial forestry organizations, colleagues, and through personal collecting trips. World politics have often made the acquisition of germ plasm difficult and occasionally impossible." In Asia during these cold-war years, only Japanese elms were easy to obtain. "We have also introduced a large number of clones from various arboreta, including the Arnold Arboretum, the Morris Arboretum, the Morton Arboretum, and others, as well as from the Dutch and USDA elm-breeding programs. Our clonal collection now contains representatives of virtually all named DED-resistant elm cultivars, as well as unnamed accessions being considered for release."

When Hansel offered Smalley a thirty-thousand-dollar grant to focus on producing a pure DED-resistant American elm, the scientist told him, "You're crazy. You can't do it." But, recalled Hansel, "I said, 'I have money and that's all I want you to work on. Elms from everywhere else, the foreign elms, they don't have that vase shape or the tremendous stature of the American elms.'" Smalley was the first to concede that "American elms possess virtually all architectural and adaptability traits demanded of a superior shade tree," but his decade-long field trials had only confirmed their susceptibility to DED. Moreover, Smalley and his colleagues were frustrated to learn that American elms were particularly fussy—they did not care to hybridize with foreign elms, whose robust genes could have shored up their DED immunity.

And galling as it was to these plant pathologists, after all their carefully controlled cross-pollinations, years of trials, and further cross-pollinations, the best elms they had identified, explained plant pathologist Ray Guries, who worked with Smalley for two decades, were "ones we found in the wild or that someone just happened to notice was surviving along a street. After thirty years, we never could really improve on them."

One of Smalley's most successful elm tree introductions perfectly illustrated this point. In 1973, after fifteen years of field tests, he patented and released to the commercial trade "Sapporo Autumn Gold Elm," a DED-resistant Japanese-Siberian hybrid with some resemblance to the American elm. This desirable tree came not from laboratory wizardry but from the free-growing seeds of a Siberian elm in the botanical garden at Hokkaido University in Sapporo, Japan. It had been naturally pollinated by a nearby Japanese elm with a pleasing shape. European nurseries eagerly introduced the tree, and those licensing fees helped support Smalley's research, as did sales of his other Asian hybrid elms.

Whatever his misgivings, however, Smalley was, like so many others, beguiled by John Hansel's crusade and grateful to have additional funding. Soon he was devoting a portion of his time to what would amount to years of research. In 1969 he selected three thousand of his most promising American elm trees, conducted thousands of controlled pollinations by hand, and then waited two years for those seedlings to grow large enough to be subjected to inoculations of DED fungus. He would see how those young elms fared and then move on to the next generation and follow the same procedure. Convinced of Smalley's abilities, Hansel retained all licensing fees for this holy grail of a tree, all potential future income for ERI.

While no American elm of sufficient promise had yet emerged from these crucibles to be a prospect for the skeptical American commercial tree growers, Hansel predicted in ERI's "Progress Report 1970" that his nonprofit would soon have a disease-resistant American elm for sale. "There is ample evidence to support our prediction," he wrote, "and if we 'go out on a limb' in making this promise of a disease resistant elm by Arbor Day 1972, not only is it the limb of an elm, but a disease-resistant elm at that."

Unfortunately no such tree materialized—not in 1972, 1973, 1974, or 1975. That year Smalley was completing his sixth year of trials, and by now only 530, or 17.7 percent, of the young trees born of his original 3,000 survived. He envisioned many more years of testing, growing, and watching most of his elms die.

John Hansel, meanwhile, was experiencing his own disappointments. Lignasan injections were not viewed by others as particularly effective, and at a treatment cost of seventy-five dollars per elm per year, the fungicide was not being adopted as widely as Hansel hoped. The Syracuse beetle traps had also proven to be ineffective, unable to weather vandals, storms, and pests; Michigan State's European Dendrosoter wasps, elm bark beetle predators, could not survive American winters; Cornell could find no common factor to account for why some elms survived and others didn't; and fungus behavior and soil fungicides proved dead ends. Hansel was also learning—to his considerable ire—that scientists he funded had been making public remarks that Hansel believed "breached ERI philosophy and recommendations."

When the ERI board gathered in 1975 at the Lodge at Crotched Mountain in Francestown, New Hampshire, for its annual meeting, Hansel complained that one scientist had told the *New York Times* that "no method of treating DED had proved more than marginally successful," which, though true, ran counter to ERI doctrine. Smalley had transgressed by instructing someone that, for Lignasan to be effective, it should not be diluted, "in direct violation of the instructions printed on the label." Richard Campana, professor of botany, forest pathology, and forest resources at the University of Maine and a major figure in DED research, had written a letter to the editor of a California newspaper that Hansel believed "negated the ERI stand on the importance of preserving the American elm as a species, denied the efficacy of chemotherapy, advised a policy of cut-and-burn, and claimed he was not an authority on DED and gave the impression that he did not thoroughly understand ERI's Specialized Elm Care plan and recommendations."

However frustrated he was at these deviations from orthodoxy, Hansel took great pride in having galvanized and brought together the scientists he viewed as "the tops. They were so excited about working on something that others thought a lost cause. They had never gotten together and talked about their work, and now they were putting all their cards on the table. The thing I saw as a challenge was how to get the interested parties working together." Unless, it now emerged, those interested parties happened to work for the U.S. government.

In January of 1970, Alden Townsend, known as Denny, a newly minted twenty-eight-year-old PhD in tree genetics and plant pathology, arrived at the USDA's Shade Tree Laboratory in Delaware, Ohio.

"When I was hired," he recalls, "my superintendent said he was open to my doing anything to improve urban trees." Townsend was born in Oklahoma, where his father worked for an oil company: "We moved around a lot, but summers I worked on my uncle Al Miller's farm in western Pennsylvania on a former strip mine planted over with trees. My uncle was a superb naturalist and farmer and he got me interested in trees." Denny was also an active Boy Scout and learned more about trees through earning merit badges.

By the time he was in high school, Townsend's family had settled on the East Coast, in a home three miles outside Somerville, New Jersey. "There was a big field behind the house with a lot of gray birches," he recalls. "For some reason, in the tenth grade I decided to start my own nursery and dug up the area and transplanted those gray birches. It was fun for me, and I'd transplant them out into people's yards. It was just a hobby. When I went to college at Penn State, I started off in geology, and when the professor told us there were twenty thousand geologists out of work, I ended up majoring instead in forest science." While pursuing his master's at Yale's School of Forestry, Townsend worked in Oregon for a year and a half on white pine blister rust, which got him interested in tree genetics. He earned his PhD in that field from the University of Michigan and soon thereafter reported to work in Delaware, Ohio, a former railroad hub that had become home to Ohio Wesleyan University.

Assigned at the Shade Tree Laboratory to "improve urban trees," Townsend started out "breeding European and Asiatic elms to improve their level of DED tolerance. They made great urban trees. . . . With the non-American elms, we would plant trees, cross-pollinate them, grow those seedlings, plant them out, and wait three or four years and then inoculate them with DED." The fenced, locked farm fields around the one-story office building were filled with tens of thousands of different species of trees, including 15,000 young American elms planted in the 1960s. In these elm plantations were even the one pair of the USDA's 35,000 American elm seedlings that had been planted back in 1937 in Morristown and subjected over the years to huge infusions of DED fungus, and had survived. It was, wrote scientists, "a dwarfish tree of doubtful usefulness, but the other had the 'vase-shaped' characteristics of the traditional American elm. This tree was grown to maturity at the National Arboretum in Washington, D.C., and developed into a smallish, spreading tree of moderate vigor." Its offspring found a home at the Shade Tree Laboratory, where it gained a name: the Delaware elm.

Not only did Townsend work on the existing American elms at Delaware, but he and his research colleague Lawrence R. Schreiber had a steady supply of new prospects. "We would get numerous cuttings from people who believed they had an ancient survivor elm growing in their yard," Schreiber said. "But we soon discovered that most of those trees were just lucky. They had simply not been attacked by the elm bark beetle."

To study those trees, Townsend and Schreiber would secure fresh shoots from high up in the elm. The shoots were dipped in growth hormone and nurtured in the laboratory greenhouses until they were several years old and could be subjected to the DED fungus. "Most of the thousands we tested turned out simply to have escaped exposure," said Townsend. He came to believe that maybe 1 in 100,000 American elm trees was DED tolerant.

Naturally, in this small world of tree pathologists and elm breeders, Denny Townsend had heard about John Hansel, the Elm Research Institute, and their work. "I never met John Hansel," he says, "but I always thought he was good at marketing and I wished him the best, as we were both in the same game. But when I first came on and we were breeding non-American elms, John Hansel thought we were breeding those to *replace* American elms. That was not the point. Our goal was to develop a better urban tree. We used elms because they had such a resilience to urban stresses—salt, drought, heat, terrible soil. You could plant elms in any kind of tough situation, and they would survive without extra care. That was the thinking."

Townsend, Eugene Smalley, and Ray Guries naturally got to know one another, being part of a rarefied world of plant pathologists seeking to develop better trees, including that most high-profile of quests: a new, improved American elm. Smalley and Townsend, however, each held very different views on how best to replicate a DED attack. "The method Gene used," says his longtime University of Wisconsin colleague Guries, "was to do the field inoculations [in twiggy crotches] in a way that mimicked how beetles fed on the crown of the tree with its young shoots. Denny Townsend would make a cut at the base of the tree and flood it with spores—a much more severe inoculation. Smalley said, 'That's not how infections happen.'

Townsend said it was faster and you got more immediate results. They always argued about it and they never reconciled on that issue."

Hansel, vocal in his scorn for government in general and the USDA in particular, viewed its scientists not as natural allies in the crusade but as unworthy rivals. Townsend had been in Ohio only a couple of months when Hansel remarked in a speech, "Even the USDA has shown new interest and activity at their lab in Delaware, Ohio. Could it be they are jealous of our progress?"

Hansel's worldview, that of a private businessman who believed all ERI's research was proprietary, also put him on a predictable collision course with the university scientists—Eugene Smalley above all. "Smalley was an academic," says Guries. "He believed in sharing his research with other researchers." Inevitably, what Smalley regarded as standard academic practice Hansel would come to view at a crucial time as an act of deep betrayal. But that was still in the future, and Hansel's vexation in these years was how very slow tree research could be. In 1980, ten years after he had promised the ERI board a DED-resistant all-American tree, Smalley was still methodically testing promising American elms.

CHAPTER TWELVE

"Having Cities Work with Forces of Nature": The Rise of the New Urban Forestry

Actor Raymond Burr and Spunky the Squirrel, the USFS Urban and Community Forestry mascot, at a 1982 tree planting in Cincinnati, Ohio. *(With permission of American Forests.)*

At 2:00 p.m. on October 10, 1982, a few hundred tree lovers gathered under a damp, overcast sky at the historic President's Grove in Cincinnati's Eden Park. The occasion was the planting of a London plane to honor President Ronald Reagan. In the hundred years since the grove had been inaugurated, no chief executive had yet graced one of these ceremonies with his presence, and no one expected Reagan, of all presidents, to be the first to show up. He was, after all, the California gubernatorial candidate who, when urged to preserve more old-growth redwood forests, infamously said: "A tree is a tree. How many more do you need to look at?"

Appearing in Reagan's stead at the President's Grove was actor Raymond Burr, who played Perry Mason on the classic television series. A longtime member of the American Forestry Association, Burr greatly esteemed trees. Now corpulent and gray-haired, attired in coat and tie, he navigated with his cane across the uneven ground of the hillside grove toward the slender London plane. Once the crowd quieted, he spoke in the famous baritone of his youthful days with the Civilian Conservation Corps, the many park structures they had built, and the many woods and forests they had replanted. Burr then shoveled the final mounds of soil onto the London plane on behalf of the president.

A century after Superintendent Peaslee had established the nation's only presidential grove, it was going strong—unlike, say, the Author's Grove and Pioneers' Grove, places now scraggly with underbrush, their markers long disappeared, their honored names virtually forgotten. At the Presidential Grove Reagan's tree was joining thirty-nine others, including defeated rival Jimmy Carter's young loblolly pine. Reagan had actually asked for an American sycamore to commemorate him, but nurseries rarely sold them, and the Cincinnati Parks Department and its new young forestry division had decided a London plane would suffice. Most present were well aware that Reagan had tried and failed to defund the Forest Service's nascent urban forestry program.

Carter's Georgia pine had become something of a minor local star, whose installation on May 21, 1979, had attracted reporters and news cameras. Dorothy Behlen, the young urban forestry coordinator for the American Forestry Association (and a Cincinnati native), had flown in from Washington, D.C., to do the official honors.

Her appearance, in fact, was helping lay the groundwork for the American Forestry Association's new urban forestry campaign. Explained Behlen to local journalists puzzled by the term "urban forestry": "Just as the president represents all the citizens of the country, his tree represents all the other urban trees. And as the Carter loblolly pine joined an urban forest instead of becoming just a city tree, so all the other urban trees become, in effect, parts of their own urban forests." In the wake of the continuing loss of the American elm, the American Forestry Association was launching a new twentieth-century crusade to rebuild the nation's urban forests. (This infelicitous but functional term had been coined by Canadian forestry professor Erik Jorgensen in 1965 to encompass all the millions upon millions of trees found in any city's streets, parks, cemeteries, campuses, yards, industrial areas, and vacant lots.)

The American Forestry Association was now hosting the Second National Urban Forestry Conference in this symbolic city. At its nineteenth-century Cincinnati meeting, the main topic had been the wanton destruction of old-growth forests and the need for trees in cities. Now the driving issue was the loss of trees in American cities in the wake of DED. "In 1970, Detroit, Philadelphia, and Dayton cut down more trees than they planted," lamented Anne Whiston Spirn, chair of the University of Pennsylvania's Department of Landscape Architecture & Regional Planning. "Detroit removed 10,623 trees in that year and planted only 2,457." As for Chicago, that city "lost 295,000 elms over twenty inches in diameter from 1968 to 1978," an eco-disaster that cost the city $24 million just in removal costs.

While Dutch elm disease destroyed innumerable trees, trees were also falling victim to the increasing prevalence of the automobile. City planners and engineers thought nothing of sacrificing mature specimens when widening streets for cars, gouging new highways through old leafy neighborhoods, and paving over small parks to accommodate more parking lots. The destruction of big trees and their shady canopies, wrote Spirn, meant that "the street, sidewalk, and adjacent buildings feel much hotter, the air grittier. Street noise echoing off building walls and unmasked by the rustle of leaves seems much louder. Seasons pass unheralded; property values of adjacent residences decline." Citizens responded by moving out.

The new generation of forestry leaders that had assembled in Cincinnati included visionaries who were scientists, landscape architects, and arborists, along with professional tree huggers with a genius for public relations. One of the most influential would be Rowan Rowntree, a U.S. Forest Service scientist and grandson of the famous California wildflower botanist, free spirit, and author Lester Rowntree. "With the loss of the elms," recalls Rowntree, "there was a tidal wave of interest in trees and the urban forest by the public. I could have been doing nothing but giving talks to garden clubs, community groups, and such about how to use trees to beautify cities."

Born in Carmel, California, in 1935, Rowan Rowntree was named by his father for the rowan tree. Rowntree Sr. eventually moved his young family to Berkeley, where in high school Rowan recalls working for a retired professor of botany: "He hired me because he knew of my grandmother Lester's work. The first day he showed me this newly planted bed, with hundreds of one-inch green things sticking up. 'Rowan,' he said. 'These are the plants and these are the weeds.' He rattled off some Latin names and showed how one had a serrated leaf edge and the other didn't. Then he spent another hour showing me other duties. Well, by the time I went back to weed that bed, I could not remember which was which and was too embarrassed to ask. So I chose one kind and pulled them all out. I had taken out all the new young plants rather than the weeds. The professor just said, 'Rowan, I can vouch for your consistency.' He was a kind man and kept me on through my senior year of high school."

Rowntree graduated from the University of California at Berkeley in 1960 with a degree in economics and an interest in human resources. His first job was with Crown Zellerbach, then the world's second-largest paper company and the owner of vast tracts of forests. Rowntree served as a recruiter of new hires and then "would teach a six-month full-time course on everything from forestry to papermaking to corporate policy. I really loved the teaching. And I learned a lot about forestry in those five years."

He returned to Berkeley for a master's degree in forest ecology and by 1973 had completed a PhD in physical geography. He was delighted to become an assistant professor in Syracuse at the State University of New York's College of Environmental Science and Forestry, an institution founded in 1912 whose alums were known as "stumpies." Its allure for Rowntree: "It was a truly ecological college that was practical."

Propitiously, as Rowntree launched his academic career, Congress allotted a few million dollars for new research on urban forests. Among those present at the Cincinnati conference was former U.S. Forest Service chief John R. McGuire, whose agency had been given the mandate in 1972 to regreen the cities but no money until 1976. McGuire was a true believer, and in a much-circulated 1975 speech he had laid out a blueprint for this new endeavor that noted the challenge of convincing "'the powers that be' of the benefits, indeed, the necessity of urban forestry . . . which by itself brings in no revenue," in contrast to the lumber that could be sold from harvesting the national forests.

McGuire viewed research as key to justifying the costs of bringing trees to where three quarters of Americans actually lived. And so had come into being the Northeast Consortium for Urban Forestry, which used the new congressional funding to support scientists at the USDA Forest Service Northeastern Forest Experiment Station and at twelve big-name northeastern universities (including Yale, Princeton, Cornell, University of Massachusetts, and Syracuse). In 1978 the U.S. Forest Service opened several new research labs; Rowan Rowntree established the Northeastern Station and then later turned his attention to the West Coast.

By the time of the 1982 Cincinnati conference, Rowntree had learned that the case for restoring city forests required a stronger argument than their mere beauty. Four years earlier Fred Bartenstein, a recent Harvard graduate and special assistant to the city manager of Dayton, Ohio, had contacted Rowntree, asking for help. The U.S. Environmental Protection Agency had ordered Dayton to clean up its dirty air and stop dumping into the Miami River or face "strict constraints on metropolitan expansion and industrial operations." Bartenstein had seen a short German documentary about how the City of Stuttgart had used trees and greenways to clear its air pollution, cool summer heat islands, and clean its water, the very problems Dayton faced.

"The idea," explains Bartenstein, "was having cities work *with* forces of nature rather than against them. I had never heard of anything like this in the U.S., but Rowan Rowntree at the U.S. Forest Service was someone who was really thinking about it. . . . Neither of us had any serious money, but we put something together with chewing gum and paper clips, and the city and the Dayton Climate Project became a hub for this widening collaboration of different agencies and academics, looking at the role and value of trees to cities."

In the world of professional forestry, foresters and lumberjacks had long known how to calculate the board-feet value of a single lodgepole pine or for tens of thousands of acres of timber. In the late twentieth-century city, though, Rowntree and Bartenstein knew that this new entity, the urban forester, could not answer a host of basic questions: What was any particular city tree worth? What were the structure and character of an entire urban forest? How many poplars, ashes, or trees of heaven were there in a given metropolitan area? How old were those trees and what size? How healthy? How did trees interact with the greater urban ecosystem? Did they really affect air quality? Anyone whose family home was shaded by large oaks or maples knew the delicious relief those trees provided on a hot summer day, but how much did they actually reduce the need for air-conditioning? How much money *did* trees save in energy costs?

When thunderstorms lashed down, how many gallons of rainwater did the leaves of a Norway maple absorb and keep out of the stressed sewer system? And what effect did tree-lined streets and tree-rich landscaping have on commerce? Or crime? Or human well-being? Did trees raise property values? How might trees improve city life and city revenues? What were, in short, the overall ecology and structure of urban forests?

If you *could* quantify those presumed benefits, believed Rowntree and Bartenstein, you had a chance of persuading city officials that trees were valuable green infrastructure and not mere ornamentation—or, worse yet, a liability. "The relationship between people and trees, in our work," says Rowntree, "would have to be described as a pecuniary one." At a time when American cities were besieged with troubles, Rowntree believed that only innovative science and hard data could reestablish trees as essential to urban life.

Fred Bartenstein and Rowan Rowntree had come to the Cincinnati gathering to report on their first real-world foray into quantifying trees as Useful and Important Citizens, part of what was gaining renown as the Dayton Climate Project. Bartenstein, who was spending part of his time commuting to Harvard as a Loeb Fellow in Advanced Environmental Studies, had by now become an impressive enough urban forestry guru to be the first speaker featured that Monday following opening remarks by the national officers of American Forestry and Forest Service. "Do you ever wonder?" he asked his tree-loving audience, "why your voice doesn't get heard at city hall?"

The reason was simple enough: Mayors and city managers rarely saw trees as essential to a city's vitality. In earlier times, Bartenstein noted, cities had cherished their trees as necessary to public health, as attested to by the 1912 Ohio Arbor Day manual: "Don't forget that an adequate number of street trees mitigates the intense heat of the summer months, and diminishes the death rate among children." Families had also found respite from crowded tenements in city parks. Then, said Bartenstein, air-conditioning and cars had obviated those compelling reasons to plant and nurture city trees, just as the elms began to die. If trees were no longer saving children and helping families survive hot summers, what *were* they doing to justify their presence in the cityscape? Why bother replacing the elms?

On the second day of the Cincinnati conference, Rowntree took up that very question in his talk on the Dayton Climate Control Study. In this novel paradigm of urban forestry, he described how the U.S. Forest Service Northeastern Forest Experiment Station used downtown Dayton as "a field laboratory . . . to understand how the urban forest system operates to modify the physical environment—air temperatures, air quality, and surface runoff." Rowntree's group deployed aerial photography to create a map of the city, divided it into smaller subunits, and determined that 37 percent of Dayton was covered with tree canopy.

Armed with the knowledge of the extent of Dayton's urban forest, could they now determine what *did* happen when it rained? As Rowntree told his listeners, "No published method exists for accurately calculating how much and how fast rainfall will run off from [such] areas. . . . So, we've developed a mathematical model that does, using only conservative assumptions." From that Rowntree could now determine that Dayton's "urban forest—that is, just the trees—is responsible for reducing surface runoff by ten percent or more."

Rowntree's graduate students, meanwhile, had investigated whether using more street trees and an experimental green parking lot could lower city temperatures, but they could not compute a comparable level of analysis. Rowntree could report only that "our model for predicting air temperatures . . . suggests that the urban heat island in most cities can be reduced by 25 to 50 percent by employing woody and herbaceous vegetation in the Central Business District." While they had not yet figured out how to translate any of these findings into actual dollars and cents, still, crude as the data was, Rowntree was taking the first steps in a revolutionary way to think about and value city trees.

As Rowntree would be the first to acknowledge, though, in Ohio the study eventually came to naught: "Honestly, in Dayton we were too grossly underfunded, understaffed and underknowing to develop the concepts we were starting to work on." When champions disappear, their pet projects are likely to follow, and when Bartenstein's boss took a job in another city, the Dayton Climate Project ceased to be. But, Bartenstein says, "It was a harbinger of what was to come."

Among those wandering the halls and lecture rooms of Cincinnati's Stouffer's Towers Hotel and the convention center was John Rosenow, the soft-spoken, boyish-looking executive director of the tiny National Arbor Day Foundation. Tall, sandy-haired, and given to wearing tweed jackets, he was neither an arborist nor a forester, but he was building his foundation into a formidable institutional power in urban forestry.

Rosenow grew up on a farm in the village of Elmwood, twenty miles due east of Lincoln, Nebraska, where he climbed the native cherry trees in his front yard for fun and fruit. A natural at math and science, he attended the University of Nebraska at Lincoln to earn a BS in agricultural engineering. While an undergraduate he discovered a talent for politics and organizing, becoming president of the FarmHouse Fraternity. On April 22, 1970, like twenty million other Americans, Rosenow joined the first Earth Day demonstrations, which, he says, "was my entrée . . . to how trees are of such importance to the environment." The brainchild of Wisconsin senator

Gaylord Nelson, Earth Day and its teach-ins prompted such an extraordinary outpouring of grassroots passion that the nation's startled political class quickly signed on to this emerging environmental movement.

When University of Nebraska administrator Stan Matzke Jr. ran for Nebraska secretary of state in 1970, Rosenow offered himself as campaign manager. Matzke lost but was appointed the new head of the state's Department of Economic Development. In spring of 1971 he recruited Rosenow to come work for a year as his assistant. Six months later, when Matzke added the state's Tourism Division to his portfolio, he appointed Rosenow, then only twenty-one, as its new director. Nebraska governor James Exon instructed Matzke and Rosenow to go all out in organizing a centennial celebration of Arbor Day in 1972, setting up the tiny nonprofit Arbor Day Foundation, seeded with ten thousand dollars.

By the time Earth Day spurred Congress to act in 1970, establishing the U.S. Environmental Protection Agency and passing the Clean Water Act, Arbor Day had been fading away for years, often dismissed as an old-timey school festivity. Its dogged supporters had struggled for decades to gain it official recognition as the nation's official tree-planting holiday, so they were jubilant when President Richard Nixon signed a bill in 1970 that declared the last Friday in April a national observance—until they learned that it was effectively a one-off, in force only for that one year.

Still, Arbor Day did retain considerable "brand power" as part of a venerable heritage of fostering respect for nature. Rosenow flew to New York City with J. Greg Smith, Nebraska's advertising consultant, and learned that the members of the Ad Council would be delighted to create a major campaign around Arbor Day and its centennial. One of their ads featured a freckle-faced, towheaded boy under a young oak next to the headline: TRY TO SAY TREES WITHOUT SMILING. The copy then gently encouraged: "On this 100th anniversary of Arbor Day . . . the Conservationist's Holiday, bring joy to your little corner of the world. Plant a tree for tomorrow." Then, further reinforcing the message—also in large, bold type—came: "Trees . . . a joy forever / they provide us with shelter / keep the air pure / make life so much better."

By the time of the Cincinnati conference, Rosenow had been full-time executive director for the now-named *National* Arbor Day Foundation for two years. While still tourism director in 1976, Rosenow had devised the tiny foundation's clever Tree City USA program, which played to every politician's and bureaucrat's love of awards, plaques, public relations, and photo ops. "DED was the driving force," recalls Rosenow, appealing to Americans' yearning for new trees to replace the lost elms. "The U.S. Forest Service saw the potential of state foresters to deliver assistance at the local level. Gene Grey, the urban forestry coordinator for Kansas, helped us craft those initial standards and draft a descriptive brochure, so it was a well-conceived effort with solid thinking behind it."

Rosenow hitched the fledgling Tree City USA program to the nation's bicentennial celebration and pitched it through a powerful network of partners: the U.S. Forest Service, the U.S. Conference of Mayors, the National League of Cities, and the National Association of Foresters. The qualifications for membership were simple: To be a Tree City, you had to have (1) a city forester or tree board responsible for city trees; (2) a tree-care ordinance on the books; (3) an active urban forestry program with a budget equal to two dollars per citizen (though that figure was soon dropped to one dollar). The fourth rule was self-serving but brilliant: "Arbor Day must be proclaimed and observed in the community"—in other words, each city would have to promote Arbor Day and plant a tree. That first bicentennial year, forty-two cities and towns in fifteen states signed on.

At the same time Rosenow understood that such promotional efforts were only a beginning. For his little foundation to make any difference in the new world of urban forestry, it needed members and money, *lots* of members whose contributions would support a substantial budget. In February of 1977, 700,000 American households sorting through their mail found a packet containing a typical direct-mail appeal: a letter from a group called the National Arbor Day Foundation, inviting them to join this tree-loving organization for a mere ten dollars annually, a certificate of appreciation, a membership card, and a window seal announcing their sponsoring of the organization to their neighbors. But then came the surprise: six actual live tree seedlings, along with a coupon that would enable the recipient either to buy more trees or to bestow gift memberships on his or her friends and relatives. That first direct-mail campaign recruited three thousand members.

By the fall of 1982 Rosenow's young foundation was growing fast. It had enrolled thirty thousand members and awarded Tree City USA status to

360 communities, bestowing upon each a large Arbor Day flag, a walnut wall plaque, and a road sign. In his Cincinnati talk Rosenow promised city arborists that supporting the tree holiday was a way to attract both attention and dollars: "The promotion of Arbor Day helps focus the media's attention on trees. The Tree City USA Award can give a local forestry program a tremendous boost. We've had a number of cities that barely met the standards the first year, but because of the interest in trees generated by the Tree City USA publicity, the funding and the magnitude of their program increased dramatically."

Urban foresters loved this award, for it gave them a yearly organizing tool that turned the spotlight on the community's trees and, equally important, local politicians. Recalls Cincinnati's forester Tim Jacob, "That Tree City USA program . . . just gained steam constantly. The National Arbor Day Foundation had simple, inexpensive ways to do hands-on things with tree planting. A lot of trees got in the ground that way."

Rosenow, a low-key sort who liked to bicycle to work long before that became fashionable, would prove to be a brilliant executive and marketer. He never did make it to graduate school, but he would credit his engineering education with giving him "a passion for facts and a problem-solving approach that is organized and methodical." By 1987 his foundation had a half million members. In coming decades he would build the National Arbor Day Foundation, with offices in Lincoln, Nebraska, into a leading player in urban forestry, with major corporate, government, and nonprofit partners and initiatives, eleven million members, ten million tiny trees dispatched annually, and the large Arbor Lodge Farm and Lied Conference Center on part of J. Sterling and Joy Morton's former estate in Nebraska City, Nebraska. More than three thousand communities claimed status as a Tree City USA.

A very different kind of up-and-coming star at that 1982 conference was Andy Lipkis, who had flown in from the West Coast to dazzle the arborists and foresters with *his* California dream, as summed up in the title of his conference session: "One Million Trees for the 1984 Olympics in Los Angeles." Lipkis, who grew up in Beverly Hills, was an entirely new kind of tree lover, a young baby boomer flower child who instinctively understood how to cast trees and tree planting as hip, glamorous, and the cool thing to do to heal cities.

Some in his Cincinnati audience had already seen Lipkis two years earlier on *The Tonight Show,* on which Lipkis looked the part of mountain-man child, sporting dark, tousled hair and full beard, rumpled work shirt, and jeans. In the wake of ferocious floods and mudslides in Los Angeles in 1980, Lipkis had mobilized three thousand volunteers through his nonprofit, TreePeople, to provide assistance.

This work undertaken in the aftermath of the floods, Lipkis told the Cincinnati crowd, had "created a tremendous amount of public goodwill and attention." A charismatic figure with a talent for logistics, Lipkis had fallen in love with tree planting at age fifteen. Alarmed by reports that local oak, mountain mahogany, and pine forests were slowly dying from Los Angeles's polluted air, he had organized the planting of smog-resistant Coulter pines at his summer camp, persuading his campmates to pull up an old parking lot to make room for this new patch of forest. As he told one writer, "It helped that I had worked with my parents on Gene McCarthy's presidential campaign as a twelve- and thirteen-year-old. I understood how to take ideas and put them into action one piece at a time. It was an incredible thing to transform that piece of earth and to have such immediate gratification."

For the hundreds of attendees at the conference, Lipkis and his group represented something completely novel: truly grassroots tree lovers who had managed, with no official governmental patronage, to become

USFS scientist Rowan Rowntree (*left*) and Andy Lipkis (*right*), founder of TreePeople. *(Left photograph courtesy of Rowan Rowntree. Right photograph courtesy of TreePeople.)*

significant local players in the world of urban trees. When former U.S. Forest Service chief McGuire envisioned a "viable urban forestry program," he imagined "very close cooperation . . . among Federal agencies, State and local governments, universities, professional associations and private industry." But a nonprofit like TreePeople was an entirely new concept, and while not all the professional tree folk welcomed these amateurs, they were already a force to be reckoned with.

The creation story of Andy Lipkis and TreePeople had a number of wonderful chapters. The first might be titled "Andy Versus the Bureaucratic Deadwood," from the headline atop a 1973 front-page *Los Angeles Times* story that was a journalist's dream: earnest, do-good college kid (Andy) approaches California Division of Forestry to ask for donation of twenty thousand tiny, smog-resistant Sierra redwoods and sugar pines needed for a big project to restore dying forests. But in a classic catch-22, the bureaucrats explain that their state agency cannot give trees away but must sell them for five hundred dollars. Andy, however, could not raise the sum, and as he told reporter Michael Seiler: "They said I had to get the money by March 16 or they'd kill the trees. They were going to plough them under to replant their seed beds so they could grow another crop of the same trees for next year." The following Monday the state foresters began the process of destroying twelve thousand of the seedlings.

Andy, all of seventeen, was savvy enough to make a few phone calls to politicians and reporters, and suddenly the embarrassed Forestry Division was delivering the eight thousand survivors, at no charge, to Lipkis's dorm at California State College in Sonoma, with a warning that the trees had to be planted immediately. Andy scrambled, persuading a local milk company to donate eight thousand milk cartons, paying sixty dollars for a truckload of topsoil, wrangling another free truckload from a local developer, and then calling out the Boy Scouts and fifteen dorm pals to help pot up the seedlings.

By the time the *Los Angeles Times* was savoring this tale of youthful enthusiasm besting jaded bureaucracy, Andy still needed four thousand dollars to rent a truck for two months, buy tools and mulch, and pay for help. In that pre–Kickstarter campaign era, the mailman delivered to Lipkis over a three-week period an avalanche of letters from schoolchildren, families, and concerned citizens. Every day Lipkis and friends would tear open the envelopes to find enclosed checks, dollar bills, and coins, totaling an astounding ten thousand dollars. "We put the money in the bank," he later told a writer. "An older cousin helped me file incorporation papers. American Motors donated Jeeps to move the trees to the camps. And TreePeople was born."

In chapter two of his new life as a tree hero, Lipkis and the three-person staff of his new nonprofit, officially registered as the California Conservation Project, showed they were not a flash in the pan. They had the wherewithal to engage all manner of people as volunteers to prepare the tens of thousands of tree seedlings, getting them into milk cartons for the large-scale forest plantings the organization orchestrated each year with kids in summer camps and schools. For Lipkis "this was one of the peak experiences of my life, working together with a group of people to create natural beauty." As he explained in his Cincinnati talk, "TreePeople developed an especially effective method of recruiting, training, and mobilizing volunteers," drawing from senior centers, colleges, religious groups, the scouts, and day-care co-ops. He also adhered to a "philosophy of cooperation," and the group remained resolutely nonpolitical, never taking a stand on public issues.

By late 1977 Lipkis and TreePeople had little trouble persuading Los Angeles to hand over a 1920s-vintage retired fire-watching station on twelve acres of wooded land hugging the side of Coldwater Canyon atop Santa Monica Mountains Park. Again volunteers from many walks of life converged and over time helped TreePeople settle into its new home, with its converted firehouse headquarters, a tree nursery, an environmental education center, and a public park. When *Los Angeles Times* columnist Art Seidenbaum caught up with Lipkis at this mountaintop redoubt full of old trees, owls, and even coyotes, he learned that TreePeople was taking up a new cause: urban forestry. "A maintained urban forest," wrote Seidenbaum, "cleans the air, cools the temperature and reduces noise, not to mention what trees do to soften the looks of a cityscape." Now twenty-two, Lipkis was also finally finishing college at UCLA in environmental studies.

Chapter three in the amazing rise of Andy Lipkis and TreePeople was set in motion by the aforementioned 1980 torrential rains. As Lipkis told it, this disaster acted to "catapult the TreePeople into a prominent position. . . . In the worst of the floods, TreePeople mobilized, trained, equipped, and dispatched 3,000 volunteers who sandbagged, laid plastic, and helped clean up mud at 1,200 houses that were either damaged or threatened. TreePeople was also called on by authorities to help rehabilitate areas damaged in local brush fires. In each case TreePeople assembled a diverse collection of volunteer groups, including members of four-wheel-drive clubs, environmentalists, ham radio operators, and church groups." Their well-respected if not well-known group became "famous practically overnight," thanks to huge media exposure and that appearance on Johnny Carson. "The flood," said Lipkis, "caused a major shift in TreePeople's thinking and direction."

Lipkis and his small staff could legitimately look with pride at the past eight years, during which they had planted 150,000 trees, a feat accomplished by energizing armies of volunteers to work in the region's threatened forests. But Lipkis had to honestly concede to the audience in Cincinnati that all this prodigious effort up in the mountains had had "little visible impact on Southern California." He had also come to realize that in Los Angeles, his hometown, "much of the city crucially needed trees." During the floods, Lipkis marveled at how the media coverage of all the TreePeople volunteers made "even the people who were sitting at home watching the work on the evening news feel involved." Lipkis pondered how to take advantage of this tsunami of publicity and the coming of the 1984 Olympic Games to Los Angeles. Why not, he said in his talk, "involve the whole city in a major challenge; to plant and maintain trees throughout the city to beautify itself or 'heal' itself before the Olympics?"

At this point the story circles back to Dayton, Ohio. For such was the renown of the Dayton Climate Project, and through it Stuttgart's efforts, that the Los Angeles City Planning Department, which had been tasked by the EPA with reducing its infamous smog, had been studying both cities. With TreePeople suddenly famous, the city planners had contacted Lipkis with a proposition: Would his group help the city plant one million trees "as an air quality improvement measure"? Reported Lipkis, "According to their figures, the trees, when 20 years old, would filter as much as 200 tons per day of particulate matter from the air, bringing Los Angeles within 80% of meeting the Clean Air Act Standards for particulate smog. They figured it would take them at least 20 years and $200,000,000 to plant all the trees." So even as TreePeople was "toying with the million tree goal," here was the city proposing the very same thing, albeit with a far more conservative time frame. Ever the optimist and dreamer, Lipkis told the city planners, "We'll do it in the next four years at a fraction of the cost."

And so was born the nation's first high-profile million-tree campaign, with Lipkis blazing the way, media savvy as ever. While Los Angeles planners were focused on satisfying EPA rules, Lipkis saw Million Trees as accomplishing far more: namely, helping energy conservation and air quality, producing fruit and nuts, and beautifying all of the city. One can only imagine what the Cincinnati audience thought as Lipkis described how Doyle Dane Bernbach, one of America's most creative advertising agencies, had developed a pro bono campaign (complete with all creative materials) for TreePeople, starting with the slogan "Turn Over a New Leaf L.A.—Help Plant the Urban Forest." Television stations, radio, newspapers, magazines, billboard companies, urban foresters, landscape architects, and local nurseries all were on board to promote Andy Lipkis's latest visionary idea. There was even an "I Brake for Trees" bumper sticker.

Citizen arborists would duly register any new tree in their yards via TreePeople's twenty-four-hour telephone hotline, while a local TV station would monitor progress with its "Treemometer." And, Lipkis explained, "It is hoped that many of the people who participate . . . will be empowered or inspired to begin actively pursuing their dreams." TreePeople would, of course, also be promoting the planting of larger trees via community plantings, targeting business groups, block clubs, schools, and youth gangs.

The several hundred tree lovers headed home from Cincinnati's urban forestry conference had much to ponder. At the poster sessions they had been able to stroll around and learn about everything from "Control of Root Damage to Sidewalks" to "Urban Forestry in Utah" to "Building Better Elms: The Wisconsin Elm Breeding Program." Ray Guries was on hand to discuss his and Eugene Smalley's work. Thus far they had not found a holy grail of elms, but there was cause for optimism. As their poster reported: "This program includes American elm, but emphasizes an Asian gene pool, especially Japanese and Siberian elms. Seasonal susceptibility testing procedures have now identified at least twenty-two elm clones and very high resistance to Dutch elm disease."

In Cincinnati, if you had had the prescience to attend the right talks, you heard the leaders who were reconceiving the way people would think about and plant trees in American cities. Rowan Rowntree's gift was translating science into data on the multiple benefits of trees, casting green as a great municipal investment even for those with no particular affection for trees. John Rosenow's talent was the methodical building up of existing institutions and ever-evolving partnerships to scale up city forestry departments via Tree City USA, even as the National Arbor Day Foundation's membership personally planted millions of young trees. And then there was Andy Lipkis, the young California dreamer, who was showing how in the biggest and most sophisticated of American cities you could make trees a powerful cause.

CHAPTER THIRTEEN

"Trees Are the Answer": John Hansel, Henry Stern, Deborah Gangloff, and George Bush

John Hansel of the Elm Research Institute in late 1969 with the dead American elm in the background. *(Used with permission of the estate of David Gahr.)*

In 1990 Michael Pollan, executive editor of *Harper's Magazine*, was delighted to learn that his father-in-law was sending him a young American elm tree said to withstand Dutch elm disease. Pollan walked out into his Cornwall, Connecticut, garden, deliberating about where to plant this promised sapling, and "imagined its trunk climbing swiftly beside my driveway and then, as is the habit of American elms, opening all at once its tall, leafy canopy." What arrived in the mail, though, was only "a leafless, 14-inch chopstick . . . from the Elm Research Institute in Harrisville, N.H., . . . a poor excuse for a tree. Swaddled in a sheet of The *Manchester Union Leader*, it had not even a single branch—just a couple of bud-eyes at one end and a straggle of root hairs at the other."

Nonetheless he installed his "dinky" elm in his perennial border, where it could "be cosseted like a delphinium." Three years later the elm was coming along, though he conceded that "a tree whose height is still measured in inches (39) and whose leaves can still be counted (73) may not be much to brag about. But I'm taking a long view. Nursing this skimpy sapling along is the part I play in the effort to restore the elm to the American landscape." Pollan was just one of the vast army of elm lovers who welcomed John Hansel's long-awaited "American Liberty" elm.

Not quite twenty years after Hansel had launched the Elm Research Institute, Eugene Smalley at the University of Michigan's Department of Plant Pathology and Forestry had finally delivered not just one but a series of six disease-tolerant American elms, all "superior survivors from our American elm screening program begun in 1958, as well as resistant individuals from the New York and USDA programs." Year in and year out, various field technicians had advanced Smalley's research with "field testing of clones, progeny arrays, and new accessions."

All told, Smalley and his colleagues had subjected more than sixty thousand American elm seedlings to concentrated blasts of DED. All six varieties released under the name "American Liberty" elm had survived this fungal assault, but in time it would emerge that one of the six trees—a cultivar with its own patriotic name, "Independence"—was more DED tolerant than others. This clone, W-510, was patented, and Hansel and the Elm Research Institute retained full rights to and control over it. However, when purchasers acquired Liberty elms, they might or might not receive this hardy Independence clone, which would only lead much later to confusion, controversy, and heartache.

Among the first to acquire one of Hansel's new trees was Pulitzer Prize–winning Boston newspaper columnist Ellen Goodman. In late 1984 she told her readers,

> I went outside this morning to pick a spot in our small urban landscape for an elm tree. . . . My task has a wonderful, even corny, edge of optimism to it. Once my entire street, like thousands of others, boasted the elegance of a dozen American elms that reached as high as three-story buildings. But twenty years ago, one by one, they were destroyed by Dutch elm disease.
>
> The blight left the brick skin of this neighborhood raw, and stripped the sidewalks naked to the sun. The elms were later replaced by Norwegian maples, which are lovely but which are not American elms.
>
> So, when I find the proper spot, there will be an elm again on my street.

Goodman, having heard about John Hansel and the Elm Research Institute, had decided she wanted to be an early adopter. "I drove up to Harrisville, N.H.," she wrote, "on one of those dismal days that come at the end of a warm November." She met Hansel in his pine-paneled office on the ground floor of a renovated nineteenth-century brick mill. The colored pins in the large map hanging on his wall, he explained, marked the 185 towns where Conscientious Injector groups were helping to save the remaining old American elms by injecting Lignasan, with eight thousand trees treated so far.

Ever savvy about publicity, Hansel was delighted to bestow one of his trees upon Goodman, a journalist whose column was featured in hundreds of newspapers. Hansel escorted her to a "misty mill room that doubles as the greenhouse." There she met her American elm. Like Michael Pollan, she was taken aback by its diminutive size. "My tree is only six inches tall, barely a treelet. If it grows a foot a year, as predicted, it will be several years before the elm reaches my height. . . . It came with a green card that bears a computer number and planting instructions and no promises. I am told that even Dr. Eugene Smalley, who cloned this tree, is not sure what it will look like. It may be majestic," Goodman quoted the scientist. "It may turn out to be a ratty dog. We won't know the answer for 20 years." Meanwhile, Ellen Goodman declared herself one who found "something wonderful in being part of a comeback story."

In the years after the Cincinnati urban forestry conference, tree advocacy had been gaining significant momentum. Although the Reagan White House had ordered the complete defunding of the U.S. Forest Service's tiny program, fifteen-term Democratic Chicago congressman Sidney Yates, a fervent tree lover, kept putting the necessary money back into the federal budget. "I was caught in a delicate situation," recalls then–USFS director of cooperative forestry Tony Dorrell. "The result was that we had a [urban forestry] program but couldn't really talk about it or promote it. It was awkward for me as Director because I wanted to provide leadership for urban forestry but couldn't say anything about it in a public way." The solution for the agency was to reach out "to our friends at the American Forestry Association."

The USFS awarded large grants to the nonprofit, still the oldest U.S. citizens' group devoted to trees, forestry, and conservation. The American Forestry Association then took the lead, organizing the Cincinnati conference and, some months after that event, hired arborist Gary Moll, who in 1976 had launched Maryland's first urban forestry program, to serve as American Forestry's director of programs, including the brand-new position of director of urban forestry. Taking on the modernized name American Forests, the venerable nonprofit also made a less likely hire, PhD anthropologist Deborah Gangloff. Like Andy Lipkis, she had no particular training in trees but a deep interest in urban populations and in the environment.

American Forests now emerged as the driving force behind a coalescing national movement. In 1981 it had established the National Urban and Community Forestry Leaders Council, and in coming years the local foresters and tree huggers running its state councils created a national network. American Forests kept them informed and mobilized through its *Urban Forests Forum* newsletter. In 1986 American Forests organized another urban forestry conference as a follow-up to the one in Cincinnati, this time in Orlando, Florida, featuring more new grassroots nonprofits like Trees Atlanta and such venerable statewide green leaders as the Pennsylvania Horticultural Society.

It was not hard to see why so many Americans were taking up the cause. In 1987, when Gary Moll of American Forests surveyed the state of street trees in twenty cities, all the news was bad. "Only one city, Lansing in Michigan, was planting as many trees as it was removing. . . . A third of the cities planted only one tree for every eight removed, and about half the cities surveyed planted only 25 percent of their losses. New York, one of the larger cities in the survey, had lost approximately 175,000 street trees, or 25 percent of its total street tree population over the past ten years. A further survey of street trees in 1989 confirmed this depressing trend."

Above all, Americans living in and around cities were mourning the continuing loss of their elms. In Milwaukee alone, between 1956 and 1988 the city cut down 128,000 American elms. Journalist Derrick Z. Jackson spoke for so many who had grown up there, describing in his case a childhood on the city's "black north side" defined by elms, whose "immense crowns arched from tree to tree, creating a deep shade that kept us cool on the hottest days. They invited long games of hopscotch, two-square and 'Captain May I?' The yard-thick trunks were used for hide and seek. When we were done playing, we plopped our backs down on grass kept cool by the shade. At night, we tracked lightning bugs as mothers talked on the porches. The trees, by promoting community with their coolness by day, provided a leafy security blanket at night." As Dutch elm disease struck, those nurturing trees sickened. "Without the trees," recalled Jackson, "summer heat reflected brutally off concrete. Children retreated from the congregational sidewalk to individual porches."

When urban foresters surveyed these denuded cities, the obvious question was what to plant instead of the American elms. On columnist Ellen Goodman's Boston street, the city had chosen Norway maples, which she judged "lovely . . . [but] not American elms." Other typical urban options were silver maples (*Acer saccharinum*), London plane trees (*Plantanus* x *acerifolia*), and littleleaf lindens (*Tilia cordata*). As USFS chief John R. Maguire declared in his seminal 1975 "Plant a Tree and Save a City" speech: "We need to know more about proper selection of trees and shrubs for urban environments. Which species will thrive in most cities; which are most suitable for shading, noise abatement, esthetic enjoyment . . . [and] we need to know . . . how to properly landscape trees on engineered sites."

In the spring of 1982 the journal of the Arnold Arboretum, *Arnoldia,* devoted an entire issue to the demise of the elm, ending with an article titled "Replacing the American Elm: Twelve Stately Trees." Authors Gary Koller and Richard E. Weaver Jr. immediately dashed any easy solutions: "At present no single tree, including all the modern elm hybrids [meaning Asian elms], has the positive architectural qualities and the environmental flexibility" of the American elm. Nor were most of the listed twelve stately trees—selected based on their performance on the grounds of the Arnold Arboretum—practical prospects for urban life or streets. Silver maples had shallow root systems that "wreak havoc with pavement." The sugar maple (*Acer saccharum*) "cannot tolerate very dry soils or roadway salt." The large nuts of the Ohio buckeye (*Aesculus glabra*) "could pose a hazard to traffic or pedestrians," while the messy husks of the black walnut (*Juglans nigra*) were a menace that "stain sidewalks and roadways as they decay." The Kentucky coffee tree (*Gymnocladus dioica*) was open and picturesque but more suited to parks and lawn areas; likewise the river birch (*Betula nigra*) with its multiple trunks. The tulip poplar (*Liriodendron tulipifera*) was too statuesque at one hundred towering feet to squeeze into standard street tree pits but could supply "grand elements of outdoor architecture" along a parkway.

So what *were* the best options for an urban forester? Advised the authors: "Red oaks (*Quercus rebra*) make excellent street trees. They are tolerant of poor, dry, compacted soils, salt, and atmospheric pollution. The thick bark and strong wood are able to withstand the inevitable impact of vehicles. . . . At maturity such a tree is magnificent." They also had praise for the Chinese elm (*Ulmus parvifolia*), deeming it the best of the elm species to "replace the American elm. . . . It will never assume the stature of its American relative, but it is highly resistant to Dutch elm disease . . . [and] they have exhibited exceptional tolerance of repeated and severe damage to their trunks from automobiles and snow-removal equipment." They likewise endorsed the still-rather-rare Japanese zelkova (*Zelkova serrata*) cultivars "Village Green" and "Parkview" as "excellent street trees." White ash (*Fraxinus americana*) and green ash (*Fraxinus pennsylvanica*) had already proven themselves "adaptable to urban conditions" and came with a fall bonus of lovely autumn foliage, "a beautiful blend of yellows, golds, reds, and purples . . . resembling a bed of glowing embers."

With so many cities facing the conundrum of what to plant, a team from the Morris Arboretum at the University of Pennsylvania also weighed in with new nominees for successful city trees. In 1965 landscape architects Patton Inc. had selected 414 trees of twenty-three different species to be planted as street trees in a twenty-two-square-block section of Washington Square East in Philadelphia. Fourteen years later the arboretum staff made detailed observations of the results and passed judgment. "Six species which stand out due to their good growth rate and general freedom from pests and diseases are ginkgo (*Ginkgo biloba*), Kwanzan Oriental cherry (*Prunus serrulata* cv. Kwanzan), black locust (*Robinia pseudoacacia*), English oak (*Quercus robur*), Japanese pagoda tree (*Sophora japonica*), and Chinese elm (*Ulmus parvifolia*)." They were also enthusiastic about willow oaks (*Quercus phellos*), as long as they were planted where "the soil is uniformly acidic."

By the 1970s many urban foresters believed they had found the ideal new street tree in the Bradford Callery pear. USDA Chinese plant explorer Frank Meyer had been the first to sing the praises of Callery pears back in 1918. He had collected several hundred pounds of seeds that year in Yichang, China (just before his untimely and mysterious death), declaring: "*Pyrus calleryana* is simply a marvel. One finds it growing under all sorts of conditions; one time on dry, sterile mountain slopes; then again with its roots in standing water at the edge of a pond." Meyer had been looking for a fruit-bearing pear tough enough to withstand fire blight, a chronic problem in American orchards.

After World War II, at the USDA's Plant Introduction Station in Glenn Dale, Maryland, scientist and station chief Frederick Charles Bradford began growing Callery pears, hoping for a tough new ornamental tree. When he was stricken with a heart attack at work in 1951 and died, his successor, John L. Creech, continued the search. By 1963 Creech had found a particularly shapely Callery pear with a winning cloud of white spring blooms, named it Bradford in honor of his deceased boss, and released it to the nursery trade. The Bradford pear would become the most popular new urban tree of postwar America, planted up and down boulevards and side streets from coast to coast.

City arborists, politicians, and tree lovers all fell in love with this fast growing charmer, with its bridal bower–white spring blooms, attractive pyramidal shape (topping out at a tidy thirty feet), and glossy foliage that in autumn turned shades of scarlet, orange, and even a deep purple. Like Eliza Scidmore's Japanese ornamental cherry, the Bradford was becoming a beloved arboreal immigrant-citizen, even if Michael Dirr, legendary professor of horticulture and author of the tree bible *Manual of Woody Landscape Plants*, warned that the planting of Callery pears was reaching "epidemic proportions."

As American cities grappled with how best to replant their depleted canopies, the trees of New York City gained their most vocal champion since Andrew Jackson Downing: new parks commissioner Henry J. Stern. Appointed by Mayor Ed Koch in April of 1983, Stern revealed his passion publicly that summer when reporter Joe Klein described in *New York* magazine how Stern eliminated one of three proposed athletic fields in Central Park solely to save two large London plane trees. These trees were neither rare nor historic but one of the most common species in the city, having been the favorite tree of city planner Robert Moses. At the park site Klein watched, bemused, as Stern patted one of the trees, saying, "Don't worry. You shall be spared." Then he leaned against it and said, "Feel the tree. It's happy." Later, in the commissioner's office, wrote Klein, "Stern was proud of his performance. 'I saved two trees today. Isn't it nice to have a job where you can save trees?'"

Stern, forty-eight, was just warming to his new role. In a city famous for large personalities, he was making a name for himself as late-twentieth-century Gotham's most eccentric and colorful public servant. He was a brilliant Harvard Law graduate and workaholic with an enviable mastery of city laws, ordinances, and alliances *and* an amazing natural talent for stunts and silliness, such as showing up in a space suit to view the lunar eclipse in Central Park, having an aide keep track of how many people petted his dog, and appending odd nicknames to high and low. Stern's own nickname was StarQuest, and his public wackiness made him a media darling—a key factor in advancing his many serious agendas, one of which was saving the city's diminishing number of trees and planting many new ones. Klein was witnessing the first act of Henry Stern's long, entertaining, and sometimes just plain odd run as New York's top tree lover, a devotion rooted in childhood father-son rambles beneath old trees in wooded Inwood Hill Park at the northern tip of Manhattan.

In late 1984 New Yorkers learned through "The Case of the Imperiled

Ginkgo" just how determined their parks commissioner was to save trees. One Matt Sabatine had purchased a brownstone for $890,000 at 110 East Sixty-fourth Street, in the heart of the Upper East Side Historic District, and wanted to cut down a beanpolelike four-story-tall ginkgo street tree to gain entrance to a garage he intended to build on his ground floor. His high-powered neighbors on the block, including JFK historian Arthur Schlesinger, were infuriated.

Parks Commissioner Henry J. Stern, whose bailiwick included street trees, jumped into the fray, championing the ginkgo: "It's a marvel. It's an urban dream. It's a sight to behold when its leaves turn to gold." Sabatine spent the next two years battling not just the neighbors and Stern but also "the Landmarks Preservation Commission, the City Planning Commission, the local community board, several lawyers, at least one architect, [and] a horticulturist." Eventually he conceded defeat, thus preserving the ginkgo—reportedly planted by John D. Rockefeller for the dedication of the Asia Society, once located next door.

Stern pronounced that he, being "the custodian of plant life in the city, could hardly allow such a magnificent specimen to be sawed to the ground for personal convenience." Stern stirred up such tempests in gilded teapots not just to raise the profile of taken-for-granted city trees but, more important, to raise the budget to plant more. He would later share his philosophy: "The dirty secret is that if . . . people with money are involved and paying attention, they naturally exert pressure and the government responds."

To keep New Yorkers paying attention to the steady loss of trees, Stern dreamed up ever-crazier high jinks. In January of 1987 he summoned reporters to the scene of a new kind of urban crime: a "vile arboricide" (a term he either coined or popularized): Tree rustlers had cut down and stolen four large empress trees (*Paulownia tomentosa*), some as tall as seventy-five feet, from the very Inwood Hill Park where the commissioner had wandered as a lad. After denouncing the thieves and offering a thousand-dollar reward, Commissioner Stern announced, "Mayor Koch has called for the return of the electric chair, but these people who cut down these trees, give them what they gave the trees." Seeing the terrific press coverage he earned in his role as the city's Lorax (Dr. Seuss's character in the 1971 book of that title, who declares himself the creature "who speaks for the trees"), Commissioner Stern became ever more vigilant.

In early 1988 Stern fined a Bayside, Queens, building contractor who admitted to ripping up twenty-six oaks, poplars, sycamore maples, and sassafras on the Cross Island Parkway to clear a better view of Little Neck Bay for his two new houses. Developer Anthony Conte acknowledged the "mass arboricide" and paid $100,000 for 180 trees to be planted in the vicinity of the crime. "The Bayside 26 shall not have died in vain," pronounced Stern, who had less luck that summer tracking down the culprits in the "Temple Square Massacre" in downtown Brooklyn. He had heard rumors that the guilty parties were local store owners who had hacked down four mature London planes so the public could see their store signs. Stern appeared with his entourage, set up a boom box playing Verdi's *Requiem*, and held a funeral for the jagged stumps.

In 1985, as part of his Cherish Our Trees crusade, Stern had revived Arbor Day, which in most places had been largely eclipsed by Earth Day. He also secured Tree City USA status for New York from the National Arbor Day Foundation. Three years after John Rosenow had launched this project, New York was among 552 cities that had signed on, each of them finding the award useful in public relations and as a political prod for greater funding for urban trees. On Arbor Day, April 26, 1985, Stern stood on Central Park West next to a 125-year-old American elm, one of the winners in his "Great Tree" search, and declared: "We in the animal kingdom give far too little thought to the other forms of life. We should be more considerate. We want them to think well of us when we are gone. Trees are for the most part largely benevolent. The worst thing they can do is fall on you, and then it's usually not on purpose."

Stern had launched the Great Trees program to celebrate 120 arboreal landmarks worthy of attention: the city's "oldest, largest and otherwise most unique trees." A *Great Tree Walk Guide*, written by Gordon Helman five years later, invited New Yorkers to enjoy the surprising bounty of history attached to especially large or rare trees. Who knew that in 1737 William Prince had established in Flushing, Queens, North America's first tree nursery, which in 1776 had a special on grafted English cherry trees? Prince's Nursery was a place of pilgrimage for tree lovers, including our first father of trees, President George Washington, who wrote in his journal in 1789, months into his first term, "I sett off in my barge from New York to visit Mr. Prince's fruit gardens and shrubberies."

In a city that loved drama and big stories, Stern showcased one of the most infamous and raffish of trees: the "Hangman's Elm" in Washington Square Park in lower Manhattan. More than three hundred years old, with a massive trunk, it towered 135 feet tall. Sometime in the chaos of the American Revolution, its sturdy branches had served to hold the hangman's noose, the sentence for traitors, or perhaps the tree had merely been an innocent witness to such grisly justice. Records show gallows were also constructed here. Three decades on, that English elm remains the biggest and most famous in New York, and no amount of debunking has managed to dispel the hanging legend.

At American Forests, Deborah Gangloff, herself a natural at attracting media attention for trees and urban forestry, so admired Stern's clever Great Trees venture that she set up her own high-profile Famous & Historic Trees program in 1986. What better candidates to launch the program than the tulip poplars that George Washington planted in 1785 at Mount Vernon? Still flourishing, these trees evoked republican grandeur: "Like Washington himself, they are stately, impressive, and tall—over one hundred feet tall. . . . The marquis de Lafayette also loved the tulip poplar; he felt it was the quintessential tree, and he took saplings home to France . . . [which he] took to Versailles to be planted and gave one personally to Marie Antoinette."

A few years later Gangloff signed on Jeffrey G. Meyer of Jacksonville, Florida, a folksy nurseryman with a deep passion for growing and proselytizing the charms of historic trees. He was soon raising and selling young saplings with beguiling names like "Moon Tree sycamores" from seeds that had flown to the moon and back, "Elvis Presley pin oaks" from Graceland acorns, "John F. Kennedy post oaks" from the tree shading JFK's grave at Arlington Cemetery, and "Walden Woods red maples" from the hills where Henry David Thoreau once wandered. They are all featured in a charming, slender volume he published in 2001. These famous historic trees proved a public relations bonanza, culminating in a PBS show featuring Meyer as a guide to some of the superstars. "I sold five thousand to ten thousand trees a year," says Meyer. "Johnny Appleseed trees were very popular, and the George Washington poplars. On Easter of 1994, I remember we planted a magnolia at the White House with the Clintons during the Easter Egg Roll. It was just chaos."

Then, in 2004, recalls Meyer, "We had three hurricanes in one summer and it just wiped me out. I had a hundred thousand historic trees in my Florida nursery. It rained thirty to forty inches during Hurricane Ivan, and everything was underwater, and then the greenhouses began collapsing under the winds. After the first hurricane I rebuilt, but Ivan finished me." He moved far away to Wyoming, where he took up wind farming. "I still plant trees. I still love trees." He had had wonderful adventures—planting trees in locales as varied as Versailles and the Kremlin, but nature had done him in. With Meyer gone, the program languished.

Gangloff also revived her organization's *National Register of Big Trees,* first dreamed up in 1940 by lumberman Joseph Stearns, and launched that September. Its very American goal was to identify the largest specimens of all the major American tree species—from the already famous redwoods like General Sherman to the soon-to-be-celebrated oaks and pines and sycamores. "If you know of a very large tree," the conservation group urged readers of its magazine, "make it your business to see that its full and accurate record is sent to the American Forestry Association; its identity as to species; its diameter or circumference four and a half feet above the ground, its height, its state of preservation, and particularly its location and ownership."

Before our primeval forests were largely clear cut, the land was thick with gigantic trees, and Americans had long been eager to take their measure—literally. When President John Adams was a young lawyer riding the circuit in coastal Maine in 1765, he came upon a felled hemlock lying across the road. "I measured the Butt at the Road and found it seven feet in Diameter, Twenty one feet in circumference. We measured 90 feet from the Road to the first Limb." He estimated its height at 130 feet.

By Gangloff's day, the once-popular *Register of Big Trees* was semimoribund. "In the mid-1980s, we realized people loved this, we got better organized, and we found it got us the most media attention." On April 20, 1986, for example, the front page of the *New York Times* reported, "The Louis Vieux, reigning king of the American elms, has been dethroned." And, as the paper proudly noted, New York state had "13 national champions . . . not always the romantic species" since one was "the nation's largest poison sumac." And yet another, that "familiar scourge of alleys and vacant lots," was the champion ailanthus or tree of heaven. Then, says Gangloff, she saw "a Davey Tree ad of a kid hugging a tree. They became the corporate sponsor in 1989, and we really pumped some life into it. These champion trees just captured people's imagination, and we began doing the *Register* every two years. And once it was computerized, it just became so much easier."

Back in New York City, even as Commissioner Stern entertained the press with his ever-more-elaborate tree funerals, he managed to anger his natural allies—the city's professional arborists. Apparently Stern's love for trees had led him to forbid the arborists from bothering the trees by actually climbing about in their upper reaches to care for them. "Hard to believe," complained one arborist in print, "but true. The commissioner does not permit anyone on the horticultural staffs in our historic parks to climb trees." Cherry pickers could get arborists no higher than fifty feet; any higher than that, and the trees' caretakers would necessarily have to clamber about them with ropes if they needed pruning or cabling. As a result of Stern's having forbidden such routine maintenance, said the arborist, the city's great trees "get cursory attention or are left untended." This made sense if one understood, as one local politician did, that "trees were people to Henry."

All of Stern's antics were aimed at getting New Yorkers to pay attention to the fact that every year the city's estimated 700,000 street trees were diminishing in number—not because of assaults upon them but because strained city budgets did not value them. "More than 16,000 trees were removed during the fiscal year that ended last month," wrote *New York Times* reporter David Dunlap in late July 1988. "Several thousand of those had been dead longer than a year, but were part of a backlog awaiting removal. In the same period, because of budget limitations, the city planted about 8,200. The city's chief forester, William B. Lough, foresees a continued annual net loss until 1991 or 1992. By that time, the average annual losses should be about 12,000 to 14,000 and the city hopes to be planting more than 12,000 trees a year." Meanwhile, Stern did what he could, holding his funerals, denouncing arboricides, enacting new, tougher tree ordinances that exacted fines, replacing trees, and even threatening jail time to perpetrators of arboreal crimes.

On January 20, 1989, George Herbert Walker Bush, Ronald Reagan's loyal, low-key, two-term vice president, became the forty-first president of the United States. In his inaugural address Bush exhorted each American to become a citizen who "leaves his home, his neighborhood, and town better than he found it." He wanted to create "a kinder, gentler nation" with "a thousand points of light." Urban tree lovers were taken aback to discover that they had in the nation's new leader a powerful ally. As one U.S. Forest Service manager marveled, "Then Bush One came along and everything changed."

Bush quickly convened a special White House meeting to inform the environmental agencies that "the President wants to plant more trees." What he had in mind was, in fact, a billion trees a year, an initiative he named "America the Beautiful."

Why, many naturally wondered, did this patrician Yale graduate who had attained his fortune via the Texas oil industry take trees on as a cause? One urban forestry activist credited the president's longtime political supporter Trammell Crow, Texas developer and tree lover, explaining, "Crow put tree planting into Bush's ear.... He was/is one of the biggest developers in the world and a very tough cookie and a good friend of the Bushes, and he's the guy." When the president made known his intentions, the U.S. Environmental Protection Agency quickly offered to take charge of "America the Beautiful."

Recalls one of those present at the White House meeting: "USFS Chief Robertson said, 'Not so fast—the Forest Service is this country's federal tree planting agency.' The White House asked both [agencies] to compete and come back with proposals. Chief Robertson came back to Tony Dorrell, handed it off and said 'Run with it!' Tony [the urban forestry chief silenced in the Reagan years] came up with a great theme for a tree planting campaign—'Trees Are The Answer' with a proposal along with bumper stickers, lapel buttons, etc. EPA also did not have a field delivery system and network for planting like the Forest Service through state and private forestry agencies. Thus the Forest Service was selected by the White House."

On October 12, 1988, just before George Bush was elected, American Forests, the U.S. Forest Service, and a coalition of city mayors had launched the aspirational Global ReLeaf program, with the goal of collectively planting 100 million trees in the coming four years, exhorting, "Plant a Tree—Cool the Globe." And now here was the new president, proposing ten times that number in a single year! By 1990 the U.S. Forest Service's Urban and Community Forestry Program won a coveted line-item spot within the Farm Bill, which ratcheted up its funding from a meager $2.5 million (its annual budget since 1978) to a respectable $21 million in 1991, rising to $24 million in 1992.

Suddenly real money (at least by the penurious standards of the tree folk) was available to fund state-level urban foresters in *every* state (where there had been but six) and state-level councils of experts and citizens to advise and work with them. The U.S. Forest Service could afford to create a National Urban and Community Forestry Advisory Council that would advise on distributing the new Farm Bill millions in the form of local grants. Over the next two years every single state would establish an urban forestry program.

In his first State of the Union address in January, Bush urged the planting of "more trees from the rural countryside to the center of our cities," a message he repeated soon thereafter when he urged a crowd of 65,000 at an Atlanta conference of the National Association of Home Builders to plant a tree to shade every home they constructed. Bush had already shown that he was practicing what he preached by planting trees himself. In the spirit of Theodore Roosevelt, who relished leaving live arboreal memorials in his political wake, the president promoted his "America the Beautiful" campaign with hands-on, high-profile tree plantings in Bismarck, North Dakota; Helena, Montana; and Spokane, Washington. On March 22, 1990, he and Mrs. Bush each grabbed a shovel and helped install an eastern redbud in the manicured back garden of the White House as members of Congress, conservationists, and industry leaders applauded the First Couple for their DIY spirit. He might also have convinced skeptical environmentalists that Republicans loved nature too, as he noted that trees helped offset global warming, then an uncontroversial assertion.

When the Bush administration came into office, John Hansel, the Elm Research Institute, and the American Liberty elms were all going strong. Even before Ellen Goodman had shown up in Harrisville to obtain one of the original American Liberty elms for her Boston backyard, Hansel had heard from a twelve-year-old Sebago, Maine, Boy Scout named Sean McNutt. Sean was proposing to nurture seventy tiny young Liberty elms until they were six feet tall, whereupon he would plant them for his Eagle Scout service project. Hansel, ever the clever promoter, soon had national scouting officials endorsing ERI's "Johnny Elmseed" program to all of its 412 councils.

By late 1989, under the "Johnny Elmseed" banner, Hansel had sent forth 75,000 one- to two-foot American Liberty elm saplings to the more than three hundred paid "municipal members" of the Elm Research Institute. In each locale Boy Scouts took charge of raising those young trees to a robust six to eight feet in new nurseries. Once the trees were big enough, the scouts organized integrating them back into the local landscape. Hansel was justly proud of this major comeback story, saying, "I figured there had to be something that somebody could do besides sit back and watch the chainsaw at work. Luckily, a lot of others felt the same way." And Professor Smalley no longer expressed any doubts in public about ERI's elms, telling the *New York Times* in a December 5, 1989, story that these much-in-demand trees "have the look of a classic American elm."

By not releasing his American Liberty elms to nurserymen, John Hansel was opting to retain control and recoup some of the Elm Research Institute's investment. This also meant, however, that American Liberty elms would be raised by amateurs—Boy Scout troops—and never at a scale or within a network that would make them widely available.

As American Liberty elms were stirring up excitement, USDA plant geneticist Denny Townsend was now hopeful that in time he could release to nurseries "disease-resistant American elm hybrids that will also be resistant to elm leaf beetles." To get American elms successfully reestablished in the urban landscape, professional tree nurseries would have to begin growing and selling them again. Townsend believed the nursery trade would be willing to take that risk only if these new American elms proved their mettle in comparative field trials.

By now Townsend had been redeployed to Washington, D.C., to head the National Arboretum's Floral and Nursery Plant Research Unit in Glenn Dale, Maryland. He had worked for fourteen years in Ohio before reluctantly moving back to the East Coast in 1984. "I had to leave behind all my trees," he says, "so the best thing I could do was to go back out to Ohio and clonally propagate the best trees, and I brought them back to the Arboretum." In 1989 Townsend had his trees moved to Glenn Dale, where he collaborated with horticulturist Susan Bentz.

As Townsend prepared to field-test all the plausible DED-tolerant American elm candidates, he wanted to include Eugene Smalley's American Liberty elms, which were enjoying a hearty welcome from the public. "We ordered trees from E.R.I., but we also got trees from Smalley," recalls Townsend. In New Hampshire John Hansel was furious when he heard that Smalley had sent the trees. "Gene didn't have my approval. He hands over our trees to what I consider the enemy. These guys at the USDA had all been Doubting Thomases. They had said, 'You're not going to get a resistant elm. This is a waste of time.' And when we're finished and have our trees, Gene sent samples to the USDA. I could have shot him."

CHAPTER FOURTEEN

"Don't *Trees* Clean the Air?": Rowan Rowntree, Greg McPherson, and David Nowak

The Chicago skyline and part of its urban forest. *(© Songquan Deng/Alamy Stock Photo.)*

When Chicago's Richard M. Daley, a self-proclaimed tree hugger born on Arbor Day, became mayor in 1989, he vowed to plant a half million trees. Daley was a man who hated concrete and was still heartsore at having seen a hundred thousand local elms slowly die. New trees would help revive his Rust Belt hometown. "What's really important?" he asked one interviewer. "It's the personal things. A tree, a child, flowers. We need to soften the cities. Neighborhoods need nature. . . . We can help people by planting trees, by putting in pocket parks that are habitat. . . . Taking care of nature is part of life. If you don't take care of your tree and don't take care of your child, they won't thrive." Moreover, as Daley, forty-seven, took over the office his father had held for twenty-one years until 1976, his city's air was the fourth most polluted in the nation, and the U.S. Environmental Protection Agency was pressing for improvements. Daley wondered, "Don't *trees* clean the air?" But no one could yet provide an answer to his question that was actually grounded in science.

Elsewhere in Chicago was a young arborist, urban forester, and self-proclaimed tree hugger named Edith Makra, who had interned every summer at the Morton Arboretum during her years at the College of DuPage. Hired by a local tree care company, she began as a climber and trimmer. By May of 1987, promoted to sales, she was calling at the seven-acre English Tudor estate of Mr. T, the Mohawked star of the TV hit *The A-Team* and World Wrestling Federation bouts. "There I was, walking his North Shore property," recalls Makra, a slender woman who wore her brown hair in a ponytail and bangs, "and he wanted to widen his driveway and take down three hundred oak trees. I just could not put my heart into that. I felt as if I would have too much sap on my hands."

Soon thereafter she joined Openlands Chicago, a venerable conservation group that had been conserving and restoring prairies. Inspired by

Andy Lipkis and TreePeople in LA, they were now looking to plant urban trees. She launched NeighborWoods for Openlands, coordinating volunteer tree plantings to regreen vacant city lots and wastelands. She ramped up Arbor Day, passing out 1,500 honey locust seedlings to schoolchildren in a big downtown ceremony in April of 1989. At a Catholic high school she helped plant "a six-foot sapling that is a direct descendant of the Stelmuze, a two-thousand-year-old, thirty-foot-wide English oak in Lithuania that is revered as a national monument."

In the summer of 1989 a *Chicago Tribune* reporter came to write a profile of Makra. "They stuck it on the obit page," she recalls, "with the headline SHE CRIES FOR THE TREES. Turned out that Mayor Daley read the obituary page that day, and I got a call saying, 'The mayor wants to meet with you.' That August I was hired to launch his new GreenStreets program, and I became known as the mayor's 'tree lady.'" She laughs. "I was all of twenty-seven and very, very green politically. . . . The mayor was always about substance. And pleasant living."

The EPA was still pushing for the city to address its air pollution issues, and in search of a solution, Mayor Daley asked Makra to investigate whether trees could actually help clean the air, as well as prevailing on a fellow tree lover in Congress, twenty-term Democratic Chicago representative Sidney R. Yates, to earmark some serious federal research dollars for the problem. (Yates, who was chair of the Interior Committee, was the same politician who had refused to let President Reagan kill the federal urban forestry program.)

Makra had met Rowan Rowntree at a recent urban forestry conference. "I remembered him talking about the Dayton Climate Project, which stuck with me. I thought it was the neatest thing." In the seven years since the Dayton Climate Project and the Cincinnati urban forestry conference, Rowntree had been actively developing an intellectual and scientific framework for urban forestry, mentoring graduate students, collaborating with a range of other scientists, and serving as editor on three issues of academic journals dedicated solely to the new research and thinking on the "ecology of the urban forest." For Rowntree the goal was still clear: understanding how trees in cities influenced "the human environment." Only this kind of scientific knowledge could allow city foresters to quantify what he was certain would be the compelling benefits of city trees, justifying "the costs of creating and maintaining an urban forest." But as he lamented in 1988 in *Landscape and Urban Planning* in the third of the issues he edited, "very little research is being done to provide these answers."

By then Rowntree had moved back to the University of California at Berkeley, where he had been tasked with establishing a West Coast USFS-funded Experiment Station. "Edith Makra called out of the blue," recalls Rowntree, and told him, "Here's the deal. We have a lot of money. Do you have the staff? Can you come to Chicago?" Rowntree was thrilled, especially when he heard further details: "The mayor will be getting us $900,000." To Rowan both the timing and the location seemed perfect.

Since Dayton, Rowntree had trained two young PhDs, Gregory McPherson and David Nowak, who were proving to be talented and innovative researchers. McPherson's dissertation revolved around modeling the role of city trees in reducing energy costs and cooling down urban heat islands; Nowak was developing a computer model to quantify air pollution removal by trees and was also measuring and quantifying the carbon dioxide absorption of Oakland's urban forest. "What Greg and Dave had just finished," says Rowntree, "was a great foundation for doing the work in Chicago." Rowntree was eager to run a study on a major urban forest, and here was Makra offering real money, a tree-loving mayor, and the nation's second-largest city, Chicago (whose motto, fittingly, was *Urbs in horto,* or "City in a garden").

Rowntree's first task was to lure his former Syracuse graduate student McPherson away from his newly tenured position at the University of Arizona in Tucson. After Syracuse, with its plentiful tree shade, McPherson had been distressed by the hot, barren, and treeless Tucson. A tall, lanky fellow who commuted to the campus on his bicycle, McPherson had teamed up in 1989 with local environmentalist Joan Lionetti to found Trees for Tucson, which was affiliated with American Forests' Global ReLeaf campaign. (Trees for Tucson's early meetings memorably opened with a standing cheer of "Trees! Trees! Trees!" while members waved their hands overhead like blowing leaves.) When the tiny nonprofit proposed planting a half million trees over time to cool the desert city and reduce energy use, recalls McPherson, "the local water utility said, 'Whoa, we're trying to conserve water, and we can't support this,' and that got me to thinking, 'Yes, trees use water, but they have other benefits, and I better include those in my calculations.'"

So he broadened his study. "I added particulate matter, because in the desert climate there's a lot of dust. Tucson's Achilles' heel with air pollution was that dust, and we all knew that trees intercepted dust. So we looked at desert-adapted trees, like native velvet mesquite, that would use less water, and I developed a method for quantifying the uptake of dust, as well as the energy savings, and compared that with the costs of watering the trees. The net benefit of the trees was greater than the cost."

In his study, which was later published in *Landscape and Urban Planning* as "Accounting for Benefits and Costs of Urban Greenspace," McPherson calculated that over forty years the average cost of planting and maintaining a mesquite tree in Tucson would be $9.61 each year. During that time period that same tree would on average provide $25.09 annually of "environmental services," each year absorbing thirty-five pounds of dust, mitigating 2761 gallons of storm water, and saving 288 kilowatts of energy. Says McPherson proudly, "This was the first peer-reviewed formal study that actually had an impact on a community. Tucson Electric Power was persuaded by this to invest in desert-friendly trees to reduce power use, becoming a supporter of Trees for Tucson. Rowan knew what I was doing and wanted that in Chicago. I saw Chicago as building on Tucson."

If McPherson had any hesitations about giving up his tenured Tucson position to serve as the USFS project leader for three years on what would be known as the Chicago Urban Forest Climate Project, Rowntree promised a plum reward: Greg would be in charge of the West Coast Urban Forest Research Center that Rowntree was organizing. In July of 1991 McPherson moved to Chicago with his very pregnant wife and two small children. He was returning to his midwestern roots, having grown up in Howell, a small county seat between Detroit and Lansing, Michigan, in the years when DED sickened the trees. "The canopy," he says, "was American elms. I remember these tubes and glass jars they were using to inject them to save them, but it didn't work, because DED was spreading from tree to tree through the roots. One tree after another died and had to be taken down. I was eight to ten years old as those trees came down, and it was pretty shocking to see the sky again, to lose that protective, umbrellalike effect softening the space overhead. Suddenly it was so hot and exposed, so glaring, with heat reflecting off the road and driveways."

After he earned a BS at the University of Michigan's School of Natural Resources in 1975, McPherson's first job had been with Goldner-Walsh, a Detroit-area nursery. He had liked the work and decided to pursue a master's in landscape architecture at Utah State University in Logan. There he discovered that design was not his strong point, but marshaling data was. In one project he and other students considered developing two sites, one with clustered housing, the other with single houses on large lots: "We calculated the amount of infrastructure," says McPherson, "estimated the cost difference, and saw how much more cost-effective it was to cluster. It was the first time I saw the power of putting numbers to things that were good for the environment."

By 1982 McPherson was teaching at Logan and considering a PhD in urban forestry when he crossed paths with Rowntree in Cincinnati. The two clicked, and the next year McPherson became Rowntree's doctoral student at the College of Environmental Science and Forestry at the State University of New York (SUNY), Syracuse. Recalls McPherson, "When I saw the program and what Rowan was doing, it was pretty cutting-edge. We just connected on multiple levels." McPherson, sensitized by the loss of his hometown elms "to the role trees play in our lives" and fascinated by data driving green policy, wrote his dissertation on how trees affect the temperature inside and outside of buildings. Having fallen in love with the American West during his years in Utah, he then moved in 1984 out to Tucson, where he taught urban forestry and horticulture at the University of Arizona's School of Renewable Resources. Now he was giving up his tenured academic job to join the USFS as a scientist and project leader in Chicago.

David Nowak, who was completing his doctorate at Berkeley, had first met Rowntree in 1983 when the younger man was still a SUNY undergraduate. "I gave a guest presentation," remembers Rowntree, "and David asked such good questions, we set him up to do a master's in urban forestry." Nowak had grown up in suburban Buffalo. Trees were fun to climb but otherwise not much on his mind. "Originally," he says, "I wanted to go into wildlife. But I realized it would be very tough to get a job and it didn't pay well. I got hooked into urban forestry. I saw it as the wave of the future. I thought, *This is where the action will be.* It would be an important area."

At Berkeley Nowak shared a big office with Rowntree, who admired the way he went about establishing the forest structure of Oakland. "To build mathematical equations," says Nowak, "you had to measure everything, so we were out in the field physically measuring local dimensions of trees—their DBH [trunk diameter at breast height], determining species, condition, crown width and height. We wrote computer code that created a 3-D model of a tree to see where the crown edge was." To understand whether a green ash better absorbed air pollutants than, say, a Norway maple, Nowak and his helpers had to figure out how *many* leaves a tree of a certain trunk size could be expected to grow, *and* the tree's leaf area index.

"We'd scan the leaves through a scanner: for American elms, green ash, hackberry, locust, Norway maples," says Nowak. "We used to do this by hand. We needed this to develop an algorithm that would work for each species to know its crown leaf area in relation to its trunk size." They further studied each of the different species' leaves under microscopes to learn to what extent they removed assorted pollutants and absorbed carbon dioxide. They then used this field data to write computer programs reflecting how different urban trees "performed" in intercepting air pollution and absorbing carbon. Says Rowntree proudly, "Dave wrote the first paper on how to measure the carbon dioxide sequestration potential of a tree." Chicago would be Nowak's first postdoctoral job, a far, far larger laboratory to further hone this work, and he and his wife headed there with their twin sons. Rowntree would commute in and out as research and work allowed.

By July of 1991 McPherson, Nowak, and a shifting cast of scientists, doctoral students, and technicians had moved into a large corner office in one of the low brick buildings of the North Park Village Nature Center in Chicago. Across the hall John Dwyer ran the USFS's new Mid-West Experiment Station, focusing on social science research. The surrounding 158 sylvan acres had been the site of the Chicago Municipal Tuberculosis Sanitarium from 1911 to 1979. "At lunch," recalls McPherson, "we used to go out and play Frisbee golf. There was this enormous park space, and we had fun just wandering around under these big old massive oak trees predating the Civil War."

At the time none of the scientists knew that before the TB hospital, the land had for decades been a famous tree nursery, owned by Norwegian immigrant Pehr Peterson. These very acres had provided "all the trees and shrubs for the [Chicago] World's Fair in 1893, most of the trees in Lincoln Park and, by some estimates, seven-eighths of all the trees for Chicago's parkways and boulevards by 1910." Peterson was also known for his "innovative techniques for moving and transplanting large trees."

The sheer physical challenge of Rowntree's Chicago Urban Forest Climate Project was daunting: It would study the urban forests not only of the City of Chicago but also of Cook and DuPage counties, 1,300 square miles that were home to six million people. "The proposed research" in Chicago, wrote McPherson, would be "comprehensive, ranging from studying the leaf area of individual trees to the aggregate effects of all trees" on "hydroclimate, air quality, energy use, and carbon cycling." The young scientists saw their goal as guiding "policy makers, regional planners, municipal/county greenspace managers," as well as local utilities, landscape architects, and homeowners.

Their first task was to learn the history of Chicago's urban forest and determine its present structure by conducting an arboreal census via sampling randomized plots. Then, using sampling and modeling, they would scientifically calculate and assign a dollar value to the ways that all these varying city trees collectively cleaned the air, saved homeowners electricity, cooled down heat islands, and sequestered carbon. Edith Makra had been delegated to serve as the Climate Project's liaison to the city, providing whatever help they needed. No longer the mayor's "tree lady," she was now on staff at the city's Department of Environment. Rowan Rowntree envisioned Chicago becoming a green pioneer, the first major American city whose leaders and citizens would showcase this new science, deploying its findings to strategically plant and manage the urban forest to "maximize net environmental benefits." These were heady days.

McPherson and Nowak settled into their cubicles, orchestrating the complicated logistics. Step one was assembling a complete set of aerial photographs. These hundreds of four-foot-by-three-foot bird's-eye views taken in 1987 hung in a thick sheaf on a rolling cart. Makra recalls filling the floor of the conference room with many of these maps, which helped determine the density of the canopy and which randomized field plots to use for sampling. As the scientists figured out the many data sets they and their colleagues had to acquire and analyze, project leader McPherson and Makra also made the rounds to explain the project to a variety of city officials and environmental leaders.

Not long after settling in, McPherson looked up from his desk to see before him "these guys wearing suits who were big and strong. And they had gold chains and kind of looked like henchmen sent by the boss to scare the bejeebers out of us. They never said exactly this, but the feeling conveyed was 'Remember, the mayor was born on Arbor Day and Chicago Urban Forestry is the best in the country.'" A bit nonplussed, McPherson chalked this up to the Chicago style of communication. Chicago Urban Forestry, its budget slashed in the pre–Daley Jr. era, was just beginning to recover.

Using the aerial photographs, Nowak selected 652 test plots (213 of them in the city proper), each plot a circle 74.4 feet in diameter and set in the real world of streets, buildings, suburban malls, schools, parking lots, highways, parks, and vacant lots. The field crews then set out to document all the trees in each plot. "At times," wrote McPherson, "we had 20 people collecting data on tree conditions, clipping leaves and branches for analysis, and measuring climate effects."

Makra was fascinated to discover how absolutely key quantifying tree leaves was to the larger enterprise. One of Nowak's first tasks was to calculate the actual total leaf surface area of typical local trees, which he and McPherson needed to determine reductions in air temperature, air pollution, and volatile organic emissions, as well as carbon dioxide sequestration. How many leaves *were* there, say, on a "typical" nine-year-old Bradford pear with a nine-inch DBH? Almost 89,000, it turned out, with a total leaf area of 1,923 square feet. Unless you knew how the leaves of different tree species of different stature interacted with air pollutants and particulates and with sun and wind and rain, you could not meaningfully calculate the collective "work" done by the urban forest.

The scientists and their teams headed into the city parks, looking for good trees to serve as baselines for their research. "To estimate leaf-surface area of urban trees," explained Nowak about this part of the project, "data were selected from 54 healthy, open-grown park trees in Chicago that were selected specifically for their excellent condition (10 American elm, 10 green ash, 10 hackberry, 10 honey locust, and 14 Norway maple)." Nowak sent out a two-man field team that mapped the "volume of the crown of each tree ... using a telescoping pole." They then measured crown height and distance from the tree base every five feet vertically and at another eight points. Nowak next had the field team maneuver their bucket truck into place up in the canopy and there among the branches methodically pick all the leaves from ten designated 14.1-cubic-foot sites. "The number of leaves per sample were counted," wrote Nowak, "and approximately 30 leaves were randomly subsampled for analysis of leaf area.... Following leaf-area analyses, all leaves were dried at 65 degrees for 24 hours and then weighed."

McPherson, meanwhile, focused on running the overall project and finding representative building types for his own energy-saving research. "I was collecting information," he says, "from local utilities on types of buildings, construction, and typical heating and cooling use for residential buildings. We did some surveys, looking at construction, insulation, thermal pane windows, so we could come up with some proto-typical buildings." A nice camaraderie developed among the staff, and Makra wanted to make everyone feel welcome in her city. "We had one another over for dinner. We went apple picking. I'd babysit for their kids. We played volleyball and had a beach picnic."

On July 13, 1992, a year into their labors, McPherson, Nowak, and Makra issued an initial report that set out their research goals and also briefly described Chicago's swift development into a major metropolis, its particular environment much influenced by the famous 1909 Plan of Chicago, which had wisely preserved some forest, created tree-lined boulevards, and designed elegant lakefront and landscaped parks. The report noted "urban forest management issues" of varying "scope and urgency" and pointed out the multiple services that trees could provide. The Chicago Urban Forest Climate Project would, they promised, translate "selected environmental benefits into dollar terms" and offer advice on how best to strategically increase these benefits. Mayor Daley had been reelected the previous November with 70 percent of the vote.

Over the next two years the group steadily gathered the necessary field data from the hundreds of circular plots, as well as myriad other data about air pollution, weather and temperature and wind patterns, urban forestry costs, and energy costs, and devised the necessary mathematical and computer modeling. In June of 1994 Rowntree and his protégés issued *Chicago's Urban Forestry Ecosystem: Results of the Chicago Urban Forest Climate Project*, a 210-page final report illustrated with hundreds of charts, graphs, and maps. In the end they had largely concentrated on the city of Chicago in order to make the most of their resources.

For the first time the city learned never-before-known details about the city's "forest structure." Chicago, home to about 2.8 million people, was also home to 4.1 million trees. Over the past 160 years its original historical landscape of prairies, wetland, and oak/hickory forests had given way to an industrialized city with an 11 percent canopy cover. This urban forest was dominated by cottonwood (13 percent), green and white ash (12 percent), American elm (7.2 percent), and Japanese cherry (6.5 percent). By the late twentieth century oaks and hickory barely registered; the city's 126,000 oaks constituted only 3 percent of the present-day tree population, not much more than the unloved ailanthus. Nowak published his formulas for estimating the all-important leaf surface area of urban trees, which could in the future inform "how managers and planners can direct urban forest structure to a desired outcome."

"The most impressive thing," says Makra, "was to be able to really *know* what we had. While there were not a lot of oaks, most were large and old. They had been here a long time. And it told us that we needed to focus on planting more to have a new generation." John Dwyer, the USFS social scientist whose group worked across the hall, remembers being amazed to learn that the city had so many millions of trees and the outsize status of the city's street trees.

Thanks to the Climate Project data, foresters now also knew that while the city's 416,000 street trees made up only a tenth of its urban forest, they also provided a *quarter* of the tree canopy or "24 percent of the total leaf-surface area." Overall, the city's 11 percent tree canopy was *half* the size it ought to be, and adding street trees was essential to expanding it. Because almost half of existing street trees were Norway maples and honey locusts, the city had to plant a much wider palette of species to avoid a potential loss comparable to that of the American elm.

In addition to Nowak's revelatory chapter "Urban Forest Structure: The State of Chicago's Urban Forest," he described for the first time "Atmospheric Carbon Dioxide Reduction by Chicago's Urban Forest." Rowntree wrote the report's introductory chapter, "The Role of Vegetation in Urban Ecosystems." Scientists Gordon M. Heisler, Sue Grimmond, Catherine Souch, and Richard H. Grant wrote two chapters on preliminary research into how the urban forest affected urban microclimates and thus energy use and rainwater runoff in residential neighborhoods. The final chapter, "Sustaining Chicago's Urban Forest: Policy Opportunities and Continuing Research," was written by McPherson, Nowak, and their mentor, Rowntree.

While everyone knew that trees cooled down buildings. McPherson's chapter "Energy-Saving Potential of Trees in Chicago" measured the *actual* energy and dollar savings involved. The shade from a large street tree growing to the west of a typical brick residence, he wrote, could reduce annual air-conditioning energy use by 2 percent to 7 percent. Three twenty-five-foot trees "located for maximum summer shade and protection against winter wind could save a typical Chicago homeowner about $50 to $90 per year (5 to 10 percent of the typical $971 heating and cooling bill)." By planting more trees in built-up city neighborhoods whose higher temperatures made them urban heat islands, and promoting utility-sponsored residential tree plantings, the city government could further curtail energy use. "Trees," McPherson wrote, "can help defer the construction of new electric generating facilities by reducing the peak demand for building air conditioning."

But what of Mayor Daley's original question: Don't *trees* clean the air? The answer was yes. Nowak's chapter "Air Pollution Removal by Chicago's Urban Forest" explained how he had calculated that the city's 4.1 million trees had in the year 1991 removed an estimated 17 tons of carbon monoxide, 93 tons of sulfur dioxide, 98 tons of nitrogen dioxide, 210 tons of ozone, and 234 tons of particulate matter from the air, at an estimated value of $1 million.

These impressive-sounding figures did come with important caveats. First, what did they represent within the context of Chicago's polluted air considered in its entirety? The city's trees had cleansed 652 tons of gases and particulate matter from urban skies, and while this total sounded huge, it constituted at best 1 percent of the pollutants spewed into Chicago's air by traffic, power plants, and millions of buildings. No savvy politician could tout that as a meaningfully significant improvement. Nowak did point out, however, that in neighborhoods where trees were large and lush, they improved air quality by as much as 15 percent during the hottest hours of midday. If the city were to strategically plant more trees and bigger trees, in time they could play a far more important role in fostering cleaner air, especially if further *avoided* pollution from energy savings from tree shade and windbreaks was factored in.

Second, Nowak cautioned, "The removal estimates in this paper are approximations based on computations that incorporate measured local urban tree canopy surface, local pollution concentrations, and local meteorology in diurnal and annual patterns. . . . The results should be considered first-order approximations of pollution removal by urban trees." So while Chicago's trees had been substantiated as cleaning the air, the actual tonnage cited at this point could only be a good, educated approximation.

What the Chicago Climate Project did establish was that trees should be regarded as essential green infrastructure—a capital investment that delivered *multiple* city services over the decades. Not only could an ever-larger urban forest steadily absorb more pollution and save more energy, but trees in the Chicago metro area could do the same for carbon sequestration. Nowak reported that the city's 4.1 million trees sequestered about 155,000 tons of carbon a year. Again, this sounded like a large amount, but as the report noted, that annual intake equaled the amount of carbon emitted by transportation vehicles in the Chicago area in just a single week.

However, as with pollution and energy, over time, if the city developed a lusher and more strategic urban forest, those trees could sequester as much as eight times more carbon. Chicago would need to invest in greater numbers of large, long-lived species such as oaks or London planes and actively nurture its existing trees to full maturity. A big tree that lives for decades or even a century or two could sequester a thousand times more carbon than, say, a crab apple with a life span of ten or twenty years. McPherson, Nowak, and Rowntree proposed that planting even 95,000 new large trees over the next thirty years would achieve this goal. The mayor had already announced he would plant 500,000.

"Are trees worth it?" asked McPherson in the report's final chapter. "Do their benefits exceed their costs? If so, by how much? Our findings suggest that energy savings, air-pollution mitigation, avoided run-off, and other benefits associated with trees in Chicago can outweigh planting and maintenance costs." The "other benefits," drawn from other researchers' work, factored in the documented higher real estate values that came with tree-lined streets and shady yards and the pure aesthetic enjoyment trees offered. The bottom line? Each tree planted on a Chicago street or in a park or backyard provided over its lifespan about $402 worth of "services"—about three times the tree's cost over thirty years. And so, concluded McPherson, "trees are worth the price to plant and care for them over the long term."

In the report's final chapter the three scientists proposed that their innovative work meant Chicago's urban forest could be reconceived as an essential element in the city's infrastructure, with a demonstrable dollar value. If properly managed, trees could serve "as a value-adding magnet for economic development." They suggested ambitious tree-planting and maintenance partnerships with the "Chicago Housing Authority, Chamber of Commerce, Openlands, Commonwealth Edison, People's Gas, Center for Neighborhood Technology, and other local, state, and federal organizations that manage public housing, energy, water, and air resources."

As the Climate Project wrapped up, it was clear to Edith Makra that while "the city of Chicago started out embracing this, once it got rolling, there really was not that much interest. We got our funding and did this great project. But, I was no longer the mayor's 'tree lady.' Had I still been in that position, I feel it would have been very different." *New York Times* correspondent William K. Stevens did write a piece about the project's findings, headlined MONEY GROWING ON TREES? NO, BUT STUDY FINDS NEXT BEST THING. But little further attention was paid to the groundbreaking research.

Rowan Rowntree, so proud of his team's original, data-rich science, was deeply disappointed at its reception. "We were scientists," he says. "Our job was to quantify and then turn it over. Then it was up to them. This was a case where it was perhaps just too much for the political system to accommodate. Maybe they thought people would expect the city to somehow follow up with a big plan."

Another scientist who was reasonably familiar with the local political system speculated, "I think Chicago Forestry worried that the study would show that somehow the urban forest was not in good shape and it would cast them in a poor light." Said Makra, "I had experience with what Mayor Daley's press machine had the ability to do. In no way did we see that press machine come out for this. I shared Rowan's disappointment." Even while continuing to transform Chicago into a world-class green city, neither the mayor nor any member of his administration ever mentioned the report. Daley pushed his 500,000-new-trees goal, but not in the scaled-up, strategic way Rowntree and his team had hoped for.

While Chicago ultimately evinced no interest in this new science of urban forestry, such was not the case in Sacramento, California's state capital, the self-proclaimed West Coast "City of Trees." For years Sacramento had been losing trees to age, neglect, and development. Tree lovers had been enraged when officials sanctioned the cutting down of huge heritage oaks on Elk Grove Boulevard in 1976, followed by further destruction on Auburn Avenue. In 1981 then-mayor Phil Isenberg, galvanized by Israeli tree planting and an "Ugliest Street" contest that showcased far too many barren city blocks, had gathered 150 citizens to take action. On California's Arbor Day, March 7, 1982, the nonprofit Sacramento Tree Foundation was born.

The goal was to plant a tree a day, and the group launched its effort with fifty shade trees along Fair Oaks Boulevard. The forty-person board soon discovered that, while it was easy to *plant* a tree, the hard part was keeping it alive and thriving in the hot, asphalt-dominated environment. As summer heated up, the new foundation hired native son Ray Tretheway, community activist and solar energy consultant, on a part-time basis to care for trees and volunteers. His first act: "I contrived two 50-gallon drums with a hose spigot, placed them in the back of my 1952 Dodge Truck and twice a week hauled hoses from the drums to water the trees." Tretheway was a graduate of the University of California at Santa Cruz who had moved east to Washington, D.C., to work at environmental think tank the Conservation Fund, where he became a convert to the power of civic engagement to improve the environment. "The trees need stewardship," said Tretheway, "and so do the volunteers. You know, it actually takes three to four years to plant a tree. Once it's in the ground right, there's the watering, staking, mulching, pruning. That's how I got my job."

Using Andy Lipkis and his TreePeople in LA as its model, the Sacramento Tree Foundation steadily grew. A local educator named Mike Weber proposed a "Year of the Oak" campaign aimed at restoring the region's original landscape and habitat using native oak trees. In 1985 Sacramento County and city officials embraced the idea and, working with the Tree

Foundation, launched a year-long campaign that collectively planted ten thousand young oaks. And so was born one of the foundation's signature programs: "Seed to Seedling." Third- and fourth-grade schoolchildren began growing little native oaks from acorns in milk cartons on the windowsills of their classrooms. After a few months, those new seedlings were all moved to a nursery to grow large enough for life on the grounds of schools and parks.

Like TreePeople, the neophyte Sacramento Tree Foundation began training a reliable corps of volunteers to properly plant and care for trees. These devoted folk provided the know-how and muscle at neighborhood tree plantings that then also drew membership from local residents. SacTree (as the group was soon known) partnered with as many interested groups as possible, drawing especially on the wealth and talents of private companies. And like Lipkis, it never stopped proselytizing on behalf of trees, stressing how planting trees could organize and slowly transform communities.

Lipkis was asked so often to speak about how others could replicate the ingredients of his success that in the spring of 1990 he and his wife, Katie, published a step-by-step how-to book: *The Simple Act of Planting a Tree: Healing Your Neighborhood, Your City, and Your World*. And to showcase the potential of TreePeople's more urban focus, earlier that year, on Saturday, January 13, 1990, TreePeople had pulled off its most ambitious street tree event ever: mobilizing more than three thousand volunteers to plant three hundred Canary Island pine trees along a seven-mile stretch of Martin Luther King Jr. Boulevard straight through the heart of South Central Los Angeles, starting at Long Beach Boulevard and ending at Rodeo Drive. For a decade Baldwin Hills resident Eudora Russell had envisioned a living memorial of trees to honor Dr. King. When TreePeople had launched Citizen Forester classes in 1986, she had been among its first set of graduates. The idealistic expectation was that each Citizen Forester would learn to organize his or her community to plant trees properly and then care for them responsibly. TreePeople would provide the initial support, but ultimately each community would create and care for its own urban forest.

Jim Hardie, also an alumnus of that original course, was serving as TreePeople's head of forestry and orchestrated this massive MLK "plant-in." "Tom Bradley was mayor then," he says. "Our first African American mayor, and he loved TreePeople and this salute to Dr. King. We trained three hundred volunteers so that each tree would have one planting supervisor, and then recruited enough volunteers to have another five or six on each team to help get each tree into the ground. The morning of the planting, we were having a light drizzle and we were all very concerned. Would the three thousand volunteers show if the rain kept up? And then, I kid you not, the clouds began to lift and a rainbow appeared over the city. We knew we would be okay. The whole boulevard had been closed down for this, and people just began pouring in from all over the city—the mayor was there, kids from schools, entire churches, all kinds of officials, tons of media." The previous January Mayor Bradley had announced that he was running for his fifth term, as well as another ambitious plan to green LA in coming years by planting two to five million trees, with help from TreePeople, part of an overall pledge to help reduce the city's greenhouse gases.

On New Year's Eve of 1988 in Sacramento, local tree lover Jeanie Shaw had written down her dream for the millennium's final decade. Inspired by TreePeople's Olympics goal, she penned the words "plant one million trees in Sacramento by the year 2000." The City Council and the Sacramento Board of Supervisors officially embraced this ambitious vision and named it "Trees for Tomorrow, a Gift to the 21st Century." By spring the chair of the "Trees for Tomorrow" campaign had joined the board of the Sacramento Tree Foundation. Many long-standing board members soon left the tree foundation, seeing no realistic way that their group, planting four hundred trees a year with volunteers, could scale up to accommodate so many more trees.

In July of 1989 outraged local voters forced the Sacramento Municipal Utility District (SMUD) to close its dysfunctional Rancho Seco nuclear plant. The following year, to reduce its peak load, SMUD's new tree-loving CEO, S. David Freeman, proposed partnering with the Sacramento Tree Foundation to plant 500,000 young five-gallon shade trees for free in the yards of SMUD's residential customers over a ten-year period. Freeman, a short dynamo who liked to wear his ten-gallon hat even at his desk, had helped get 20,000 trees planted in his previous job in Austin, Texas. SMUD would pay $1 million a year for Sacramento Tree to run this new program, called Sacramento Shade. Ray Tretheway, the Sacramento Tree Foundation's new first paid executive director, says: "Overnight, we literally grew from three part-timers to 30 full-time and a $2 million budget."

In 1991 Tretheway attended an urban forestry conference in Los Angeles, where he listened avidly as Greg McPherson spoke about the newly launched Chicago Urban Forest Climate Project. "He blew me away," Tretheway recalls. "Afterwards, I went up and said, 'Hi, I'm Ray from Sacramento. I can't believe it, you've clarified what I'm doing. These tree benefits, I'd never heard of this before.' So Greg said, 'Let's get some lunch,' and he introduced me to Rowan."

After meeting with the Chicago pioneers, Tretheway prevailed on his new congressman, Victor F. Fazio, to press the USFS to open what would be known as the Western Center for Urban Forest Research and Education at the University of California at Davis, not far from Sacramento. McPherson would soon take charge. In coming years Tretheway and Sacramento would acquire a wealth of studies and new data from McPherson and other tree scientists, who developed detailed portraits of the city's six million trees and their multiple benefits.

By the summer of 1993, when McPherson moved his family to Davis, the Sacramento Tree Foundation had already planted 111,500 trees in the yards of the city's homes, and SMUD wanted to assess whether these young shade trees were actually starting to reduce energy use. McPherson and colleague James Simpson gathered information from 326 sites on tree mortality, location (seventy-two different real-life shading scenarios), species, and size, as well as all relevant information about each house included in the study. McPherson's number crunching revealed that a tree planted to the west of a house saved three times more energy than the same tree planted to the south ($120 versus $39).

Those findings, wrote SMUD's Misha Sarkovich, an economics PhD who monitored the program's impact evaluation (and went from skeptic to convert in the process), caused the SMUD shade program to undergo "a paradigm shift. . . . Instead of tracking program performance in terms of number of trees planted, the program is now evaluated in terms of Present Value Benefit of each tree planted (as expressed in dollar terms)." Sarkovich would become SMUD's long-term program director, and under his aegis, trees would be strategically planted around customers' houses to save kilowatts.

McPherson's graduate student Qingfu Xiao began conducting pioneering research on the impact of trees on storm-water dispersal, measuring how much rainfall trees of various species and sizes intercepted. Storm water was an emerging (and expensive) problem for many cities as they were increasingly paved over. With rising standards for clean water, cities also faced new federal mandates to upgrade their sewer and water systems. The results of this new arboreal science, says Tretheway, "changed everything. Trees were no longer just a decorative item for our streets, parks, and homes. Now we had scientific information and it opened a whole different audience for us. We could talk about the cost benefits of the trees."

In the post-Chicago years, both McPherson and Nowak would further develop their science and models, engaging in increasingly ambitious studies. McPherson began systematically studying a reference city in each of sixteen climate zones to expand his database. As this new research became known, city foresters and nonprofit city tree groups alike drew on it to advocate for trees.

CHAPTER FIFTEEN

"We Stand a Great Chance of Seeing a Return of the Stately and Valuable American Elm": Rebirth of an Iconic Tree?

USDA horticulturist Susan Bentz with USDA scientist Denny Townsend with his American elms at trial fields in Glenn Dale, Maryland. *(Courtesy of the U.S. National Arboretum.)*

In 1989, at the seventy-acre National Arboretum facility in suburban Glenn Dale, Maryland, USDA plant pathologist Denny Townsend was finally ready for the great American elm survivor contest. He was preparing to test how the few, rare American elm clones and cultivars—survivors of the original waves of Dutch elm disease and years of further fungal assault in field trials—now stacked up against one another. "We were not trying to advocate any particular tree," he says, "we were just doing a scientific experiment."

One journalist driving into the USDA Glenn Dale facility gazed around and at first glance admired "tree-shaded lawns bearing resemblance to a posh country club's grounds." But after a closer look, he likened the facility, with its acres of semiabandoned test fields and hodgepodge of decaying greenhouses, wooden storage barns, and concrete-block offices, to "an obscure research station in Ukraine." He had come to meet Townsend, who ran his elm operation out of a spartan stucco lab and office with "the ambience of a small-engine repair shop, its drabness redeemed only by red celosia flowers blooming at the doorway." In fact, the building had originally served as "a cold-storage building—just concrete blocks lined with six inches of cork."

David Fairchild had established this then-rural outpost in 1919 as one of four national Plant Introduction Stations, a place of quarantine for the ornamental and commercial foreign plants sent home by plant explorers like Frank Meyer. Suburbs and highways now encircled the USDA fields, many of them cleared and fallow, while others were still filled with rows of old fruit and nut orchards. A few of Fairchild's original Japanese cherry trees—relocated from In the Woods and more than eighty years old—still bloomed here each spring, while Meyer's memory lived on in a row of old conifers and, of course, his most influential introduction, the Callery pear, which still held pride of place. Glenn Dale even had a towering dawn redwood from the 1948 China expedition.

Townsend's great American elm contest would include nine different cultivars. Four of these cloned specimens had emerged from his own decades of research and included a tree he and a colleague had first spotted on a 1976 trip to Springfield, Ohio. "We noticed this American elm on Interstate 70 that looked pretty great," he recalls. They kept an eye on it for another four years, watching in admiration as it flourished while all the nearby American elms withered away from DED. "We got permission from the landowner to propagate it. It was a wild native tree." This notable survivor was dubbed Clone 680.

Townsend also had been growing on his Glenn Dale plantations the fifth of the nine clones in the trial, this being R 18-2, the most tolerant of the Cornell University–Boyce Thompson Institute trees dating back to the 1930s. Another, which came to be known as "Delaware," was the last survivor of the USDA elms dating to the 1930s. Townsend had then secured as the sixth candidate tree one known as a Princeton elm, a tree first admired in a cemetery in 1922 and placed into the nursery trade by Princeton Nursery of Allentown, New Jersey. "When I was at Delaware, Ohio," says Townsend, "Bill Flemer of Princeton Nursery sent me a tree. It had been bandied about in the nursery industry that the tree was disease resistant. I decided it was time to include it in this test." Townsend also included as a control an eighth clone, 57845, a tree grown from random American elm seeds, as well as two Asian elms, Prospector and Frontier, a pair of his own successful new trees intended for city streets. But despite this wealth of test subjects, as he himself acknowledged, "nothing matches the grandeur of the American elm."

Of course, there was also Eugene Smalley's American Liberty elm, which had now been available for almost six years, launching the "re-elming" of America and garnering the Elm Research Institute much favorable publicity. Townsend recalls that while Smalley readily sent along his trees, "they would not identify the six different elms that made up the American Liberty. That's where the science faltered with John Hansel. Everything was secret. He was attached to his trees and he was convinced they were better." Townsend, however, would base his own judgment on the trial outcomes.

In March 1989 he and his two colleagues had started planting the test plots in Glenn Dale with the nine varieties of young American elms, as well as his two Asian types. His plan was to plant twenty-eight one-year-old trees of each clone or cultivar, but he only managed twenty of the Delawares, sixteen of the Princeton elms, and six of Clone 3.

By May of 1990, 287 test trees were growing on about ten acres of land. Townsend would allow the elms to grow for another two years and then subject them to "an extremely severe" injection of the fungus *Ophiostoma ulmi.* Of course, this was the very method of inoculation that Townsend and Smalley had so disagreed about

John Hansel, needless to say, was not happy about any of this. As far as he was concerned, he already knew that the American Liberty elm was DED resistant, and he did not need the USDA becoming involved in his affairs. At around the time that Townsend had planted most of his test fields, Hansel told the *New York Times,* "The future of the elm looks bright. We will start seeing some beautiful streets again in 10 years." Since 1983 his Elm Research Institute had placed 75,000 Liberty elms. Now he had three hundred mayors in forty states signed on as "municipal members" of his institute and in 1990 anticipated a ramped-up delivery via the Boy Scouts of 100,000 saplings.

Even as Townsend had been launching his elm trials, the search for a viable American chestnut had been revived in Minnesota, almost four decades after a scientist named Jesse Davis had made an unsuccessful attempt to develop a blight-resistant American chestnut at Glenn Dale. The leaders of this new effort were Charles Burnham, a retired septuagenarian University of Minnesota corn geneticist, and Philip Rutter, a middle-aged zoologist who had retreated to a log cabin and apple farm to raise his family and live off the land in southeastern Minnesota.

In the mid-1930s Burnham had happened to be teaching at West Virginia University when the chestnut blight rampaged through and killed all its stately giants. He moved to the Midwest, outside the chestnut's natural range, and assumed the USDA was plugging along at the slow work of developing a resistant American variety. Upon retirement he had time to read more widely and was astonished to discover that the USDA not only had abandoned its program but had not employed backcross breeding as a strategy to save the American chestnut. In 1981, when he decided to take action, he was advised to call Rutter, known in state tree circles for his obsession with the American chestnut. Two years later they and a handful of other scientists established the American Chestnut Foundation, whose mission was "the preservation and restoration of the American chestnut through funding a scientific breeding program and related research." Its strategy: Create through backcrossing (a process that would take several decades) a tree that looked like the American chestnut but that had one sixteenth Chinese genes to fortify it against the blight. Burnham was delighted to discover that the USDA's program had, it seemed, produced one chestnut hybrid known as the Clapper tree, which survived from 1946 until it finally succumbed to the blight in 1976. It was seventy-six feet tall at its demise, but its scions had been grafted onto Chinese chestnut trees and done well.

The scientists who oversaw the American chestnut breeding project enlisted members in the fourteen states where the trees once flourished, both to underwrite their research and to serve as citizen growers by planting, raising, and documenting their results with the American chestnuts entrusted to them by the foundation. By late 1988 Burnham was declaring in an article titled "The Restoration of the American Chestnut" in the journal *American Scientist* that the group was advancing steadily with its program of backcrosses and that application of Mendelian genetics "is still a powerful tool in solving a problem that has not yielded to other approaches. What it appears to offer here is the recovery of the American chestnut with all its desirable qualities."

In 1989 the American Chestnut Foundation established headquarters on a donated old twenty-acre farm at the southern end of the Shenandoah Valley in rural Meadowview, Virginia, where scientists could pursue the breeding program in the tree's natural range. Fred Hebard, the foundation's lead scientist, moved in and "diligently filled the fields . . . with American, Chinese, hybrid, and early backcross trees." Born in Chestnut Hill, Philadelphia, to a family wealthy from timbering, Hebard had completed a year at the University of Virginia when he joined the army and landed in Vietnam at the height of the war. He briefly returned to Columbia College before dropping out to work on a dairy farm in Connecticut.

When Hebard was helping his boss find stray heifers in the woods, the older man pointed out an American chestnut sprouting from an old stump and told of the billions of magnificent trees laid waste by the blight. Described as "taciturn, almost shy . . . not given to talking much about matters other than the science of chestnuts," Hebard said, in what would become a standard line, "I thought it would be nice to go back to school and learn about biology and try to do something about chestnuts. Little did I know it was a lifetime proposition."

For the next two years Townsend nurtured his 287 young elms in their well-arrayed rows in the Glenn Dale test plots, pleased to see them growing strongly. Finally, in late spring of 1992, he prepared the first round of DED assaults. On May 18 and then again on May 27, he and two colleagues set forth among the three-year-old trees to administer the *Ophiostoma ulmi* to each. As Townsend reported in a subsequent paper, "This inoculation was designed to be extremely severe in order to insure symptom expression on even the most disease-tolerant clones. The percentage of the crown showing wilting or death of the foliage, and the percentage of the crown's branches showing dieback (a lack of foliage), were visually estimated at four weeks and one year, respectively, after each inoculation date."

In the subsequent weeks and months of that summer, Townsend was thrilled to witness the hardiness of some of the elms subjected to this fungal assault. By the summer of 1994, two years after the inoculation, Townsend could declare the winners of this first-ever scientific American elm face-off: "Results of this study show a wide and promising degree of tolerance to O. *ulmi* among American elm clones. American clone 3 ['Valley Forge'] appeared to show the highest tolerance to O. *ulmi*, followed as a group by clones 180, R 18-2 [Delaware 2], 680 ['New Harmony'], 'Princeton,' and perhaps 'American Liberty.'" (Note the diplomatic "perhaps" used to describe the performance of this latter also-ran.)

Townsend was well aware that Smalley, and especially Hansel and his Elm Research Institute, would not be pleased with their tree's poor showing in these trials, so he tried to cushion the bad news. He acknowledged that Liberty had in previous studies (by Smalley) shown "significant disease tolerance." He offered "two possible explanations . . . for this high degree of dieback in this cultivar. . . . First, 'American Liberty' is comprised of six different clones, and it could be that the most disease-susceptible of these clones was more frequently represented in our experimental plot. Second, 'American Liberty' ramets [independent members of a clone] were significantly taller at the time of inoculation." He wondered if those factors, in combination with the "unusually potent inoculation technique," might have dealt the Liberty elms an especially powerful blow.

Still, the hardiness of a handful of these other American elms was exciting news. "We were quite surprised!" Townsend recalled. "It was eye-opening." As he modestly stated in his paper, "The evidence presented here offers new hope for the reestablishment of the American elm. Even under the most strenuous of tests, several clones showed a significant tolerance of DED. This tolerance will allow for the future release of American elm clones." Moreover, these hardy trees could also serve in further "breeding programs designed to maximize disease tolerance in American elm."

Townsend continued the Glenn Dale field tests but now felt sufficiently confident about two of the cultivars—Valley Forge and New Harmony—that on December 27, 1995, just three months after his paper was published, the National Arboretum officially released these varieties to the nursery trade. Townsend chose the cultivar names as tributes: "Valley Forge," like George Washington and his revolutionary troops, had survived brutal testing. (Also, Townsend had a particular affection for Valley Forge itself, as it had also been the site of a huge 1957 Scout Jamboree that he viewed as a highlight of his teen years.) With "New Harmony," Townsend aimed to honor the nineteenth-century utopian village of New Harmony, Indiana, a community dedicated to education and science.

Fittingly, the USDA celebrated Townsend's dendrological triumph by planting the first public Valley Forges on the lower grounds of the U.S. Capitol, east of Third Street NW, there to join a veritable living museum of historic and memorial trees.

On Thursday June 6, 1996, favored with a cool clear morning, Senator Christopher S. ("Kit") Bond of Missouri, a moderate Republican and self-described "tree junkie," presided over the ceremony to honor Townsend, the Agricultural Research Service, and the National Arboretum. The senator commended Townsend as a "scientist who has spent a lifetime studying and developing new trees for cities and towns. . . . Now, with the help of Dr. Townsend, and the National Arboretum, we stand a great chance of seeing a return of the stately and valuable American elm."

While John Hansel had opted to propagate Liberty elms through the "Johnny Elmseed" nurseries partnership with the Boy Scouts, federal scientists saw as their raison d'être finding, developing, and releasing better plants and trees to the nation's arboretums, garden clubs, and professional nursery trade. As the products of government research, trees like Valley Forge and New Harmony were effectively in the public domain, and American taxpayers were entitled to enjoy the fruits of these USDA programs.

These were heady days for American tree lovers. After an absence of decades, young elms would in coming years again be planted in cities. And as Townsend reminded any who had forgotten: "Elms represent a superior group of trees which can withstand a multitude of environmental stresses, including air pollution, deicing salts, soil compaction, drought, and flooding."

Not only was there the prospect of new American elms, but surviving large elms could now be saved through fungicide treatments. While dedicated "sanitation" techniques had minimized losses, the systemic fungicide Arbotect had established itself as a near-guaranteed treatment for saving and maintaining giant old American elms. "The problem with Lignasan," says Tom Prosser, whose Minneapolis company Rainbow Treecare Scientific Advancements pioneered swift injection methods, "was it had a short half-life. Arbotect was the only thing that gets into the new wood and gives you three years of protection. This gives the trees the time to outgrow the old wounds." (In time, Prosser's company became the sole distributor of Arbotect.)

Still, without the development of young trees able to withstand DED, the American elm had no future. Among those Townsend favored with rooted cuttings of his long-awaited trees was Keith Warren, then chief horticulturist of the nation's biggest and most innovative wholesale shade tree nursery, J. Frank Schmidt & Son Co. of Boring, Oregon. In 1946 J. Frank Schmidt Jr., twenty-six, son of an Oregon nurseryman, had planted his first trees—maples and birches—with his bride, Evelyn, on ten acres near Troutdale, a place of rich alluvial soil just an hour northwest of Portland. With the post–World War II building boom driving great demand for new trees, Schmidt's soon stood out as a company committed to finding, growing, and shipping superior trees—maples, oaks, ash, and ornamental cherries—able to thrive in the real world of city streets or suburban developments and backyards. With success came expansion, and by the 1970s Schmidt's was raising one-, two-, and three-year-old saplings on more than two thousand acres in the Willamette Valley.

By 1974, when Keith Warren arrived at the company's Hood Acres directly out of college to work as one of its farm managers, J. Frank Schmidt Jr. had become, he says, "the Henry Ford of trees. He had perfected the mass production of very high-quality bare root shade and flowering trees." (Bare root trees could be cheaply shipped in refrigerated trucks from January to April while still dormant.) His boss's remarkable eye for promising new varieties gave substance to the firm's slogan, "We're Growing New Ideas!" Frank Jr.'s legend had been solidified in the 1960s when the sunset maple (*Acer rubrum* cv. Franksred) became "the single most planted ornamental tree variety in America" while also selling "well in the rest of the world." Soon Schmidt's was shipping hundreds of thousands of bare root saplings to smaller wholesale nurseries, mainly in the Great Lakes region, New England, and Mid-Atlantic states.

When Warren started at Schmidt's, he recalls, no sane commercial tree grower was still planting or selling American elms, for every tree person knew that "the fate of the American elm was sealed." Two years later, in July 1976, he flew to the East Coast for the first meeting of the new Metropolitan Tree Improvement Alliance (METRIA), which had been established as a forum where nurserymen could network and collaborate with city and state foresters, landscape architects, and academic and government scientists like Denny Townsend—all passionate about somehow finding better ways to grow, plant, and manage city trees.

"METRIA was like a phoenix that rose out of the ashes of the disaster of the American elm," says Warren, "a wonderful coming together of so many important names in what became urban forestry." It was here that Warren met Denny Townsend, who was serving as METRIA's first secretary-treasurer. "We became just great friends," Warren recalls. "He was a real tree geek, like me, and a heck of a nice guy, with no ego. I went to visit Townsend when he was in Delaware, Ohio, and was impressed. He was very sharp and his inoculation tests were very rigorous." When the USDA released Townsend's Valley Forge and New Harmony, Warren soon had Schmidt's raising them in commercial quantities with an eye to shipping to the local wholesale nursery trade by 2001.

While enthusiasts like John Hansel were interested only in the American elm, others had aspired to breed a tree that *looked* like an American elm but was partly Asian and thus harbored an innate immunity to the fungus.

George Ware, the other major researcher in the elm field, took this approach. Ware, who had a PhD in forest ecology from the University of Wisconsin at Madison, was chief dendrologist at the 1,600-acre Morton Arboretum.

In 1968, as the arboretum neared its fiftieth anniversary, it had committed to a greater amount of scientific research, wooing Ware, forty-three, away from Northwestern State University in Louisiana. Four years later Ware, who specialized in developing urban trees, noticed from his office window a lovely old elm with a splendid vase shape and glossy, deep green leaves that was unaffected by Dutch elm disease. The Morton's own records listed the tree as an American elm that had been planted in 1925, but how was it still thriving? The arboretum staff referred to it as the "mystery elm"—grown from seeds sent by the Arnold Arboretum in 1924 with no information about them. Ware began sleuthing about the grounds of the Arnold, finally finding in a far corner two elms—a Chinese (*U. wilsoniana*) and a Japanese (*U. japonica*)—that looked very much like his own tree's "mother" and "father." Overjoyed, Ware named (and trademarked) this promising Asian elm "Accolade."

As others had learned to their chagrin, wrote Ware in 1992, "American elm cannot be crossed or bred easily with other elms because the chromosome number is double that of other elm species, rendering them reproductively incompatible." So Ware felt the Morton Arboretum could best fulfill its mission to help midwestern urban foresters find tough but beguiling trees by pursuing "American elm 'lookalikes' using Asian elm species . . . as alternatives for producing attractive arcade-forming trees."

In 1990 Ware took the first of five plant-hunting trips to China, following in the footsteps of Frank Meyer and Ernest "Chinese" Wilson, and another three to Russia, in search of elms that could survive the stressful life of American city trees. On this inaugural expedition he journeyed north to a nature preserve in Manchuria close to the Russian border, looking for a particular David elm (named after Father Armand David, a nineteenth-century French missionary in China who first wrote about the tree) that was reputed to grow in trying climates and poor soil. When he saw the David elm flourishing in the "rugged, open terrain" of Manchuria, he said, "It was exactly what I was looking for. I was euphoric." Ware believed that the "David elm's tolerance for bitter winters, open spaces, and barren soils made it a perfect candidate to thrive in Chicago and other cities and towns across the American Midwest."

Like Townsend and Smalley, Ware was growing hundreds of elms—in this case solely Asian hybrids—not only in test fields at the arboretum but also throughout Chicago in the real world of parks, college campuses, cemeteries, golf courses, and nature centers. Both Denny Townsend and Eugene Smalley had also successfully hybridized a number of European and Asian elm cultivars that were already commercially successful, but there was always that yearning for the genuine American elm. Ware knew this as well as anyone, but he was optimistic that among the arboreal riches of Asia, and especially China, with its twenty-four native varieties of elm, resided genes that researchers at the Morton Arboretum would in time hybridize into the new, improved city elms of the future.

For the first twenty years of Keith Warren's career at Schmidt's, elms barely registered in the company's growing plans, catalog, or sales. But the release of Valley Forge and New Harmony galvanized growers. "Ten years ago," Warren, now director of product development, told those at an International Elm Conference hosted by the Morton Arboretum in 1998, "our nursery catalogue listed two elm cultivars. Today, we have 18 cultivars in some stage of production or stock increase, and we are evaluating about a dozen more."

At the same conference Townsend was certainly entitled to take a victory lap. Colleagues like George Ware lauded the reintroduction of his American elm cultivars as "milestones in the return of elms to urban landscapes." But Townsend continued his research in order to reassure the conservative nursery trade that they could once again grow and sell American elms. He had pressed on with the Glenn Dale trials and in June of 2001 published a second paper describing how the respective cultivars fared seven years after the administration of Dutch elm disease fungus. Once again, Valley Forge, New Harmony, and Princeton took the honors, appearing "able to respond and recover over time from fungal inoculation, expressing a true tolerance."

The news for John Hansel, Gene Smalley, and the American Liberty elms, however, was not as favorable. While Townsend wrote that "results of this study confirm the promising level of tolerance to *Ophiostoma* shown in earlier, 1-year evaluation of American elm clones," he was no longer hedging his assessments: "[O]ther selection, such as 57845, 11, and the cultivar 'American Liberty,' did not express such resilience." Not only did these cultivars show "the poorest survival rate" but they barely grew any taller in the seven-year span of the trial. "This 'negative' growth," Townsend reported, "most likely was due to severe crown dieback sustained by these two biotypes after inoculation."

Hansel was enraged at this government disparagement of his trees, and he vented his ire in his newsletter, *Elm Leaves*: "Elm Research Institute believed in the idea of disease-resistant purebred American elms at least ten years before anyone else jumped in. We became actively engaged in sponsoring research that resulted in the tree that led the way, the American Liberty Elm, introduced in 1983. . . . Today, it is the only street tested, purebred, native American elm *with a lifetime replacement warranty against Dutch elm disease*. Now, because a few new elms have appeared on the market, consumers should know what to look for."

As far as Hansel was concerned, the USDA and Denny Townsend could criticize Hansel's trees all they wanted, with their artificial field tests aimed at the nursery trade. But how could they deny the reality of the ERI's "250,000 American Liberty elms (some over 40 feet tall) thriving in 1000 communities; loss to Dutch elm disease—less than 1%"? Hansel was similarly unimpressed by Townsend's peer-reviewed, field-tested science, warning: "Only after years of growing in communities, where elm bark beetles provide a natural inoculation with the disease fungus, can true chances of survival be determined." "We suddenly have imitators," he wrote, "some of whom have little or no genetic research or field tests to back up claims for resistance."

For Townsend the elm trials were not a personal matter, as they were for Hansel. "We were doing a scientific experiment," Townsend said years later, "and we had to publish our paper. Unfortunately, when the publicity came out, I had to tell people who interviewed me about American Liberty." This was small consolation to John Hansel, who a decade afterward was still lamenting the entire episode: "The USDA has the power to pan a tree. They decide our tree is no good." Townsend had a broader perspective: "I see John Hansel as an innovator, and he deserves a lot of credit for capturing the excitement of the American elm and doing what he could to get trees released and into the landscape."

CHAPTER SIXTEEN

"I Never Saw Such a Bug in My Life": Attack of the Asian Long-Horned Beetles

The Asian long-horned beetle. *(Photograph by Anson Eaglin, USDA APHIS.)*

In early August of 1996, developer Ingram S. Carner was finishing up eight town houses on McGuinness Boulevard in Greenpoint, Brooklyn, when he noticed several large finger-size holes in his thirteen-year-old street trees, and piles of fine sawdust at their bases. Why, he wondered angrily, would vandals be attacking his Norway maples with pocket knives or drills? "A few days later," he says, "I noticed more holes, but now up at seven or eight feet, too high for boys to reach. It was then that I realized something else was going on."

The next morning, when Carner came to supervise construction, he set up a plastic beach chair and table in the adjacent parking lot. He kept glancing up from his paperwork to check the Norway maples, a species that had been heavily planted throughout New York City when the American elms began dying. "Around one p.m., this beetle emerged from one of the holes. I have about sixty trees on my property where I live in Whitestone, Queens. I'm a gardener and amateur horticulturist, and I never saw such a bug in my life."

The beetle was an inch and a half long, black with white spots and inch-long black-and-white-banded antenna that waved slowly back and forth. "I made a drawing of it. Then I started calling around. The city parks department, their attitude was 'This is some seventy-five-year-old guy who says a beetle with horns is coming out of his trees—we'll get back to you.' Over the next ten days, I was just calling desperately around. One entomologist sent me a picture of a similar beetle but not quite the same. . . . I tried the U.S. Department of Agriculture and they brushed me off. Finally I called the Parks Department again and just screamed and insisted they get someone down here."

The parks department finally sent a young intern, but no beetles appeared. Carner, however, was persistent, and a senior foreman now came to investigate. "Just as he's getting ready to drive away," Carner recalls, "the beetle emerges. When he sees this thing, he calls his office." The insect was captured and dispatched posthaste to Cornell University upstate to be identified. "When I first laid eyes on this beetle," recalls Cornell entomologist E. Richard Hoebeke, "I had the feeling it could be bad news. I checked it against our collection."

On August 19 Hoebeke confirmed the worst: The bug was the Asian long-horned beetle (ALB), the first ever identified in the United States. Dutch elm disease beetles target only elms. The voracious Asian long-horned beetle was far, far less selective. While it might prefer maples, given the opportunity, these beetles would as happily dine on (and therefore kill) about half of New York City's 5.2 million trees—the Norway and sugar maples, elms that had survived Dutch elm disease, horse chestnuts, birches, hackberries, black locust, and mimosas.

The beetle, explained Hoebeke to city parks and forestry people and the relevant federal officials at the U.S. Department of Agriculture's Animal and Plant Health Inspection Service (APHIS), had long been a minor pest native to the forests of China, Japan, and Korea. But the Chinese Maoist government planted millions of poplar trees in the 1960s and 1970s as windbreaks in the northern provinces, and when these monoculture forests came to maturity in the 1980s, the population of the "starry sky beetle," as the Chinese call it, exploded, destroying the trees in those plantations. The dead poplars proved handy as wood for pallets and crates to hold China's mushrooming exports—televisions, umbrellas, plumbing fixtures—which filled eight million shipping containers in 1980 and the ten million annually that by 2000 were flooding U.S. ports.

For the first time since 1933, when the USDA's Curtis May confirmed Dutch elm disease, America's cities were facing a new, potentially catastrophic alien insect invasion, one that could wreak havoc on urban forests. The most alarming aspect of the threat was the fact that the preferred food of Asian long-horned beetles was maple trees. Only a few hundred miles north of Brooklyn were the New England woods, where maple syrup and the tourism that brought visitors to leaf-peep at the maples' brilliant autumnal colors were major industries. Moreover, just as real hope was developing for reestablishing the American elm, here was an insect scourge that would eat an elm if a maple was not easily available.

Hoebeke had further dire news: The only known way to combat this pest was to first cut down and remove every single infested tree and reduce it to mulch and then cut down any nearby potential host trees. In a city so heavily planted with the beetles' favorite arboreal foods, officials could expect to be almost clear-cutting entire blocks around Greenpoint and taking down many a private tree in family yards behind brownstones. And in a worst-case scenario, should this beetle horde conquer Brooklyn and press on to other fronts, entire American forests and ecosystems could be at risk. The question remained: Was this a relatively minor attack that could be quickly rebuffed or the first skirmish in another drawn-out but doomed war to avert ecological and economic disaster?

So began year one in the war against the beetles. After a muddled start, the USDA dispatched Joseph P. Gittleman, a longtime APHIS employee, to take charge. A veteran of the agency's Kennedy Airport operations (where he dealt with matters like smugglers secreting rare venomous pygmy snakes), Gittleman was a big, rumpled guy with aviator glasses and mustache, attired in the agency's uniform of khaki shirt with epaulets and silver badge affixed above the left breast pocket, dark green cargo pants, and sensible black brogue shoes. A biologist who spent his youth outdoors and on the water at Rockaway Beach in Queens, he was that rare native New Yorker who grew up fishing and hunting. "Frankly," he says some years later in a heavy New York accent, "I originally took the job because it cut my commute in half. Also, sometimes you just need a change of pace." In early 1997 he assumed command of a tiny, ragtag ALB corps whose mission was to identify the beetles' position, estimate their strength, and rout them from their strongholds—all preferably on the cheap.

"When we first started," he says, "we weren't operating on a shoestring, we were operating on dental floss. We had six or eight people, we were working out of our cars, we didn't have our own binoculars, you didn't have your own anything." Despite the dire warnings from scientists, the beetle's presence in a gentrifying Brooklyn neighborhood set off few alarm bells in high places. The ALB "doesn't have the same kind of political clout as something that affects citrus or wheat," said Gittleman, as damage to food crops wreaks obvious economic harm on the private sector. For the moment, at least, only urban trees were immediately at risk.

In 1996 Henry Stern was thrilled to be back for an encore as parks commissioner, this time under Republican mayor Rudolph Giuliani. A master of publicity, Stern had been taking his act as the Lorax to new heights. The previous July he had paraded before the press one Andrew Campanile, the employee of a billboard company who was under arrest and in handcuffs. "Mr. Campanile's crimes," said Stern, "constituted arboricide in the first degree, premeditated arboricide": He had destroyed five honey locusts and two London plane trees in three different locations, all so his company's billboards could be better viewed. Possible jail terms for tree death were, under the law-and-order Giuliani administration, now one year rather than ninety days, while fines could reach $15,000 per tree. Campanile's employer paid $34,000 to plant sixty-nine saplings. Tree lovers everywhere could only marvel at this high-level city official meting out harsh punishments to the philistines who failed to love trees.

And yet somehow Commissioner Stern was largely absent when it came to the struggle against the Asian long-horned beetle, the biggest threat to the city trees since Dutch elm disease. On September 12, 1996, Stern was present for the first unhappy public announcement of the beetle's arrival, brandishing aloft a specimen with tweezers. Compared with his elaborate antics on behalf of the East Side ginkgo or park trees, Stern's response seemed muted, almost perfunctory. Years later he did declare the Asian long-horned beetle "the single greatest threat to our trees. It was a war." Perhaps the matter was simply all too serious for jokes and street theater (though he did bestow the nickname "Beetle Buster" on Gittleman).

How had the Asian long-horned beetle made its way to Brooklyn? Some joked that the insects had heard in their travels that a tree grows in Brooklyn and had cruelly come to kill it. The answer was far more prosaic. "We believe," says Gittleman, "that during a big Brooklyn sewer treatment plant project heavy pipe and materials were coming in from China in wooden packing crates and that was the foothold. From there it spread, because people were carrying away that packing, trucking it away to use it in other projects."

Whether on the street, in parks, in backyards, or on balconies high up on apartment buildings, every host tree within a quarter mile of Carner's maples had to be inspected. With each new sign of beetles, the perimeter of blocks to be checked was pushed out farther. All that winter Gittleman and his small force scrambled to get on top of the situation. They often felt they were searching for needles in haystacks as they stood with their binoculars and scanned the branches and trunks for enemy positions in and around Greenpoint. The dime-sized exit holes were the most visible signs of longtime occupation. Far more subtle were the slight round indentations of gnawed bark that meant a female beetle had laid her eggs.

By early in year two of the beetle war, Gittleman's commandos had reconnoitered and condemned 470 trees as currently infested or potential breeding grounds. That spring the APHIS corps launched its first active offensive. On Humboldt Street in Greenpoint, Patricia Ferris watched, heartbroken, as arborists dangled from the eighty-foot-tall horse chestnut that had long shaded her backyard. "I raised six children here, my grandchildren's playpens were under it," she said, mourning the giant tree that had grown old with her, had brought nature so close. Now young men in hardhats were dismembering twenty-six years of memories with chainsaws, carting away what had been her beloved tree in three-foot chunks through the house and out to the street. "It's difficult to describe," she said. "The whole yard just doesn't look like it belongs to me anymore."

Elsewhere in Greenpoint neighbors watched that March and April in similar stunned sorrow as Gittleman's troops ruthlessly felled trees just budding out with pale green leaves and then fed them into giant, roaring chippers. McCarren Park was nicknamed "McBarren Park" when its decades-old Norway maples—replacements for the elms lost to the last scourge of alien insects and the fungus they carried—became more collateral damage. "People come here to get a break from the heat, sitting under the trees," said one resident. "What is going to shade them now? Not a skinny little new tree."

Among those working to remove the besieged trees was Fiona Watt, a graduate of the Yale School of Forestry and Environmental Studies and the 1995 Parks Department Rookie of the Year—an honor she had earned by spearheading the city's first-ever street tree inventory, something of a bureaucratic trial by fire. She was a country girl from Vermont whose China scholar father and PhD-in-education mother had opted to raise their three girls on a farm in a "back to the land" 1970s manner. "I was a tomboy," Watt told a *New York Times* reporter of her "idyllic" childhood. "Every tree I saw I had to climb: I loved to scale the heights and look at the world from the top of a tree." After stints as an Aspen waitress/ski bum and Random House PR person, Watt had attended forestry school and then signed on with

Henry Stern, who nicknamed her "Treetop." The Brooklyn ALB invasion was one of her worst job moments: "On one street, with a wonderful leafy canopy, we had to dismember maybe 30 trees. I watched the starkness of the block emerge. It was profoundly distressing." Within two years she had become director of central forestry, where she began making a name for herself as an articulate advocate who knew how to win allies and dollars for urban trees.

But in that terrible spring of 1996, time and ruthless tree removal were of the essence, for the beetles would reemerge with June's warm weather to mate and renew their attack. As he came to better know his foe, Gittleman lamented its reproductive prowess: "There is no synchronous emergence. If they all emerged in two weeks, you could hit them with everything you've got. Instead, the adult beetle emerges starting in late June and continues till frost, mating, laying eggs in eighty to ninety places on the bark, one per site."

When the rice-sized white larvae hatch after eleven days, they begin the attack, eating into the tree's cambium layer, intercepting the nutrient flow, before gnawing through the sapwood and heartwood, where they create a cozy pupal chamber. Twenty-one days later, the now-adult beetle chews its way back out of the tree, boring a perfect round exit hole three-eighths of an inch in diameter. All summer long this invading army added numbers to its ravenous ranks. Nor would frost and winter put an end to it, for the larvae overwintered (as could the eggs), ready to attack anew the next fighting season. Gittleman, other officials, and entomologists holding war councils in these early months were grateful to observe that the beetles seemed slow-moving, not particularly inclined either to stray too far from trees they occupied or to advance precipitously into new territory. Less reassuring, though, was the fact that the beetles had easily infiltrated Brooklyn, their sleeper cells silently proliferating (scientists determined) for almost a decade.

When Gittleman worried about the beetles' conquering new terrain, he suspected that most often it was people who provided unwitting transport. A month after Carner's discovery, the beetles were spotted in Amityville, Long Island, where Mike Ryan Tree Services had mistakenly disregarded the Greenpoint quarantine zone (no tree wood could leave that had not first been ground into mulch) and escorted the beetles out into leafy Long Island, where a new battlefront soon opened.

In the summer of 1998, year three of the beetle war, the situation took a most ominous turn. On the Fourth of July weekend, a suburban parks worker named Barry Albach stopped by his friend Gary Luka's two-story brick house in the Ravenswood neighborhood of Chicago. Luka, annoyed by his neighbor's failure to respond to polite requests, had taken action and himself sawed off a huge box elder branch from the neighbor's tree overhanging Luka's above-ground pool in the backyard. Luka had offered some of the cut-up branch as firewood to Albach, who tossed the pieces in the back of his covered pickup and headed home to Morton Grove. A few days later Albach's mother pointed out a black-and-white beetle crawling on his truck's rearview mirror. Curious, Albach went inside to look it up on the Internet. When he clicked on a photo of what looked like the beetle, the screen flashed: "Wanted. Be on the look-out for the Asian Long-Horned Beetle Attacking Trees!" Fortunately he had not yet unloaded the firewood in his covered truck.

Chicago city forester Joe McCarthy (soon to be known as Beetle Joe) remembers being deeply skeptical about the identification of Albach's beetle, but days later entomologists at the Smithsonian Institution confirmed it. Until now U.S. officials had viewed the Asian long-horned beetle as a small, isolated outbreak in an obscure Brooklyn neighborhood that spread to Long Island through careless arborists. But when alarmed USDA personnel arrived in Ravenswood, a community of modest brick apartment buildings and homes set back on trim front lawns, their spacious rear yards shaded with big trees, they quickly spotted dozens of trees riddled with the telltale holes.

If you stood in Gary Luka's backyard next to the infested gray-leaf maple (another name for box elder), you could look directly across Montrose Avenue and see a beige cinder-block building with a side parking lot. This was the home of Polar Hardware, which like millions of other American companies now imported from China certain industrial components—in this case for its door and truck body hardware. Those components arrived on wooden pallets. The beetles did not have to fly far to find their first American tree.

In contrast to New York's top officials, who left matters largely to Gittleman, Chicago's Mayor Richard Daley personally took charge when he learned that the beetles had appeared. Marshaling his considerable political might, on Thursday, July 16, 1998, Daley showed (as he had not for the Climate Project) what his press machine could do. Flanked by the U.S. Secretary of Agriculture, U.S. Senator Carol Moseley Braun, numerous officials from the U.S. Forest Service and state agencies, as well as a clutch of ward politicians, the mayor stood before the press to declare war. "The potential destruction by these dangerous pests goes far beyond Chicago," he warned, "and that's why these agencies are deeply involved." The Chicago story made *Good Morning America, Nightline,* and CNN and received continuing front-page coverage in the local papers.

That September the USDA pushed through new shipping rules first proposed after Ingram Carner found the original beetle. As of December 18, 1998, all wooden pallets coming from China had to be kiln-dried or fumigated. The Chinese were not pleased and even sent a delegation to Ravenswood to tour around. "They came," recalls forester McCarthy, "but they refused to accept any responsibility." APHIS scientists had, meanwhile, begun collaborating with their counterparts in China. There, said one American official, the Asian long-horned beetle population was so out of control that it had become "an ecological disaster of biblical proportions, and wholly manmade [as Communist officials had created the forest monoculture]." The Chinese farmers who watched the insects kill forests in the north had their own name for the "starry night beetle": the "forest fire without smoke."

As Mayor Daley's ALB ground troops began surveying Ravenswood for beetle damage, scanning the trees with binoculars, they found that the first wave of invaders from the wooden pallets and their many offspring, which had gone undetected for at least several years, had already established strongholds in several hundred trees. As year four of the beetle war began, the USFS proposed an innovation in reconnaissance techniques: foresters in aerial bucket trucks who could more closely inspect the higher reaches of the winter-bare canopies. "They showed us in a matter of minutes how many [infested] trees we missed from the ground," said one official.

But trucks could not get into people's backyards, where many large trees might be serving as beetle redoubts. The Forest Service suggested training its forest firefighters—known as smoke jumpers—to serve as tree-climbing beetle scouts in their off-season. "The next thing we know," said a Chicago official, "we've got Forest Service smoke jumpers coming from everywhere. They were the lifesavers."

With the bucket trucks and the smoke jumpers scouring the canopies, the Chicago ALB corps detected hundreds more stricken trees. Many a family watched anxiously as the lean young men and women in hard hats scaled their trees and looped expertly from branch to branch. When they landed back on the ground, if they took up a can of purple or green spray paint and marked a circle around its trunk, the family knew that their tree would be gone come spring. Day by winter day more trees were tagged. The community requested that their trees not be removed before Christmas, as the loss would be so dispiriting.

Eighteen years earlier Nancy Nagler's then-fifth-grade son had brought home from school a little maple tree he had grown in a coffee can. Planted beside their two-story brick house, it had matured into a beautiful shade tree. No one had to check its canopy, for as Nagler observed, "Now, it looks like it's been machine-gunned." On another block Gary Dickerson stood on the back lawn under his family's towering sugar maple, a tree his father, now eighty-five, had rescued and transplanted when another neighbor was going to cut it down. "I've watched that tree grow up," said Gary. "It was here for my birthday parties and for my kids' birthday parties. We'd hang decorations off it and play under it. Even when we didn't notice it, it kept us cool." His three children had been planning to build a tree house in its lofty branches. But the beetles had tunneled through, and its trunk now sported a blaze of spray paint.

February 3, 1999, dawned cold, gray, and drizzly. By now all the infested trees in the neighborhood, as well as possible host trees, had been marked with spray paint. Residents of Wolcott Avenue, the hardest hit, had parked their cars elsewhere, and an army of officials and tree-removal crews roared in early in the morning. On one block all the doomed trees had poems attached to them, written by grieving families. Resident Gina Bader was asked to photograph and document this first Chicago offensive. "There were two trees on my street that I was really attached to," she said. "One was an elm tree that was maybe 50 years old. But it was amazing—within minutes they cut that tree down and it was really powerful—the sound of it falling across the street and just how long and how much space it took up. It just crashed. . . . It was a very sad day."

All around Ravenswood families gathered to bear witness to the destruction and loss of hundreds of their trees. "It was like a war zone, trucks everywhere," said one man, "people everywhere, people clutching each other, looking at the trees. It was like family members being destroyed . . . huge trees just one after another were being cut down and branches and the sound of trees being chewed up in whatever they call that machine. . . . It was just really hard." All told, 453 trees were felled. Notably exempt were a few big old catalpas, whose cambium and heartwood did not entice the beetles.

When the sun came out the next day, the neighborhood looked shorn, the harsh light revealing every crack in the walls of some of the houses. The winter beauty of bare trees, their soaring, dark branches normally silhouetted against the sky, and their presence a familiar balm in an industrialized world, were brutally absent. All that remained were piles of sawdust and scatterings of twigs from the giant chippers. That spring and summer the city planted sizable five-inch-caliper replacement trees, but as one woman said: "The grandeur is gone." The streets felt hot, forlorn. Neighbors missed the sounds of the leaves high above riffling in the breeze, the restful cool of the overarching branches dense with green, and the trill of birdsong. Some of the local churches held services just to mourn the loss of the trees, to assuage the community's grief at being caught in the midst of an ecowar that was just beginning.

As the weather warmed, Mayor Daley again exhorted Chicago's citizens to keep vigilant watch for the beetle armies: "Our urban forest will not be safe until the Asian beetle is totally eliminated from the city. It's important to find these trees quickly because where one beetle has emerged, there likely are larvae in the same tree that would come out later." As beetles were sighted in additional Chicago neighborhoods and suburbs, the mayor secured all the funding his ALB team needed. Thankfully, no other area was as badly infested as the site of the original outbreak. By late 1999 Chicago forestry officials in Ravenswood had cut down and ground into mulch another 712 trees.

In New York City Joe Gittleman quickly adopted Chicago's pioneering use of bucket trucks and smoke jumpers for inspections, which generated far more reports of infested trees. As APHIS employee Clint McFarland recalls, "I still remember climbing two Norway maples in New York City in front of a woman's house that had been previously cleared. We found when we looked up high that they were infested. She started to cry. She had planted them decades earlier to honor her son who died in Vietnam. It was just especially hard in those cases."

Gittleman did receive one particularly unwelcome surprise. "They really only know pine trees," he said of the western smoke jumpers. "So though London planes were not on the host list, one of the jumpers was checking one right at the edge of the sewage-treatment plant [ground zero of the original invasion] and he found it infested with Asian long-horned beetles. *That* was a shocker. Just a *huge* extra expenditure, because it meant we had to add the London plane to the host list. Do you know how many London plane trees there are in New York City?" Fortunately, beetles were found in only a handful of the city's more than 100,000 London planes, but every one did have to be checked.

Year four of the beetle wars was a discouraging one for the New York warriors, now a hundred strong. In February beetles had been detected in Bayside, Queens, and in July in Flushing, Queens. The quarantine area kept expanding, and the number of trees that had to be inspected and cut down mounted. Nonetheless, New York's top officials still viewed the beetles as a not-so-pressing problem affecting the outer boroughs. In August of 1999 the ALB team was meeting to discuss how to deal with impending budget cuts when the next round of bad news arrived: The beetle had struck in Manhattan.

An alert resident of the Upper East Side's Yorkville neighborhood had spotted the black-and-white beetle on a tree at the Ruppert Playground on the east side of Third Avenue at Ninety-second Street. Closer inspection of the Norway maple revealed the dreaded "bullet" holes. "When they heard about that tree, everything changed," says Gittleman. "There was no more talk about defunding." He adds, "The beetles have a tendency to show up at playgrounds. Kids like bugs. They stick them in a jar and their mother says, 'Get it out of my house.' So it could have been a kid visiting from Greenpoint. I think kids are vectors."

Gittleman's team arrived promptly to chainsaw and grind up the playground's twenty-eight handsome Norway maples. "I cried when I opened my blinds and saw it," said one neighbor. "That shady area was our only enclave, and a lot of seniors and children used it." U.S. Secretary of Agriculture Daniel Glickman, former congressman from Kansas, showed up in this wealthy neighborhood for the replanting of Ruppert Playground, remembers Gittleman. "'Why?' I asked, and he looks at me like I'm crazy. It's the Upper East Side, end of story."

Gittleman now publicly voiced everyone's worst fear: "This is red alert time for Manhattan. It's very possible the beetles could hit sensitive areas like Central Park." If they did, many of the park's trees—some hundreds of years old, soaring, majestic, irreplaceable—might have to be felled to thwart the expansion of the beetle. As Parks Commissioner Henry J. Stern said, "You have to kill the trees to save the forest." The tree cognoscenti were immediately concerned for the quarter-mile-long Literary Walk, lined with its irreplaceable quadruple row of elegant mature American elms towering above expanses of velvet lawns. This was "Central Park's most important horticultural feature," declared its public materials, "and one of the last and largest stands of American Elm trees in North America." No matter the season or weather, locals and tourists alike flocked to photograph this famous site, especially when it snowed. Then you could drink in the full architectural beauty of the dark-barked elms with their sinuous limbs arching skyward, etched in white. Could beetles have made their way to this irreplaceable landmark?

CHAPTER SEVENTEEN

"On That Branch Was a Four-Inch Green Shoot with Leaves": Ground Zero Survivor Trees

Rebecca Clough poses with a badly damaged tree she helped save at Ground Zero in Manhattan. *(© Mike Browne, RLA.)*

In the period after the attack on the World Trade Center on September 11, 2001, the last thing on most people's minds was the fate of nearby trees. At Ground Zero in Manhattan, "Everything had been very chaotic at first, surreal," recalls Rebecca Clough, an administrator at the city's Department of Design and Construction handling the logistics of clearing the southwest quadrant. "To me it looked like the beginning of the world, so cataclysmic and yet serene in many ways, with sound but no sound. Somehow I knew everyone who had perished here was at peace. It took me two weeks to realize that the four or five inches of mud covering everything was pulverized concrete." Suited up in protective garb, Clough directed the debris removal via heavy trucks.

With the passage of time growing numbers of spontaneous memorials began to adorn nearby walls and fences: photographs of the dead and missing, flowers, flags, poems, and stuffed animals. One October dawn Clough was navigating the area near Church Street when a few fresh green leaves caught her eye amid the gray ruins. A landscape architect by training, she had earlier worked for two years for the Parks Department. "I saw this tree, almost completely destroyed, laying on its side in its planter. There was debris on it, and the bark was scorched and burned. The tree still had one branch about five feet long coming out of the trunk. And on that branch was a four-inch green shoot with leaves." She stopped to clear away some of the dust. Aside from that one long branch and its few leaves, all the other branches had been blasted off, reduced to jagged stumps splayed out of the eight-inch-wide trunk. The tree was a Callery pear that had been about thirty feet tall before its decapitation, growing there since the towers opened in 1976. "Anything that's trying that hard to survive, I thought, we should save it. It was that simple. Later, when I told my boss, he said, 'I am not going to spend money on that.'"

Clough, who grew up north of Rochester, New York, says, "Trees have always been a big part of my family—knowing what they are, seeking them out, climbing them. I have traveled extensively and sought out and seen some great trees. I would rather have a huge pine cone on my credenza than a sculpture." Ignoring her boss, she contacted an old colleague, Michael Browne, another landscape architect and assistant chief of design at the New York City Department of Parks and Recreation: "I was well familiar with [Commissioner] Henry Stern and I knew he was passionate about trees. Mike said he was going to see if he could get the troops moving."

In mid-October Parks Commissioner Stern dispatched Browne (nickname "Greenman") down to Ground Zero to see what trees might be salvaged. Nothing Browne had seen or read prepared him for the devastation: "It was a massive mess on an incomprehensible scale. The fires were still burning all over the site. It was smoky and treacherous, just frightening. I thought the air smelled like a combination of burnt plastic and Earl Grey. People were risking their lives just working there."

He set out with Clough to survey the surviving trees they felt could and should be saved. At the northwest side of the plaza along Church Street, they identified six obvious candidates—three Callery pears and three lindens, all about twenty years old, that were not badly damaged. Browne and Clough inspected them, and he took photographs and made notes. Then Clough guided him toward what had been the space between World Trade Center buildings 4 and 5, where the blasted tree was still lying on its side in what had been a very long tree pit with low pink marble walls. By now it had sprouted even more leaves.

"It was unique," says Browne. "It had sprung back to life despite losing most of its branches and all the trauma around it. We suspected that the extreme heat generated from the pile fires had fooled it into thinking it was spring again. That leafed-out tree inspired Becky and myself, especially with all the ruin around it, the extreme loss of life, and the fact that there were so very few survivors from the attack."

Commissioner Stern appointed Bram Gunther (nickname "Dogwood") of the city's Forestry Division to organize the tree-salvage operation at Ground Zero. Gunther was making his way back from examining the six Callery pears slated to be moved when he stopped to have a look at the tree that Browne and Clough wanted to save. "I felt, there is no way this tree is going to survive. It was maimed, mangled. I felt I'd already helped rescue the six healthy trees and they would get a new start on life in a park near City Hall by the Brooklyn Bridge." To Gunther the tree was a hopeless corpse, one not worth saving.

On the morning of September 11, Deborah Gangloff, by then executive director of American Forests, had been sitting in her office at 910 Seventeenth Street NW in Washington. Two days earlier, the National Urban Forestry conference at the nearby Omni Shoreham had wrapped up. "At about 8:45 a.m. on 9/11," she says, "my bookkeeper came in and said that a plane flew into the World Trade Center." Gangloff had been hosting two of the conference guests, Sergey Ganzei from the Russian Academy of Sciences in Vladivostok, and Zane Smith, a retired USFS forester from Oregon. They all got into the car with Gary Moll, her colleague and carpool partner, and joined the lines of traffic crawling slowly out of town. "As we left D.C., we could see the smoke from the Pentagon," she recalls.

When Gangloff drove back to work the following morning, she was shocked but reassured to see four large tanks and soldiers in a vest-pocket park near her offices. "Honestly, when I got back in, I knew the world had changed. I struggled to think what that meant for American Forests. I went to an already scheduled meeting at the U.S. Forest Service, and that got me back into the tree swing. By the next week, on September 20, I sent out my first e-mail to the network of tree-planting agencies and organizations to gauge their interest in tree planting for those who gave or risked their lives on 9/11." By then, says Gangloff, "I realized that the woman from Hawaii who had been at the conference in D.C. and decided to go sightseeing in New York had ended up being on the plane that came down in Pennsylvania."

As for the largely forgotten tradition of memorial trees, Gangloff says, "I always felt it was part of my duty as executive director to be up on our rich history. I used to buy old American Forests memorabilia on eBay. I was absolutely familiar with World War I memorial trees and groves promoted by American Forests and knew how we had worked with First Lady Florence Harding after World War I planting memorial groves. All that just resonated with me." One weekend in 1996 American Forests and the Congressional Medal of Honor Society had planted 250,000 trees nationwide for memorial plantings. "Now," she says, "I thought, *This is a great thing to revive.*" Within weeks Gangloff had unveiled American Forests' new nationwide tree-planting program, the "Memorial Trees" campaign. "We contacted our planting partners," she said, "and all agreed that planting memorial groves was a positive response."

Among the thousands of rescue workers toiling to clear the wreckage at Ground Zero were five hundred dispatched by the U.S. Forest Service: firefighters, smoke jumpers, Geographic Information System technicians, and "incident management specialists." Two days after the attack one of the agency's New York–based landscape architects, Matthew Arnn, found himself, like Michael Browne and Rebecca Clough, noticing the surviving trees. "I saw rescue workers in Battery Park City sitting on benches under the linden trees," he said. The trees were coated in thick gray ash, "but they still offered solace and an opportunity to get away from the pile."

Arnn told *New York Times* garden writer Anne Raver that this scene brought to mind E. B. White's essay in *Here Is New York* about an old willow tree in Turtle Bay: "It is a battered tree, long suffering and much climbed, held together by strands of wire but beloved of those who know it. In a way it symbolizes the city: life under difficulties, growth against odds, sap-rise in the midst of concrete, and the steady reaching for the sun."

Over the next few weeks Arnn sampled soils near Ground Zero and collected debris to assess the environmental damage. He found that the two- to four-inch cement dust coating every surface contained varying levels of calcium oxide, lead, fiberglass, glass, and asbestos, as well as pulverized paper clumped into piles. Arnn came to see as he walked along these downtown blocks that the trees had functioned as a buffer, and their living green and shade now served as a refuge and refutation of so much death and destruction.

Like Gangloff and American Forests, Arnn regarded trees, groves, and gardens as ideal memorials. Not long after American Forests' public announcement, he was appointed to lead the USFS's $1 million Living Memorials Project. Aimed at communities in and around New York City, Washington, D.C., and southwest Pennsylvania, the Forest Service would collaborate—often with American Forests—to create "sacred spaces" where "the resonating power of trees" would bring people together and "create lasting, living memorials to the victims of terrorism, their families, communities, and the nation." For the many people who had suffered loss, the memorial trees, said Arnn, would serve as "emotional symbols of strength and revival."

Michael Browne, meanwhile, was still determined to save that one mangled Ground Zero tree. "I spent a lot of time cajoling and negotiating and coordinating to figure out who could help," he says. "Everyone was so heartsore and the situation was so sad and chaotic, it was a tough time to really get people to focus and cooperate on a tree." "Mike was rebuffed by the forestry and horticulture staff," recalls Clough, "because their tree contract was limited to salvaging healthy specimens." Browne and Clough could see that if they did not get the Callery pear out soon, it would get removed with the surrounding rubble, dumped in a tractor trailer, and hauled away.

Then, says Browne, "I was lucky to find an ally." Bobby Zappala, director of the NYC Parks Department Van Cortlandt Park Nursery, located in the north Bronx, was eager to help. "Bobby also immediately understood the significance," recalls Clough, "and with Mike arranged for a truck to be sent to the site to retrieve the tree. I arranged contractors on-site to provide a piece of equipment and laborers to dig it out, B&B [bag and ball the root ball] it, and load it on the truck."

On the afternoon of November 9, 2001, a green New York City Parks Department rack truck with a lift drove up to the perimeter security entrance to Ground Zero. Clough and Browne were inside getting things organized. "The tree was huge," says Browne, "The ball had to be more than a ton. The contractors did the best they could to B&B the tree and lift it with what they had—a construction strap and a backhoe." With the sun setting, the green Parks Department truck maneuvered into place, the backhoe roared to life, and the tree, which had sprouted many more leaves on its battered branches, rose in the air and was deposited gently on the bed of the truck. Clough and Browne were relieved to watch the truck head out with the rescued tree. "Now," says Browne, "it was up to Bobby Z. to perform his magic."

As landscape architects and people who loved trees, Clough and Browne had felt the power and solace of that blasted tree with its hopeful green leaves, tangible signs of new life in the smoking wasteland of Ground

Zero. As it turned out, they were simply ahead of the rest of America in rediscovering the essential healing power of trees, their incomparable ability to symbolize myriad emotions and serve as a kind of living calendar, reflecting the passage of time. The first public manifestation of those sentiments was the 9/11 "Miracle Sycamore" of St. Paul's Chapel, a historic chapel (where George Washington worshipped when first president) and cemetery across the street from the World Trade Center. A seventy-five-year-old sycamore, part of a grove of trees shading the cemetery grounds, was felled by a beam crashing down from the collapsing Twin Towers. Hailed as an arboreal martyr, the tree was credited with shielding St. Paul's from destruction.

"Inside St. Paul's Chapel," related an article for children, "from the arched ceiling fourteen colonial crystal chandeliers swayed but did not fall. Outside, the uprooted sycamore tree lay sprawled across the historic graveyard. Miraculously, St. Paul's Chapel remained perfectly intact, saved by the giant sycamore tree." While it was true that the falling towers did shower down large debris near the chapel, it was actually sheer luck, and not the intervention of the tree—which had been felled at the far end of the cemetery—that preserved St. Paul's.

But the Miracle Sycamore became a powerful fable that no amount of debunking could dispel. Sculptor Steve Tobin, forty-four, hearing the story in the weeks after 9/11, made it his mission to honor the downed tree. Initially officials of Trinity Church, which owns St. Paul's Chapel, paid Tobin little heed, for their chapel had become "the site of an extraordinary, round-the-clock relief ministry to the more than 14,000 volunteers offering assistance in the recovery effort."

But Tobin persisted, and on September 11, 2005, *Roots*, his towering orange cast bronze of the sycamore's gnarled stump and roots held aloft by powerful root legs, was installed in front of Trinity Church, the first "substantial permanent memorial" to the attack. The four-ton sculpture was strong enough to welcome clamberers of all ages and became an instant Ground Zero attraction. The sycamore's stump, professionally preserved by Tobin, returned to the chapel.

Before 9/11, it is probably safe to say that few New Yorkers knew much about faraway Oklahoma City. Now they connected, two American cities traumatized by devastating terrorist attacks. On April 19, 1995, Timothy McVeigh, an antigovernment army vet, detonated a 4,800-pound bomb in a Ryder rental truck at 9:02 a.m. in front of the nine-story downtown Alfred P. Murrah Federal Building in Oklahoma City. The massive explosion killed 168 people, including 19 children, and destroyed or damaged 324 buildings within a sixteen-block radius. Across the street in a private parking lot, the force of the bomb blasted dozens of cars into a jumbled pile of smoldering wreckage under an eighty-year-old American elm, long the lone source of coveted shade in a small sea of asphalt.

"That parking lot with that tree was right across from the Ryder truck," says Gary Marrs, the Oklahoma City fire chief in charge of rescue and recovery. "That explosion blew that tree bare as if it was wintertime, and the parking lot, the morning of the bombing, instead of asphalt, was just covered in green with all the leaves blown off that tree. I remember thinking, *That poor tree is not going to survive.*" Every car parked in the lot had pieces of McVeigh's truck, plastic barrels, and explosives embedded in it, as did the tree, which not only had lost major limbs and many branches but also was badly scorched.

Bombing investigators from the FBI and ATF came close to chopping down the tree (which had been notably uninfected by Dutch elm disease) to recover potential evidence from its mangled branches and blackened bark. "As time went on," Marrs says, "when people could come back on the site again, people who'd been working there, the tree became a gathering place for them. 'Let's meet at the tree,' they would say." By summer a 350-member Memorial Task Force had begun to wrestle with how best to create a physical space to remember and honor those who had died and those who survived while also offering "comfort, strength, peace, hope and serenity." By then, recalls Marrs, "what we assumed was a dead tree had started growing leaves again and that started people talking about 'Let's find a way to save this tree.'" Originally planted about 1920 in the front yard of a family home, this American elm was spared when the house was demolished after World War II and turned into a parking lot. In the ensuing half century, it had endured its inhospitable setting, its sturdy double trunks leaning low and northward, creating a V shape between them. This was not the elegant classic American elm shape of New England but a tough "prairie" version, smaller in stature and with a scraggly, spread-out umbrella of a crown. Now, as this tough old elm recovered from the bombing, it became known as the "Survivor Tree."

Kari Watkins, who became CEO of the Oklahoma City National Memorial Foundation, says, "Immediately we knew when we got the site that the tree was special. When we wrote a mission statement, one of the requirements was the saving of the tree and the placing of the tree on the memorial grounds. The tree is at the highest point of the memorial, and you feel it's keeping guard over all of the site. It's my favorite spot." On January 16 the Memorial Task Force Advisory Committee's second resolution made the Survivor Tree a key component of the memorial, whatever overall design prevailed.

Urban forester Mark Bays of the Oklahoma Department of Agriculture, Food and Forestry had arrived at the remains of the Murrah Building soon after the blast. With the Survivor Tree now central to the future memorial, the Memorial Foundation's first volunteer executive director, Rowland Denman, asked Bays for help. About a year after the attack, Bays, a seasoned arborist in his midthirties, found himself standing in the parking lot, directing a work crew as they removed by hand a wide circle of asphalt that had for years covered the tree's roots and come right up to its trunk. When the site was officially acquired for the memorial, Bays and the same crew pulled up all the rest of the parking lot's asphalt. "Then we covered it all with organic compost," says Bays, "and brought in a big watering truck. As we watered it, I could just imagine what that tree was feeling. Like a big cement block was off its chest. We put some special blends of fertilizer in and then we just wanted to monitor and see what happened."

Bays had worked with many trees over the course of his career, but he had developed a deep reverence for the old elm. "The power of this tree is like nothing else I've ever been around," he says. "It's just tangible. It never gave up on living. It has such a spirit." He made sure that the long-neglected elm was fortified against Dutch elm disease with injections of Arbotect. He also cloned the tree so its genetic offspring (their locations kept secret) would be ready when the Survivor Tree died—though it is likely to outlive us all.

Like Bays, mourning Oklahomans came to cherish the Survivor Tree, its arboreal burst of life amid the charred blast ruins serving as a rallying spot and inspiration for all. The old elm was emerging as the most emotionally resonant symbol of Oklahoma City's journey back from tragedy.

Recognizing this, Bays teamed up with Steve Bieberich, a nurseryman devoted to bringing back American elms, to create the Survivor Tree Sapling Program. In the spring of 1996 Bays collected all the tree's papery seeds, and Bieberich nurtured several hundred into small saplings at his Sunshine Nursery in western Oklahoma.

In 1997, at the two-year anniversary of the bombing, Bays presented these year-old treelings to families of the victims. And so began a tradition. At subsequent anniversaries he went on to distribute several hundred year-old Survivor Tree offspring at the memorial site to the survivors, to the rescue workers, and then to whoever got in line early enough. "With those saplings, the tree's spirit would live on," says Bays. "It's such a message of hope." Survivor Tree saplings were also dispatched as arboreal "ambassadors," including one to New York mayor Rudolph Giuliani, which was planted near New York's City Hall.

When Hans and Torrey Butzer were selected to design the Oklahoma City Memorial, they viewed the Survivor Tree as "such a powerful character in this story" that they enshrined it on a promontory overlooking the entire site. Bays personally led the hand-digging of a system of piers that supported the Survivor Tree's new place of honor: a large, circular flagstone plaza-promontory above a series of cascading grassy terraces. "We didn't damage any of its roots," he says. "It's really a model of how you can build around a living tree while respecting the tree." At the same time, he oversaw installation of an underground aeration and irrigation system around the piers.

On April 19, 2000, the fifth anniversary of the bombing, when President Bill Clinton dedicated the Oklahoma City National Memorial and Museum, the American elm was thriving. Where the Murrah Building once stood, 168 stylized bronze-and-glass chairs are arrayed, one for each victim, on a manicured grassy slope. All around the perimeter of this greensward, plump native loblolly pines serve to enclose the site of the blown-up tower, its destruction the nation's bloody introduction to terrorism. The loblolly branches are vibrant, swaying, and alive in the breezes. Mourning doves wing from tree to tree, as do loudly cawing starlings. A few other visitors wander quietly, stopping here and there to touch a chair, read the names. It is a poetic place.

As you emerge from between two of the pines, on the far right stands the first Gate of Time, a thirteen-ton dark bronze monolith with "9:01" incised on top, the moment when ordinary life in Oklahoma City was violently upended by the bombing. From there a long, placid reflecting pool flows imperceptibly along the 318 feet of what was North Fifth Street (where McVeigh parked his truck) to the second Gate of Time, whose inscription of "9:03" marks the moment when life after the bombing began. The gentle susurration of those dark waters mingles peacefully with the wind in the pines.

On the other side of a reflecting pool rises a series of grassy terraces, culminating in a round, low-walled patio where the Survivor Tree lives, its two thick, leaning trunks forming a *V* shape topped by an umbrella-shaped cloud of leaves. Even under the pampering care of urban forester Mark Bays, this hallowed centenarian looks a bit blowsy and scruffy, an odd arboreal occupant for its elegant flagstoned circular patio. Below are more trees—the Rescue Orchard of redbuds, Chinese pistache, maples—honoring all those who helped in the horrific aftermath.

One survivor of the blast, Polly Nichols, recalled walking by the Survivor Tree for years as she headed from her car to work in the Murrah Building. "I used to think the tree was just so ugly, such an abomination. And wonder, 'Why is that tree still standing?'" Now she marveled at how the old elm had not only regrown its branches but had turned into a late-life celebrity. The tree's will to live and its vibrant recovery were to her a "wonderful lesson."

Michael Browne kept tabs on the Callery pear he and Rebecca Clough had gone to such lengths to rescue from Ground Zero. In December of 2001 he paid a call to the tree in its new Bronx home, the Arthur Ross Nursery, with its twenty-seven plastic hoop houses, heated greenhouses, and plant beds. Browne was startled to see how drastically it had been pruned, but Bobby Zappala reassured him, explaining that otherwise the tree would be structurally imbalanced. He was giving the injured Callery pear maximum coddling, having planted it in an extra-large hole with a special rich soil mix, and was feeding its roots with regular infusions of nutrients.

"It came back strong the following spring," Browne recalls. "I was pleasantly surprised when I visited that time because the tree had grown new branches and had formed a canopy." One day that summer, when Browne came by the nursery, "I was amazed to discover a Mourning Dove had taken nest in the branches. I was even more awe inspired when hatchlings were found in the nest and thought how fitting it was to have baby chicks in the reborn tree. The wonders of nature's ability to heal." Richard Cabo had been working under Zappala for about six months when the tree arrived, and he helped with its care before leaving to work in Manhattan parks. A former corrections officer who had been badly shot in a street robbery, Cabo noticed that as time passed, "the tree started to attract fans. Underground word got around and firemen would come and police officers who wanted to see the tree."

On the first anniversary of 9/11 Americans embraced plans for a new generation of memorial trees. On January 31, 2002, Deborah Gangloff of American Forests had organized kickoff press conferences in Washington and New York. Politicians, dignitaries, sponsors, and survivors participated in ceremonial tree plantings to announce that American Forests and the U.S. Forest Service were spearheading the planting of memorial groves in New York City, at the Pentagon, and in rural Somerset County, Pennsylvania. All would serve as living testaments "to the police, firefighters, emergency workers, and citizens who risked their lives to aid their fellow citizens," with the trees intended to represent "bravery, life, and the hope that we will leave our children and their grandchildren a more peaceful world."

Gangloff found herself attending meetings at the Pentagon, where, she heard, "Donald Rumsfeld [then secretary of defense] was personally involved. He wanted to be the first to get a memorial completed. The military was going to get this thing done sooner than anyone else. Every day commuters were driving by, looking at the whole side of the building gone. He contributed $100,000 personally of his own money to jump-start the work. We decided very early on not to have one tree for each of the 184 people who died, because trees die. And they would be living in a sea of concrete, not a grassy park, and it was a very difficult site."

A design for the Pentagon site by Julie Beckman and Keith Kaseman was selected from more than a thousand proposals, and American Forests agreed to acquire the ninety paperbark maples that would grow alongside a series of 184 spare cantilevered benches, each hovering above a lit oblong pool of flowing water. These particular maple cultivars, expected in time to grow to thirty feet, were "chosen for their vibrant red foliage in fall as well as their distinctive peeling bark. They also hold their leaves longer into winter than many other species, which the designers hope will give the sense that time within the memorial stands still."

On Monday, November 25, 2002, Matt Arnn dedicated the very first of New York City's 9/11 memorial groves on behalf of the city's Department of Parks and Recreations at Sunset Park, a hilly twenty-four grassy acres in the heart of the Brooklyn neighborhood of the same name. Parks Commissioner Adrian Benepe, who had succeeded his longtime boss, Henry Stern, arrived with a bevy of local politicians and members of the NYPD, fire department, and Port Authority. On an elevated corner of the park, the forty-five new white-flowering memorial trees were already planted in an oval—an attractive grouping of yellowwoods, two-winged silver bells, and white-flowering redbuds.

With a forever-changed lower Manhattan skyline in the distance, Commissioner Benepe began the ceremony by saying, "After the September 11th attacks, many New Yorkers searched for ways to help the city heal. This grove will serve as a site for remembrance and reflection. The tree, one of the most enduring symbols of life, will stand as a memorial for generations to come." Others delivered brief remarks, and a dozen officials gathered for a photo op and, wielding shovels full of ceremonial dirt, collectively tossed it around one of the young yellowwoods.

Then Arnn spoke on behalf of the U.S. Forest Service, praising trees for "working so hard for so many people, every day," adding, "When we donate or plant a tree, its benefits extend far beyond our own physical and mental well-being. It's an unselfish act." He pointed out that none other than that quintessential New Yorker, Andy Warhol, had admired the city's "hardworking" trees, and he quoted the artist: "In the city everything is geared to working. . . . In the city, even the trees in the parks work hard because the number of people they have to make oxygen and chlorophyll for is staggering. . . . In New York you really do have to hustle, and the trees know this, too—just look at them." And so with this benediction, the first 9/11 Living Memorial was inaugurated.

CHAPTER EIGHTEEN

"I Was Surprised It Was So Aggressive": Waging War on the Emerald Ash Borer

Death by emerald ash borer. Belvedere Drive in Toledo, Ohio, in June 2006 *(above)* and June 2009 *(below)*. *(Photographs by Daniel A. Herms, the Ohio State University.)*

On June 25, 2002, David Roberts, a plant pathologist at Michigan State University, met up with four concerned federal and state entomologists. Serving as chauffeur and tour guide, he wheeled his 1982 Jeep Wagoneer to several sites where stands of large ash trees were visibly sick and dying amid the subdivisions, office parks, and shopping malls of Novi, a small city on the outer edges of greater Detroit. Roberts, fifty, a solid man with a close-cropped brown beard, had been for fifteen years the director of the university's Plant Pest Diagnostic Clinic, and as such helped arborists, nursery owners, and greenhouse growers solve woes as varied as spruce canker and poinsettia fungus. A farm boy from Delaware, Ohio, he had been introduced to plant pathology by his girlfriend's father, a researcher at the USDA facility there. For a science fair project, Roberts had grafted an American elm scion onto Asian rootstock to thwart Dutch elm disease. For decades that tree grew in his parents' front yard (until it succumbed to DED). Roberts liked this world of plant pathology and had written his PhD dissertation on golf course turf grass bacterial wilt.

Roberts had been puzzling over ailing ash ever since arborist Guerin Wilkinson of Green Street Tree Care in Ann Arbor had alerted him a year earlier to eighty unhappy-looking trees at the Bradbury Parkhomes Condominium in nearby Plymouth Township. "The branches were all suffering from dieback, with sparse or stunted foliage," noted Roberts. "One consistent and quite conspicuous symptom was the extreme proliferation of shoots on the trunk, sending out very large, dense foliage. Some of the shoots have already exceeded 4 feet of growth this year."

When the American elm disappeared as a shade tree option in the postwar decades, Midwest developers and landscapers turned to the American ash, particularly favoring the "Marshall Seedless" cultivar of green ash (*Fraxinus pennsylvanica*) and the "Autumn Purple" cultivar of white ash

(*Fraxinus americana*). Like the American elm, the ash was a towering, hardy native as likely to thrive in a swamp or on a busy street as in a forest. Nursery managers and urbanites liked its quick, solid growth, indifference to road salt and other insults, handsome compact-enough shape, and striking autumn foliage.

At first, when Roberts had assessed the sickened trees in June of 2001, he diagnosed excess herbicide as the issue. As he and a friend, retired state agricultural botanist Carl Dollhopf, began making further informal windshield surveys of ashes in the surrounding communities, Roberts was stunned: "I don't think I would be exaggerating by suggesting that 1000's of trees are affected," he reported on his Tree Doctor Web site. Now he began to suspect a bad wave of the bacterial disease ash yellows, but a definitive DNA diagnosis required a six-thousand-dollar polymerase chain reaction (PCR) test, funds not in any budget. He also wondered if a bad drought that had begun in 1998 was playing a role.

That September Mike Meyers of Shade Tree Mechanic Tree Care in Canton had directed Roberts to a client's infected ash. When they stripped off its bark, Roberts took photos of the distinctive wood borer serpentine galleries and half-inch-long beetle larvae, cream-colored and flat with brown heads. At the end of the larvae's ten-segmented bodies, they sported little pincers known as urogomphi. Such borers typically colonized unhealthy trees. Roberts gathered some in a vial and submitted them to the MSU Plant Diagnostic Services laboratory, whose entomologist could "identify the larva only to the family *Buprestidae* (metallic wood-boring beetles) and probably to the genus *Agrilus*." Their best guess was the larvae were specimens of the two-lined chestnut borer, which had preyed on American chestnut trees until they disappeared, at which point it had switched to oaks. Had it now taken to eating ash? MSU entomologist David L. Smitley advised Roberts: "Cut some branch sections with larvae, bag 'em up, and see what emerges next spring."

On January 25, 2002, Roberts convened a public forum on dying ash in western Wayne County at the MSU Extension Service facility. Eighty local officials, foresters, arborists, master gardeners, and nursery owners showed up, alarmed by the spreading malady. "I presented my findings up to that point," recalls Roberts, "and ash yellows seemed the most obvious explanation. Most borers are secondary on weakened trees." After the meeting the Michigan Arborists Association, as well as the cities of Canton and Livonia, agreed to underwrite the PCR analysis.

As soon as the ash trees started leafing out in April and May, Roberts began driving out around Wayne County, knowing that he would obtain the best results from PCR testing by using the fresh first growth of leaves. All told, he gathered sixty-five different samples. The PCR results eliminated ash yellows as the culprit; only three leaves tested positive.

During the winter, meanwhile, Roberts had begun noticing "intense woodpecker injury" on many ash trees, telltale bare patches where the birds were scraping back bark to get at the succulent white larvae. He followed through on Smitley's advice to raise the ash-eating larvae into beetles. "In early April," he explains, "I started collecting logs with larvae from trees that arborists were cutting down. I wanted to be sure as a scientist I wasn't looking at just one tree. I gathered three or four logs each from about a dozen locations." He stored each location's logs in a separate large clear plastic bag and laid them all on his long lab bench.

One day in May when Roberts was in his laboratory, he looked down and realized, "There were all these dead beetles on my lab floor. They had just chewed through the plastic bags, and a couple of dozen were dead on the floor, probably from eating the plastic. I wasn't all that excited. To be honest, I didn't know exactly what they were—same shape as chestnut borer, just not the same color." Unlike the dull bronze of the two-lined chestnut borer, these winged beetles were a subdued iridescent gold green, slender of body and about half an inch long.

"Since late May of this year," Roberts would write on his *Tree Doctor* blog, "adults of this larval stage have been emerging. Thanks to my contacts around the SE region, we've been able to collect these insects from various geographical areas." Roberts eventually passed the beetles on to Gary Parsons, a coleopterist (beetle specialist) in MSU's Department of Entomology. Parsons could identify them to genus level (*Agrilus*) but not to species and sought help from a colleague at the Oregon Department of Agriculture, Richard Westcott, a specialist in the family Buprestidae (metallic woodboring beetles). Westcott responded that only by examining male beetle genitalia could he render an opinion. Roberts sent him two males and at the same time others submitted specimens—per protocol—to the Systemic Entomology Laboratory at the USDA Research Service in Washington, D.C.

Even after studying the male beetle's genitalia, though, Westcott could not identify which of the fifteen thousand species of Buprestidae the beetle might be. The federal entomologists at the USDA in D.C. were equally baffled, as were those at the Smithsonian's Museum of Natural History. The bronze green specimen did not precisely resemble anything in their world-class collection. Consensus was emerging, though, that it was probably a foreign intruder.

Westcott now contacted Eduardo Jendek, the world expert on European and Asian wood-boring beetles, at the Institute of Zoology at the Slovak Academy of Sciences in Bratislava, Slovakia. After Jendek studied Westcott's e-mailed images and description, he made a preliminary identification of *Agrilus planipennis*. Actual specimens arrived in early July, and by July 9, Jendek had conclusively declared the beetle to be *Agrilus planipennis Fairmaire*, an Asian beetle observed near Peking by French entomologist Léon Fairmaire in 1888 and duly classified.

MSU forest entomologist Deborah G. McCullough, part of a small, loose-knit group of Michigan forest entomologists, remembers how she learned the news. "The phone rang and I picked up," she recalls, "and it was the Michigan Department of Agriculture saying, 'We've got an ID. It's exotic and we're having a meeting with the nursery and plant people. We'll have to have a quarantine. People will lose millions of dollars.' After that, everything just exploded. Southeast Michigan was a big nursery area for ash trees, and they would be stuck with all those trees. An exotic thing like the emerald ash borer, it's just huge."

Raised in Flagstaff, Arizona, McCullough had grown up surrounded by forests and earned an MS in forestry at Northern Arizona University and a PhD in entomology at the University of Minnesota at St. Paul, writing her dissertation about jack pine budworms (*Choristoneura pinus pinus Freeman*) on young jack pines in burned and distressed forests. An enthusiastic outdoorswoman, she was an avid hunter, especially of wild turkeys in the spring. She wore her reddish hair short cropped and had the powerful shoulders of one at ease wielding a chainsaw.

This was the first known sighting of this beetle *anywhere* outside Asia, adding to the mystery of its presence in Michigan. "Little was known about it," McCullough and collaborator Therese M. Poland would later write, "beyond taxonomic descriptions . . . and a few paragraphs published in

Chinese reference books." While awaiting Jendek's opinion, the Michigan State University academics had agreed on a name for this alien invader: "emerald ash borer," with the handy acronym EAB. The Entomological Society of America conferred its official blessing.

No one ever could pin down exactly how the emerald ash borer arrived in Michigan. But molecular evidence "suggests China was the likely source," later wrote Ohio State University entomologist Daniel A. Herms and McCullough. "EAB was probably imported into North America via crating, pallets, or dunnage made from infested ash."

Now that the beetle had been identified, McCullough and others joined the Michigan Invasive Species Task Force as the state grappled with how best to address this new ecothreat. On Tuesday, July 16, the state's Department of Agriculture formally announced the discovery of the emerald ash borer in five southeastern Michigan counties—Wayne, Oakland, Macomb, Livingston, and Washtenaw—and issued a quarantine of "shipments of ash trees, branches, logs and firewood from five counties around Detroit in an effort to stop the unusually aggressive insect." It also provided a phone number for a new EAB hotline.

The following day the state held a press conference in Northville. The "experts" could offer little reassurance or specific knowledge. As McCullough told a reporter from the *Grand Rapids Press*: "Literally, nobody in the country had seen [the EAB]. . . . We don't know how fast it's going to spread yet. I guess I was surprised it was so aggressive. I was surprised at how fast trees were being killed. It makes me sick, to be honest." The same front-page news story continued, "The pest is so new to Michigan that state foresters and entomologists are unsure how to fight it. MSU entomologists are trying to contact Chinese agriculture officials for advice." Dave Roberts, the MSU plant pathologist who had originally sounded the alarm and provided the first live beetles, now estimated that as many as 200,000 ash had already died in the five quarantined counties.

While this new battlefront against invasive insects had opened in Michigan, the war against the Asian long-horned beetle had been grinding steadily on in both greater New York and Chicago. Since 1996 Joe Gittleman had supervised the removal and destruction of almost ten thousand mature trees. The great fear was that the Asian long-horned beetle would make its way into Central Park. On January 30, 2002, an APHIS contractor in an aerial bucket truck was surveying park "host" trees, poking around in the upper canopy of a sugar maple in the Hallett Nature Sanctuary, when the crew spotted the telltale "bullet" holes on the upper branches of first a sugar maple and then a Norway maple.

"This is a serious thing for parks," newly elected Mayor Michael R. Bloomberg told the *New York Times*. "The only thing that you can really do when you find an infested tree is cut it down. Chop it up and burn it." Gittleman's crews quickly checked the elms of Literary Walk, the allée of extraordinarily beautiful intertwining American elms whose urban isolation had largely preserved them against Dutch elm disease. Commissioner Benepe, who began his career as a forester, said of this latest conquest of the Asian long-horned beetles: "Imagine what happens if this [beetle] makes it up to the maple sugar plantations in upstate New York and Vermont."

Few but the beetle authorities realized how kind the Fates had been to Central Park. In 2000, year five of the beetle war, APHIS scientists working in China had announced a prospective game-changing new weapon: the pesticide imidacloprid, a nicotine-based systemic neurotoxin injected into a tree's vascular system and absorbed all the way to the outer twigs and leaves. If this stealth chemical warfare was successful in poisoning the beetles in situ, it would eliminate one of the most painful aspects of the scourge: the killing of all the nearby uninfected potential host trees.

In Chicago's Ravenswood, which had lost so much of its tree canopy, foresters and residents were ecstatic. In 2001 the city began testing imidacloprid, injecting at-risk host trees. At the end of the season, APHIS scientists confirmed imidacloprid's efficacy. The following year foresters removed only 18 infected trees and injected 92,000 potential hosts. For the first time since chestnut blight and Dutch elm disease almost extirpated their respective victims from the urban landscape, scientists and arborists had a plausible weapon to fight back against the latest tree killer.

In New York City the ALB troops now swarmed through the treetops of Central Park, inoculating the park's nine thousand at-risk trees with imidacloprid, including the irreplaceable American elms of Literary Walk. Not another beetle was found, and as Neil Calvanese of the Central Park Conservancy observed, "This has been a real conundrum for us. Why did the Asian long-horned beetle not move on and how in the devil did it get into the Hallett Nature Sanctuary? Did it fall off a building? Come off a tree truck? We just don't know."

Had the Chicago beetle warriors, led by Mayor Daley, not pioneered imidacloprid in the United States the previous year and proved its efficacy, thousands of Central Park's maples, birches, horse chestnut, poplars, willows, elms, and ash would have been under a swift death sentence, doomed to execution before warm weather and beetle emergence. So it was fortunate that Gittleman's troops had failed to locate the infestation sooner; the beetles had been lurking undetected for five years.

The logistics of the ongoing battle against the beetles were mind-boggling. In 2002 alone Joe Gittleman's New York ALB team would survey 284,700 trees, with bucket trucks or smoke jumpers making three quarters of those inspections up in the canopy. They inoculated a total of 134,101 trees throughout the city. As new infestations were determined, the quarantine area steadily expanded from the original ten square miles around Greenpoint, Brooklyn, to encompass 124 square miles in all the city's boroughs but the Bronx. New Jersey had discovered infestations as well. The total cost—federal, state, and city monies—of fighting this war in New York, New Jersey, and Illinois had reached close to $200 million.

New York City faced a much bigger battle than Chicago, as its insect foes were entrenched on far more difficult terrain. In 2005 infested trees had been found in the private, hard-to-access yards behind a handful of Upper East Side brownstones. It was likewise not easy to gain entrée to such aerics as Donald Trump's penthouse, with its forest-sized birch trees on an outdoor patio. It took three years to obtain permission to get access to a private property in Long Island City that had a gigantic weeping willow. Foresters had been admiring this magnificent specimen, its cascading branches swaying over several lots. It was, as it happened, also riddled with holes harboring 140 live adult beetles, the most ever found in one New York tree. It took crews three full days to cut it down and grind it up.

Controlling the spread of the emerald ash borer was shaping up to be a very different story. Unlike New York City or Chicago, southeastern Michigan had more than 9,500 nurseries that grew or sold ash trees. Logging companies in that area harvested ash, especially white ash. More than two thousand local sawmills cut it into lumber; manufacturers in turn used that wood to produce railroad ties, pallets, tool handles, pulp paper, furniture, and, most famously, baseball bats. None of those ash trees or the lumber harvested from them could exit the quarantine zone. The economic repercussions just in the five affected counties would be painful.

Unlike the American chestnut or the American elm, the American ash was not what is today called a "charismatic" tree. It had long been a solid, reliable arboreal citizen, but it had hosted no revolutions or famous events beneath its boughs, and whatever mystique it possessed involved snakes' purported aversion to the tree. "How many thousand-thousand of untold White Ash trees are respected companions of our doorways, kindliest trees in the clearing," asked Donald Culross Peattie in his classic *The Natural History of Trees*. "No one can say. But this is a tree whose grave and lofty character makes it a lifetime friend. White Ash has no easy, pretty charms like Dogwood and Redbud; it makes no over-dramatic gestures like Weeping Willow and Lombardy Poplar. It has never been seen through sentimental eyes, like the Elm and the White Birch. Strong, tall, cleanly, benignant, the Ash tree with self-respecting surety waits, until you have sufficiently admired all the more obvious beauties of the forest, for you to discover at last its unadorned greatness."

Black ash was a somewhat different story, explained Deborah McCullough, for that species, a "country" tree found mainly in swamps and bogs, had "cultural and spiritual significance for many American Indian tribes from Minnesota to Maine. . . . Some tribes even trace their origin to a black ash tree that split—one fork became a man and the other became woman. The art of black ash basketry has been handed down from generation to generation in many tribes. Basket-making families have traditional harvest grounds, where they carefully select and harvest a few black ash each year."

As the first spring and then summer of the EAB attack unfolded, the arborists and entomologists were hopeful that they could contain the beetles. However, they did not yet know when the first wave of emerald ash borers had established strongholds. During that initial occupation, local nurserymen in the now-quarantined counties had been the unwitting accomplices of the beetles lurking in their trees, as they had been selling and shipping perennial ash favorites to customers far outside the quarantine zone. The same was true of harvested ash lumber and especially firewood, which had only expanded the emerald ash borer killing fields. One APHIS employee said, "Here in Michigan, many of us have lake cottages and when we go up, we throw everything we need into the van or the pick-up, including firewood. Or if we go camping somewhere, we'll bring along that firewood." In September, Monroe County, which bordered Ohio, was added to the quarantine zone.

In October APHIS convened a Science Advisory Panel. To vanquish the wood borer, scientists had to learn its basic life cycle and habits: How quickly did it reproduce? How far could it fly? How fast and far might it spread on its own if the quarantine stemmed the tide of its wood products? Equally important, how could the advance guard be detected as it was invisibly feasting and killing trees? Just figuring out how far it had already spread was needle-in-a-haystack work. How long had it been living and reproducing in Michigan?

What scientists did know was that one tiny female could lay as many as one hundred eggs on crevices in the outer bark; when those eggs hatched into larvae, the borers chewed their way into the inner bark and phloem, and the labyrinthine channels they made slowly cut off the flow of water and nutrients from the canopy to the roots. Those channels also scored the outer ring of sapwood, interfering with the tree's ability to transport water from there to its leaves. As the larvae multiplied, they left in their wake galleries and channels that ultimately circled the tree, causing its death by starvation in two to three years.

By November 2002, McCullough, a member of the panel, was warning: "If we don't get this thing stopped here, we might see ash removed from the landscape in our lifetime, especially in urban areas." The scale of loss was mind-boggling, for not only were midwestern cities and suburbs rich with planted ash on streets and in yards, but the tree easily self-seeded and grew wild in urban woods and wetlands. Experts estimated at least five to seven million ash trees were already dead or dying in just the six-county quarantine area of southeastern Michigan.

In late December Michigan issued its eradication plan: Remove and chip (into one-inch pieces) all ash trees three miles beyond the 2,400-square-mile quarantine zone. "That's a monumental task," Roberts told a reporter for the *Toledo Blade,* "but if we don't do it, it would definitely spread throughout Michigan and throughout the country. That could mean billions of dollars in damage." David McKay, the APHIS official in charge, explained, "Like a forest fire, we need to have a firewall of prevention to ensure that it doesn't spread beyond southeast Michigan." The theory was that EAB populations within the quarantined areas would eventually die off once deprived of live ash trees.

The problem: Where to clear-cut that fire wall? As arborists and foresters combed the streets and woods, they kept finding more ailing ash. As one APHIS official observed, "Every time they drew a circle, then they learned that the beetle was well beyond that circle." And could pesticides like imidacloprid, which was now saving trees from Asian long-horned beetles, protect ash trees as well? Only time would tell, but to those now surveying the situation, "it was becoming apparent that EAB had the potential to devastate ash on a continental scale."

The year 2003 was a litany of bad EAB news. In February the intended Michigan fire wall was breached before even a single noninfested tree was preemptively cut down. The buyer of a 1.5-acre tract near Whitehouse, Ohio, directly south of Michigan's quarantined Lower Peninsula via I-75, noticed bark falling off his ash trees. He telephoned Amy Stone, an Ohio State University extension agent at the Toledo Botanical Garden, who discovered beetles in several ash trees on the land. Daniel Herms referred to the date of discovery, February 6, as "The day everything changed."

At that time, researchers optimistically viewed a place like Whitehouse as a unique infestation—an outlier—that could be contained, and did not consider it an expanded front per se. Ohio authorities mobilized swiftly, and Department of Agriculture inspectors confirmed almost 100 infested ash trees on just five parcels with 1,500 ash trees. By March 13 a quarantine was declared on just those properties. In May, scientists thought, the beetles would emerge, mate, and seek new trees to colonize. While Michigan estimated it had 700 million ash trees at risk, Ohio had five times as many: 3.8 billion.

Time was now of the essence. In late March, Tom Harrison, plant pathologist with the Ohio Department of Agriculture and head of the state's EAB task force, announced the defense strategy: "Clear-cutting as many as 2,700 privately owned trees in the Whitehouse area, then applying an insecticide to about 300 trees around the perimeter. The plan is akin to a biological firebreak, with the chemical treatment used to ward off any stray beetles that might make it through." Melanie Wilt, state agriculture spokeswoman, defended the measures: "We have few options." Nor was Ohio going to merely chip the condemned trees: It would burn them.

Residents were not pleased. Gary Lowry, owner of 5.6 acres along Reed Road, had 122 small, medium, and very large ash among his 3,000 trees. "I'm not going to let them clear-cut my trees. If they are infected, cut them down," he said. "Once they clear-cut this, you'll see right through my woods. I have ash trees that take two people to get their arms around. I'll never see ashes like that on my property again in my lifetime." When the question was raised of trying imidacloprid first, it was noted that it was not yet a proven solution, as it was for Asian long-horned beetles. "By the time we wait and see," said Wilt, "it's too late. We believe we're taking the appropriate action to prevent a widespread problem."

Deborah McCullough scotched any hopes that the brutally cold winter might have diminished the numbers of emergent borers: "I can assure you the larvae are alive and well." What little McCullough and her entomological colleagues had been able to glean thus far about the emerald ash borer's habits and predilections did provide some reassurance. While the insects did spend a year or so becoming adults, most did not appear to venture far from the tree where they emerged. This suggested that containment and eradication were possible if the gathering forces of state and federal beetle warriors waged a timely and ruthless ground campaign.

Farther north, conscientious township arborists were racing to cut down all their dead and dying ash before the new generation of beetles spread that spring. Residents of Sterling Heights, Michigan, knew yet another towering tree would soon disappear when a big spray-painted pink *X* appeared on its trunk. By February 18, tree removal crews had dismembered 200 ash trees, and 450 more sported the telltale pink *X*.

On April 19 emerald ash borers had been discovered at Ed's Plant World in Prince George's County, a Maryland suburb of Washington. A state plant inspector had noticed the insects in a shipment of 121 ash trees. By then 27 trees had been sold. Officials believed that because they could identify how and when the infested trees had entered Maryland, they could eradicate the EAB. Nonetheless, it constituted yet another outlier, calling for arduous extermination of all the ash trees in a half-mile buffer zone around the nursery.

In June of 2003 Craig Kellogg of APHIS took charge of the EAB war in Michigan. "In the beginning," he says, "there was just very little information. The only scientific literature was a half page in Chinese. There was just so much we did not know." Tom Holden, an arborist with the Davey Tree Expert Company in Canton Township, the heart of the infestation, recalls:

> By summer of 2003, I have this vivid memory of going to a customer's backyard: There were so many emerald ash borers flying about—they were like mosquitoes in a cloud around the tree, all homing in on it. They were just everywhere. If we would spray a tree and you went back the next day, you would find just dozens and dozens of dead beetles on the ground.
>
> Many people were just heartbroken. It was really tough. The majority of specimen ash trees we saw were just too far infested to bother with trying to treat them. I had one customer who had fifty ash trees in his backyard, green ash mainly, and each forty to fifty feet tall. His whole backyard, that woods, was going to be gone. And there were subdivisions where every street tree was an ash and they took those down and there was not a single tree left.

The estimates of tree loss were staggering in the six afflicted Michigan counties: half of the 650,000 landscape ash in decline or dead (and the other half doomed in coming years), with another 10.6 million forest ash also dead, dying, or doomed.

As survey crews—some up in planes and helicopters—scoured Michigan cities and woodlands contiguous to those six quarantined counties, they found more afflicted ash. By October 8 Kellogg and his army of EAB scouts had confirmed that the beetles had infiltrated another seven counties, expanding the potential clear-cut zone immensely. Later, "as we were chasing this beast," says Kellogg, "we found from the dendrochronology studies that EAB had been around for some years. It wasn't that EAB was spreading that fast; it was that we were finding infestations that had been there for years. . . . We were spending millions, because you have to locate every tree, mark it, grind it up, burn it. At first there was plenty of money because the politicians in Washington—the U.S. senators, the representatives—as it moves from county to county, there's more alarm from new people. But you do not want to keep throwing money down a rabbit hole. We realized we needed better tools to find this thing."

The year 2004 yielded no better news. On April 21 Indiana announced that the EAB had been found at Yogi Bear's Jellystone campground on Barton Lake in Steuben County, forty miles north of Fort Wayne. Two days later the state imposed an ash quarantine in Jamestown Township. A month later Asplundh Brush Control trucks from Michigan rolled into the huge Jellystone campground, with its multiple swimming pools, water slides, and mini golf, and headed into the woods to begin three days of chainsawing and grinding up 423 ash trees. This Indiana infestation added another outlier to the tally. Ohio, meanwhile, continued to discover more of its own outlier infestations, first at the Toledo Express Airport (10,000 ash trees), then in Swanton (2,500 ash trees), Hicksville (3,600 ash trees), Rossford (424 ash trees), and Easton (17,000 ash trees).

Two years into the war against EAB, though, the quality of intelligence was steadily improving. By the fall of 2004 the teams of EAB scientists had an answer to one key question that would determine how to proceed: How far could these insects fly? Just as with elm bark beetles decades earlier, humans had again underestimated their foe. Researchers tethered mated EAB females to computer-monitored contraptions—"flight mills"—that measured the bugs as they flew around in circles at the end of a wire thread. A few ambitious endlessly circling borer females—about 1 percent—could (and presumably did) wing forth as far as three miles, much farther than anyone predicted, to lay their many eggs. "This discovery," noted the scientists in their report, "is rather alarming as it suggests females are programmed to make a dispersal flight."

Determining how far the emerald ash borers had actually spread was proving to be a key challenge. Locating the beetles hiding in the ash trees—and thus delineating the outer perimeters of the infestation—was a hit-or-miss proposition. In the winter and spring of 2004, the Michigan

Department of Agriculture created a statewide grid of so-called trap trees—girdled ash (that is, with a strip of bark removed from their trunks to slowly kill them) slathered with a sticky substance. That summer the anti-EAB troops fanned out to visually inspect each of these trees (two to nine trap ash per thirty-six square miles) for bugs caught in the adhesive traps. That fall and winter crews arrived to cut down each trap tree and peel off the bark in search of larvae. Several new outlier infestations were detected, and once again the EAB sleeper cells were found to be "much further advanced in southeastern Michigan than previously thought."

Scientists had also learned that the EAB was attracted not to any beetle pheromones but rather to the particular odors and colors of ash trees. It was after they arrived on a given tree that they found mates. Devising a trap sufficiently alluring to the beetles, therefore, posed a challenge. Early versions experimented with varying sizes, colors, shapes, and scents. As one group reported when all the scientists convened in October of 2004 in Romulus, Michigan, for a two-day meeting called "Emerald Ash Borer: Research and Technology Development": "It is unknown if low EAB trap recoveries were due to ineffective traps or low EAB populations at our trapping sites. . . . EAB may not distinguish trap colors in low-light conditions."

At that same gathering the first of many entomologists from the U.S. Forest Service, MSU at East Lansing, and the Chinese Academy of Forestry in Beijing reported on the predators of EAB, noting: "Should eradication fail, however, conventional biological control will be needed to suppress populations of this invasive buprestid. To this end, we are studying the natural enemies of EAB in Michigan and in China." In the provinces of Jilin, Tianjin, and Liaoning, the scientists found native wasps that devoured as many as half of the EAB larvae. "Based on those results, we established our 2004 study sites . . . in cooperation with local foresters to determine the species composition and seasonal abundance of EAB natural enemies."

In 2004 the federal government spent $45 million on EAB-eradication efforts, almost triple the 2003 budget, but Ohio had found EAB in another township in October, and Indiana confirmed a second EAB infestation a half mile west of the first in Jellystone. At the end of the year Kellogg added another seven Michigan counties to the quarantine.

As scientists completed more sophisticated dendrochronological studies, Kellogg and his team now understood another critical piece of data: EAB had been mating and multiplying in Michigan's Lower Peninsula since the early to mid-1990s.

As officials assimilated this new data, any real hopes of eradication disappeared. "The reality is that success in eradicating any invasive species," Kellogg said, "is directly related to when you find it. We could see it was not realistic to continue along this path. . . . So the strategy in Michigan was no longer going to be eradication, but containment."

As of December 2004 Kellogg declared a new "gateway policy" that would focus on keeping the emerald ash borer confined to the Lower Peninsula of Michigan, while still eradicating it in all other states. The gateways were fifty-mile-wide bands subject to intense surveying and regulatory activities. Wherever the beetle warriors located an infested ash tree, all ash within a half-mile radius would be extirpated. The Michigan anti-EAB forces would defend new strategic battle lines: in the north, the Straits of Mackinac between the Lower and Upper Peninsulas, and the St. Clair River into Ontario, Canada. The other critical gateway would be along the state's southern border with Ohio and Indiana.

Firewood had by then emerged as *the* main means by which EAB spread. The rare super-determined female emerald ash borer could fly several miles into untouched acres of ash, but she was far more likely to arrive in new territory thanks to campers heading out for a weekend in the woods, lugging along firewood full of EAB. Officials carried out "firewood blitzes" along major highways, checking cars at rest stops for ash logs heading toward lake cabins, or scanning pickups as they crossed the Mackinac Bridge. In Michigan, Ohio, and now Indiana, ominous television ads warned against aiding and abetting the EAB by transporting firewood, a message repeated on huge billboards, in public service ads on the radio, and in statewide mailings: Moving firewood was now a crime.

"The fines," warned a press release, "range from a minimum of $1,000 up to $250,000 and/or up to five years in jail." Yet out of ignorance or lack of civic responsibility, citizens continued to transport those convenient logs.

The EAB situation did not improve in 2005. By April, Michigan's regulated area had expanded from thirteen to twenty counties in the Lower Peninsula, with another twenty-five outlier outbreaks, with firewood always the chief suspect as vector. Like many Michigan cities, the college town of Ann Arbor learned in 2003 that its ash—which had long graced its streets and parks and composed 17 percent of its trees—were badly infested.

Ohio, which in 2004 had removed 33,524 ash trees, now faced far larger infestations. While the original outliers had all been close to the Michigan border, in the spring of 2005, Russ Heilman of North Baltimore, a Columbus suburb in central Ohio, showed up at the state agricultural office with one of the tiny green beetles. He and his wife had first noticed that their huge old ash trees were sickly in 2003 and the following year had informed officials, who responded with skepticism. But when he produced an actual beetle, "they got excited then," Mr. Heilman told the *Columbus Dispatch*. The suspect had come from nursery stock planted at a nearby housing development.

It was by no means just homeowners who suffered from the loss of the mature trees. What of the birds who nested in those vaulting branches and the many harmless insects and small creatures who depended on their bark and leaves for nourishment? At first no one could really say what ecosystems encompassed the stalwart and little-studied ash. One of McCullough's technicians, David Cappaert, noted, "Ash trees are generally prolific seeders and a variety of ducks, song birds, game birds, small mammals, and insects feed on ash seeds. In many ecosystems, ash trees provide browse, thermal cover, and protection for wildlife, including white-tailed deer and moose." One of Daniel Herms's students, J. K. J. Gandhi, determined that more than 250 arthropod species could be seen feeding on ash trees.

"It was all just really depressing," says Nate Siegert, who was working as a postdoc on a variety of EAB studies with McCullough. "One day you'd be in a neighborhood with nice large trees and next time you came through they were gone—it was just bare. What was even worse was areas where they were not able to take the trees out. The streets were a pretty high priority to get the trees down because of the safety, but in parks and small woodlots, you'd just be driving all around greater Detroit and there were all these dead trees, all brown in the summertime. And it just kept getting worse and worse in a larger and larger area. By 2005–6, it was just highly visible."

In these areas spring no longer brought signs of new life, just more dead ash trees silhouetted against the sky. As the ash ecodisaster expanded to new states—on June 9, 2006, Illinois confirmed EAB in the Windings subdivision north of Chicago—the federal funds to fight it were contracting. In 2004 the budget allocated was $45 million; the following year it had fallen to $24 million. By 2006 it was hard to ignore what time and experience had shown: the futility of firebreaks and massive tree removals aimed at eradication. The emerald ash borer was known to occupy almost 25,000 square miles in five states. In 2006 EAB funding plummeted to $10 million.

"Now we are in containment mode," Craig Kellogg acknowledged. The nursery industry had taken its losses and largely stopped growing ash trees, eliminating that vector. "We still are actively looking for EAB through better detection and then regulating," Kellogg said. "We don't want to help any movement of firewood or lumber." The focus would now be on slowing and containing the menace, with the hope that this would buy time for research and the development of better control mechanisms—whether pesticides to save specimen or street trees, or biocontrols like wasps.

The U.S. Forest Service estimated that there were eight billion ash trees in the United States, or about 225 trees per American. Midwestern cities and landscapers had been especially enthused about ash as the ideal city replacement for the American elm. Whatever lessons might have been learned about the dangers of arboreal monocultures had somehow not been absorbed by the arborists, city leaders, and developers. Now, once again, they had lost or faced losing another mature canopy. And every place with ash knew it was just a matter of time.

CHAPTER NINETEEN

"Putting in an Urban Forest Instead of a Storm Drain": High-Tech Meets a Million Trees

Big Bird watches as Mayor Michael Bloomberg plants the first tree of the MillionTreesNYC campaign on October 9, 2007. *(Courtesy of the New York Restoration Project and © 2016 Sesame Workshop. All Rights Reserved.)*

By 2005, says U.S. Forest Service scientist Greg McPherson, "I was focused on street trees." Not only did city trees lining roadways and boulevards deliver the most "services" to the humans around them, but they were the main preoccupation of urban foresters. Says McPherson, "We worked on developing sampling methods using a thousand trees in any city that would give you a good enough estimate of the total number of street trees, their species, their size, what the management issues were—how many were dead, how many needed to be pruned, the number that should be planted—and then figuring out all the benefits from these street trees. That would serve as a basis for going to the city council, giving you a legitimacy and road map for educating the public. . . . It was relatively easy to learn what a municipality was spending per year per street tree and then compare that to the benefits those trees provided."

Starting in 1998 McPherson's staff of fourteen scientists and technicians at the University of California at Davis, working in concert with a team from Davey Tree, had been laboriously gathering a set of standard street tree data for sixteen "reference" American cities, each representing one of the nation's sixteen climate zones. The data for a particular city reflected local conditions that influenced tree growth, building energy use patterns, and rainfall.

Paula Peper, a USFS support scientist, had joined McPherson's lab in 1995, and she was a key player in gathering and analyzing these mountains of data. "One of my first assignments from Greg," she recalls, "was to find a way to figure out the LAI—the leaf area index—on urban trees. I did not know what the heck LAI was, and I didn't want to look like an idiot with my new boss." She quickly got up to speed. "David Nowak had developed an equation for LAI in Chicago and we were testing our findings against that. The most accurate [approach] was a digital method where you filled the frame of the camera with the whole tree from top to bottom and then excluded the background with Photoshop so that you showed just the crown. But then you had to test that against the actual tree, which meant taking down a twenty-foot-tall tree and four or six of us working to pick all the leaves off and feeding the leaves through an LAI meter using a conveyor belt."

Over the course of two years, using hired student helpers and prison crews, Peper stripped the leaves off the full canopies of almost one hundred mature specimens representing fourteen different common species of urban trees. "The first couple of years, I had permanently sore thumbs," recalls Peper. "I really hated cherry trees; they did not want to get rid of their leaves. And oak trees—it is phenomenal the number of leaves an oak tree has."

Peper, a native of Sacramento, had spent years as a traditional forest ranger in national forests, burned out, and left to run a farm and happily teach high school English. In 1992 she was laid off, along with most of the state's other new teachers. As she recalls, "Dr. Phil Barker, a forest service researcher at UC Davis, called and asked me for breakfast. We met, and he said, 'You can operate a backhoe, drive a tractor, are red-carded on fire, and then you were an English teacher, teaching AP writing. You're perfect to help me.' He was in urban forestry—something I thought was an oxymoron—and he needed someone to manage his research site where he was looking at the effect of barriers on tree roots. I could cut the trees down and use the backhoe to remove barriers. Then we'd excavate to see where the roots were growing, note any damage to sidewalks, and we'd cut out and weigh everything. And then I could spend winters writing papers with him. In those days, when I'd go to a U.S. Forest Service meeting and meet other forestry people, they'd say, 'You're doing what? The U.S. Forest Service has an urban group?' Like me, before I met Phil, they'd never heard of urban forestry."

When Barker retired, Peper joined McPherson in his triple-wide trailer office/lab, which housed the Pacific Southwest Research Station's Center for Urban Forest Research. McPherson had been building on his Chicago climate research, working up a Sacramento Urban Forest Ecosystem study that he viewed as "the son of Chicago" while actively expanding the realm of measurable benefits delivered by city trees. "I had a PhD student in urban hydrology, Qingfu Xiao, and he developed the methodology to measure how tree crowns reduce storm-water runoff, a new component we could add to our list of benefits," he explains.

In the fall of 1999 Scott Maco joined McPherson's group at UC Davis as a doctoral student in environmental horticulture. "I had an urban forestry degree from the University of Washington," Maco says. "In the 1990s there had been an explosion of urban forestry programs, and for a while there was a huge demand for graduates.... When I came on the scene to work with Greg, we got this idea to make this [street tree] data available to a broader audience of municipal foresters. That was our approach—to make this easy for people to do, to have a minimal amount of information that could establish street tree benefits." Maco began developing software (STRATUM, an acronym for "Street Tree Resource Tool for Urban [Forest] Managers") for foresters who were eager to convince municipal authorities that urban street trees were hardworking arboreal citizens.

In the years after Chicago, David Nowak had been operating out of the U.S. Forest Service's Northeastern Research Station in Syracuse, New York. While McPherson was focused on amassing street tree data in each of his sixteen climate zone reference cities, Nowak was expanding on how best to measure and depict the big picture of an urban forest—the overall structure and species makeup, the health of all those trees, and their collective effects on air quality and greenhouse gases. "All the work since Chicago," says Nowak, who had become a project leader in 1997, "has been how to improve the efficiencies. After Chicago we were pioneering the sampling of vegetation in some East Coast cities like Baltimore and Philadelphia. In each city we did two hundred random samples, and each of these plots was a tenth-of-an-acre circle."

Nowak was slowly working toward automating his model. He and his team were analyzing and honing all aspects of the urban forest data as it flowed in, with the goal of improving its collection and interpretation, expressing this information via mathematical equations, and developing applicable software. "You can't do all this stuff by hand," he says, "so by the mid-1990s we started the UFORE [Urban Forest Effects] model and were testing the software, and it could just drive you nuts. So many little things can kill you in the code." To make his data usable by a larger audience, David Nowak began collaborating on the programming for the first iteration of UFORE with Gregory Ina, manager of Geographic Information Systems and IT for the Davey Resource Group, a division of the Davey Tree Expert Company.

In November of 2002 Mark Buscaino became the director of the USFS's Urban and Community Forestry Program. He had graduated from the

University of Maine at Orono in 1983, served three years in the Peace Corps' Forestry Program in Benin in West Africa, earned an MS from SUNY's College of Environmental Science and Forestry at Syracuse, and then held urban forestry jobs in New York City and the suburbs of Washington, D.C., before becoming chief and state forester in D.C.'s Urban Forestry Administration. Buscaino, boyishly handsome with tousled brown hair, was an easygoing, congenial leader.

Once Buscaino familiarized himself with McPherson and Nowak's nascent but impressive field of tree science, he was truly perplexed. "I saw huge potential," he says, "which is why I started looking into the tools they were developing. What I found made no sense to me. Why were all these folks—most approaching rocket scientist status and all working for the forest service and all working to improve urban forest understanding and management—why were they all working separately? Much of the base data and algorithms were not even compatible." In addition to McPherson and Maco, scientist David Bloniarz was developing a Mobile Community Tree Inventory (MCTI).

Buscaino began discussing with Nowak the possibility of merging all these individual efforts into a single user-friendly free software platform, offering $150,000 from the Forest Service for development. Some other entity would have to match that sum, and in the penurious world of urban forestry, that was not a simple matter. Moreover, a year into his job, Buscaino had concluded that his agency "didn't have the resources or the mind-set to understand that all products need to be continually refined and maintained or they will deteriorate and disappear. It would be as if Dell sold you a computer and if you called and got a message that said: 'We'll get back to you in a few weeks, but we don't offer customer support.'"

In fact, the U.S. Forest Service had earlier recognized that it needed help promoting this science. A few years after the 1993 Chicago Climate Project, the agency had approached its longtime partner in urban forestry, American Forests, and, as Buscaino explains, "the forest service said to them, 'Your strength is in publicity and policy. We have this model of ecosystem benefits and we want to get it out there.'" American Forests executive director Gangloff had embraced the opportunity, focusing on UFORE: "The science that McPherson and Nowak did was fabulous," she says. "Our program—CITYgreen—was using GIS, remote sensing, and Landsat satellite [mapping] imagery to look at the canopy from above. This was Gary Moll's version, and it really showed how much tree canopy cities had lost in recent decades. He had his own interpretation of the ecosystem benefits."

Moll, by then a vice president at the Urban Forest Center of American Forests, had begun working closely with Dave Nowak, who ran pollution models for a series of cities. When American Forests published its first CITYgreen study in spring of 1996, the results made front-page news in the *Atlanta Constitution*. The "before" (1972) and "after" (1993) satellite photos starkly illustrated the headline: HOTLANTA GETS HOTTER AS TREES FALL TO DEVELOPMENT, while the subhead lamented the consequences: "The increase in heat has boosted the cost of keeping cool and added to air pollution problems." The statistics were startling and alarming: "Atlanta's urban forest has declined by about 65 percent, says the new study by American Forests." All those lost trees also meant far more polluted storm water pouring into local waterways—enough to fill seven thousand Olympic-sized swimming pools during a major twenty-four-hour rainstorm. "The new study says that properly planted trees could reduce residential cooling costs in the city of Atlanta alone by as much as $4.6 million annually."

A year and a half later, when Moll released a more ambitious CITYgreen study of the larger metro Atlanta region, the satellite photos showed red urban heat islands, and an *Atlanta Constitution* story titled THE UNGREENING OF ATLANTA reported that an estimated "30 acres of tree cover is being lost per day in metro Atlanta and surrounding counties." Over the next several years, Moll worked with Nowak, targeting twenty major cities for CITYgreen treatment, and major headlines and attention followed whenever American Forests released its work. A PIXEL WORTH 1,000 WORDS, declared a 1999 headline on a *U.S. News & World Report* story: "Satellite images reveal startling tree loss in American cities." Wrote the reporter: "Without a scientific method to measure tree loss and its economic consequences, tree advocates were hard pressed to make a case for preservation." Moll was quoted saying the new science was all about making the "planting and maintaining of trees as integral a part of city planning as decisions about buildings, streets and sidewalks."

In the fall of 1999 American Forests released its CITYgreen study of Washington, D.C., with 1973 versus 1997 satellite images graphically revealing a 64 percent loss of canopy in the "City of Trees." The blame lay with the usual culprits: Dutch elm disease, new urban development, the loss of trees to age, and few new trees planted to replace them. On November 16,

D.C. mayor Anthony A. Williams held a press conference under a tulip poplar in Pierce Park in Adams Morgan, deploying these powerful CITYgreen images and data to argue for the importance of restoring the capital's urban forest. "If we can't take care of our trees, how are we going to take care of our schools?" asked the mayor. "How are we going to take care of our roads? Trees are a metaphor of public space. That's what we're fighting for—vital, robust, public spaces."

Mayor Williams announced hikes in the city's tree budget and the resumption of a level of tree planting—the new goal was about four thousand annually—that would at least replace the number of trees lost each year. Fiscal constraints prohibited anything more ambitious; the city would have to secure funding to save its remaining elms, plant new trees, and somehow prune twenty thousand long-untrimmed trees.

Philanthropist Betty Brown Casey was among the many *Post* readers who studied the satellite images. The third wife and widow of real estate developer Eugene B. Casey, she was the sole controlling trustee of a very rich foundation named for her late husband. Mrs. Casey approached the Garden Club of America with an initial offer of $1 million to "get started" on somehow solving the problem. After long negotiations the Garden Club agreed on December 6, 2001, to accept $50 million as an endowment to finance the newly created nonprofit Casey Trees. Its mission would be to help restore the capital's canopy and its trees to their former lush glory.

The birth of Casey Trees was the first major demonstration of the potential power of the new tree science. Not only was CITYgreen proving to be an influential public policy tool, but American Forests' Gangloff viewed it as a promising source of revenue. "As a nonprofit, we took the government research," she says, "and leveraged it with private-sector dollars. We had costs that needed to be offset, and this allowed us to be paid for our interpretation and format."

Around the year 2000 or so, though, David Nowak, whose science was key to Moll's work, recalls, "We noticed errors in a CITYgreen outcome. 'Something is off,' we told them. We wanted to review their model and their numbers. That is when they told us CITYgreen was proprietary. We saw problems, but they did not want to work with us to fix them. For us that was a turning point. If American Forests was no longer going to partner with us, then I felt we had to build a more open forum for our science."

So Nowak introduced Buscaino to Greg Ina of Davey Tree, which was happy to put up the crucial matching sum of $150,000 and eager to serve as a partner in creating a tree-benefits software suite with the U.S. Forest Service. For Ina, i-Tree fit well into his company's overall goals: "It was a great way to support our brand and mission. It was a good way to give back. And it amplified the importance of trees." By the fall of 2004 Buscaino and Ina had agreed to "a simple framework of development, dissemination, support and refinement." Buscaino was especially proud that the refining would "use the ideas and thoughts of the users themselves to make this public domain product as good as it can be." At this auspicious juncture, Bloniarz coined the catchy i-Tree acronym, short for "Inventory of Tree Resources: Economic and Environmental." Ina credits Buscaino as being "the Godfather of i-Tree, [as] without him it would not have happened."

In January of 2005 Davey Tree hired Scott Maco, who had been developing the STRATUM software, to mastermind this new tech project. "I understood the models," says Maco, who left his PhD program, "and the user base. And the beauty part was I could live anywhere I wanted—so I got to move home to Seattle."

Buscaino began wooing additional partners, and soon the Society of Municipal Arborists and the National Arbor Day Foundation joined the i-Tree Working Group. Buscaino's great regret was that American Forests did not become part of i-Tree. "This was truly a tragedy," he said. "If there ever was a partnership that could have changed the way the country thinks about trees in and around human settlements this could have been it. We had research that was on the cutting edge. On the other side, American Forests was very politically connected and had enough savvy to connect to markets. If these groups had joined forces, they could have done amazing things. Instead, it turned into rancor, resistance, a struggle for dollars and visibility, and disputes about what was and was not 'right' in terms of the science."

By early 2005 Fiona Watt had been chief of forestry and horticulture in the New York City Department of Parks & Recreation for three years, and it was time to conduct another street tree inventory. "One of the great things about working for Henry Stern and then Adrien Benepe," she says, "was that as an agency we prided ourselves on always bringing in new technology, reinventing ourselves." She was sure that if she could plug her inventory results into McPherson's STRATUM program, she could impress the politicians and public with how hard street trees "worked," as it was exactly this kind of data that might persuade New York's technocrat mayor, Michael Bloomberg, to raise her budget. In late March, when Watt learned that McPherson and Peper had not yet chosen a "reference city" for their NorthEast Zone, she urged them: "Use Queens!" and sealed her case by delivering some funding and the requisite sample of one thousand Queens street trees, located and identified by address, courtesy of her own staff.

By mid-July Paula Peper, a native of Sacramento who had never set foot in New York, was standing in the muggy heat of the city studying multiple maps. She was making the first of many visits to train the two-man Davey Tree team in the special challenges of collecting the requisite thirty pieces of data on each of the one thousand sample trees in Queens. "All I knew," says Peper, "was that you heard people in New York did not want to give you the time of day, but my personal experience was they were the most helpful people around. Maybe it was our orange vests and hard hats, but folks were so interested in what we were doing. So many people had a favorite tree in their neighborhood."

As Peper and her group toiled away, so did the 1,100 volunteers Watt and her staff had unleashed to conduct the "Trees Count!" street tree census. "Equipped with maps, clipboards, tape measures, and tree identification keys," says a Forestry Web site, "surveyors enumerated trees by species, size, location and condition in neighborhoods across New York City. . . . The [thirty-thousand-hour] effort represents the largest participatory urban forestry project in any city in the United States." Beginning in April 2005, the volunteers had spent the spring and summer walking every block in the five boroughs, counting and assembling data on what turned out to add up to 595,130 street trees, confirming an increase of 93,000 street trees (19 percent) since Watt's first trial-by-fire census.

As the Davey Tree data rolled in and Peper compiled the Queens report, Watt persuaded her to take subsamples in New York City's four other boroughs, enabling the New York City Department of Parks & Recreation to place a value on all of New York City's street trees. As Peper recalls, "Fiona called and said, 'We've just completed our new [street tree] inventory. And we'd really rather have the latest information in the report.' My heart sank, because that meant everything changes, every detail, every graph. I talked to Greg and he said, 'Go ahead and do it.' We were very willing to work with this amazing city, because it was just an outstanding opportunity to promote good management purposes."

In April of 2007 Peper and McPherson and their team produced a full-scale "Municipal Forest Resource Analysis" for all five boroughs using the Davey Tree team data combined with that of the new 2006 street tree census data. Energy savings: New York City's street trees annually saved roughly $28 million, or $47.63 per tree. Air pollution: Each street tree removed an average of 1.73 pounds of air pollutants per year (a benefit of $9.02 per tree), for a total of more than $5 million. The report also calculated that street trees reduced storm-water runoff by nearly 900 million gallons each year, saving the city $35.6 million it would have had to spend to improve its storm-water systems. The average street tree intercepted 1,432 gallons, a service worth $61, a figure large enough to impress cost-conscious city managers.

McPherson and his colleagues were also by now able to tally various benefits associated with aesthetics, increased property values and economic activity, reduced human stress, and improved public health, which were estimated at $52.5 million, or $90 a tree. These findings drew on straight-up economic studies of real estate prices as well as social-science research, which showed, for example, that hospital patients who could see a tree out the window of their room were discharged a day earlier than those without such a view. Other studies demonstrated that shopping destinations with trees had more customers than those that didn't, and leafy public-housing projects experienced less violence than barren ones.

The bottom line was that New York City's street trees delivered annually $122 million in benefits, or about $209 a tree. The city's parks and forestry officials received $8 million a year to plant and tend street trees and spent another $6.3 million to pay personnel. The net benefit was therefore an impressive $100 million: For every $1 spent on a New York City street tree, it was generating $5.60 in benefits, quite a handsome return. However, while foresters had planted 168 different tree species in recent years, five were dominant: London plane (*Plantanus acerifolia*, 15.3%), Norway maple (*Acer platanoides*, 12.7%), Callery pear (*Pyrus calleryana*, 10.9%), honey locust (*Gleditsia triacanthos*, 8.9%), and pin oak (*Quercus palustris*, 7.5%). Almost a quarter of the street trees were some variety of maple.

USFS scientists Greg McPherson (*left*) and David Nowak (*right*). *(Courtesy of USFS.)*

Had Joe Gittleman and APHIS not discovered that they could protect tens of thousands of trees—above all, maples—against the Asian long-horned beetle with imidacloprid injections, most of those would have long since been cut down. Moreover, the most hardworking trees—the large-stature London planes and Norway maples planted in the era of city master planner Robert Moses—were "nearing the end of their natural lifespan." While Watt had managed to ramp up new street tree plantings to about 8,000 a year, the large cohort of aging trees was changing the equation. Peper wrote, "Annual tree removals average about 9,300. Nearly 1,300 fewer trees are being planted than removed." Peper's report also included another seldom-expressed reality: "Given current mortality rates, only 41% of the 8,000 newly-planted trees will live to 40 years, leaving only 3,280 of the original 8,000."

The new urban forestry science finally had the data to drive public policy. In 2008 Mayor Michael Bloomberg quadrupled the city's forestry budget from $8 million to $31 million and joined mayors in Denver, Salt Lake City, Sacramento, and Los Angeles in declaring million-tree goals. MillionTreesNYC was launched in 2007 as a partnership between the city's

Forestry Division and entertainer Bette Midler's greening nonprofit, the New York Restoration Project. The MillionTreesNYC planting goal was part of a larger sustainability plan, PlaNYC, designed to address climate change.

While many U.S. cities and tree groups had launched urban forestry programs by 2007, no other American metropolis matched the scale of New York City's efforts, thanks to high-powered support (Bette Midler's fund-raising power, plus $5 million each from the Bloomberg and Rockefeller foundations); the innovative deployment of technology to map, plan, plant, and monitor trees; and sophisticated research on many aspects of urban trees. New York, unlike many cities, had a full, up-to-date tree inventory in a database that drove and informed almost all planning as Watt and her staff aimed to create a "maximally functional urban forest."

In 2007 the tree canopy covered 24 percent of the city; the goal was a steady increase in coming decades to 42 percent. The plan was this: City forestry workers would in the next decade plant 220,000 new street trees and 380,000 trees in city parks and green spaces. As for the remaining 400,000 trees: "The city is hoping nonprofit groups [above all, Midler's group], community organizations, businesses, developers and ordinary New Yorkers will plant another 400,000 trees, in backyards and other open spaces at a cost of about $200 million." On October 9, 2007, Mayor Bloomberg, Midler, and a glamorous entourage gathered on a shabby block of Teller Avenue in the Bronx and launched MillionTreesNYC with Big Apple brio: "Cast members from 'Wicked' [no costumes, just black T-shirts] read tree poetry written by third-graders," reported the *New York Times*, and "Big Bird led a procession down the block to deliver the tree to its new home." The mayor, wearing a navy suit and red tie, introduced "The Divine Miss Elm," who was dressed for planting in jeans and a leather jacket. His Honor flung the first shovelful of dirt at the foot of a Carolina silver bell. Midler wielded her shovel and conjured up a "cool, green and refreshing" New York. As for raising the requisite millions, she flashed that signature thousand-watt grin and, with twinkling eyes, suggested, "If just one hedge-fund guy contributed his bonus, we'd meet our goal."

To plant 100,000 trees a year—a quantum leap—Fiona Watt and her veteran arborists had to swiftly scale up. "It was a huge responsibility to execute and do this right," she said. "The scale became so enormous. We had to double the staff and have better procurement." Matthew Stephens, director of street tree planting and thus the arborist in charge of guaranteeing the availability of so vast a quantity of healthy young trees year after year, explains how things had worked in the old days: "We as foresters would spend hundreds of hours visiting streets, figuring out the best trees for particular spots. We ordered our list of new trees through a middleman. When you went to the nursery with him, it was 'Oh, so sorry, this tree is not available as I thought it would be. But here's this linden instead and it'll be fine.' Or you'd see a wonderful rare tree you wanted, but it wouldn't be big enough or shaped right or would have a scar on the trunk. Many of the trees we might desire the nursery didn't grow or have in enough quantities. To get the quality of trees we wanted and the variety of species, we cut out the middleman." New York City Department of Parks and Recreation now contracted directly with three respected wholesale tree nurseries (Moon in Chesapeake City, Maryland; Schichtel's of Springville, New York; and Whitman of Aquebogue on Long Island) that were thrilled to have years of huge orders in hand.

When Stephens, a lanky, soft-spoken fellow with gelled brown hair and a fashionable four-day beard, roamed the loamy "aisles" of Moon's eight hundred acres one January, choosing trees for spring street tree plantings, he reveled in knowing that every tree he wanted was there and of the highest quality. Stephens explained that "99 percent of these trees come from Oregon. They arrive at Moon's as bare roots in liners from Schmidt's." After two or three years at Moon's, once the trunks of these country-bred trees had achieved a healthy 2½ to 3 inches caliper, they were sturdy enough for rough-and-tumble city life. On that winter day Stephens and three other forestry staff members were "tagging" more than four thousand young trees of seventy-four different cultivars, from hedge maple (*Acer campestre*) to chestnut oaks (*Quercus prinus*) to zelkova (*Zelkova serrata* "Mushashino"). Where once New York arborists had about 15 choices for street trees, they would now have 120.

On April 1, 2008, Mayor Bloomberg stood in Seward Park in lower Manhattan to officially launch MillionTreesNYC (now undergirded with an advisory committee of seventy city agencies, corporations, and nonprofits) with a whirlwind of tree-planting events, pruning workshops, ranger-guided tree-identification park walks, and all-around publicity. The mayor and David Rockefeller touched down briefly at the Housing Authority's East 112th Street Thomas Jefferson Houses to bring fleeting fame to a young redbud and announce that each was donating $5 million—a sum that could fund the planting of 18,300 trees. Longtime forestry veteran Jennifer

Greenfeld parlayed that occasion to reassure David Dunlap of the *New York Times* that the redbud was just a symbolic beginning—another 10,700 trees would soon shade other Housing Authority grounds.

On April 12 MillionTreesNYC cosponsored a huge early Earth Day celebration with New York Cares, which signed up 1,050 volunteers to plant "the most trees ever (20,000) in New York City in a single day" during its seventeenth annual "Hands on New York Day." In all five boroughs the young and old spread out as assigned to plant trees—not always the large-caliper ones, but also many smaller, easier-to-handle specimens—in and around local parks, gardens, playgrounds, community centers, homeless shelters, and schools. A week later two hundred employees of JetBlue Airways' and Bette Midler's New York Restoration Project descended on East Harlem to plant 270 street trees on Pleasant Avenue for "One Thing That's Green Day." The following week Midler herself and a gaggle of celebrities (always a hallmark of NYRP-led tree events) from *Saturday Night Live* and *Lipstick Jungle* planted another 155 trees with two hundred NBC Universal employees at the Thomas Jefferson Houses. On Arbor Day, April 25, major MillionTreesNYC sponsor BNP Paribas fielded two hundred of its staff to volunteer with two hundred students in a tree planting. In its first full fiscal year, MillionTreesNYC could justifiably boast of having planted a remarkable 110,754 trees, and having done so in the midst of a collapsing economy.

Equally important was the greening of city laws. On May 1 the New York City Council passed zoning changes that would oblige real estate developers to plant more trees in and around new buildings, dictating larger tree pits—ideally long tree lawns fitted with "bioswales" to slow and hold rainwater to nourish the trees. Developers could no longer just plant but also had to care for their trees—watering and maintaining them for two years, "the most critical period for a tree's survival in the urban environment." The new zoning required greener parking lots, mandating one tree in a planting island for every eight parking spots. "These zoning changes," announced NYRP's executive director, Drew Becher (a former aide to Chicago's Mayor Daley), "will account for approximately 10,000 new trees annually throughout New York City." Such was the power of ordinance to remake a city at no cost to taxpayers.

Across the country things were not going quite so well on the arboreal front. On July 2, 2005, the new, tree-hugging mayor of Los Angeles,

Antonio Villaraigosa, had promised to make his the greenest city in America with a Million Trees LA campaign. An older and wiser Andy Lipkis, still the leader of TreePeople, was skeptical. "Campaigns like this are enough to give you early ulcers," he had warned in his 1990 how-to tree-planting book. Back in 2000 Lipkis had told a reporter that his original 1984 million-trees campaign "didn't work, because there was no follow-up care, and young trees need a lot of attention in order to survive. There was this incredible amount of energy and money and effort spent on getting trees planted, and then it would all go to hell."

In many ways by 2005 Andy Lipkis had embarked on a far more ambitious and visionary green journey. The 1992 Rodney King riots and escalating gang violence in Los Angeles had forced him to reconsider his belief that tree planting alone could heal a community. "A sociologist I know said what was really needed was jobs," he explains. "And I had a thought cloud about how you could both use nature to heal our city and create real green jobs." A year after the riots the Army Corps of Engineers had proposed to deal with the city's escalating storm-water problems by again raising the concrete walls of twelve miles of the fifty-one-mile-long Los Angeles River (and the twenty-seven bridges that spanned its banks) to better funnel now-polluted rainwater off city streets and into the Pacific Ocean. At the same time, the city was importing 90 percent of its drinking water.

"A half-billion dollars to raise the [Los Angeles River's] walls to dump the water we so badly need—that seemed the height of insanity to me," said Lipkis. Why not capture all that rainwater where it fell via a retrofitted system of green infrastructure that would also create parks and trails, new greenery, and wetlands within the city? As he told Sara Catania of *LA Weekly*, this new kind of natural urban infrastructure could address multiple environmental problems while adding beauty and recreation. "I began to believe we could retrofit L.A. to be a working urban forest," said Lipkis. "Everything is connected, and trees are the way in." He dubbed his new initiative T.R.E.E.S., for Transagency Resources for Environmental and Economic Sustainability.

Lipkis knew that shifting from a gray to a green urban paradigm would require dollars-and-cents persuasion. In May 1997, partnering with the U.S. Forest Service, he used all his charm and guile to gather at the unfinished Getty Museum seventy-five top engineers, landscape and building architects, environmentalists, and agency and community leaders for an intensive four-day "Second Nature" workshop. The goal was to agree on how to environmentally retrofit five typical LA properties: a single-family home, an apartment building, a mini mall, a school, and an industrial site, with an eye to dealing with flooding, potable water, water and air pollution, energy use, and green waste. Lipkis emerged from the workshop with what the industry refers to as best management practices for each retrofit design and, equally important, cost-benefit analyses for each.

By August of 1998 Lipkis had his first pilot project ready to go: Mrs. Rozella Hall of West 50th Street in the Crenshaw neighborhood in South Central had agreed to a Second Nature retrofit of her typical 1920s single-family bungalow. A construction crew had arrived and completely excavated her front and back yards, lowering the lawns (which were replaced with drought-tolerant St. Augustine turf) by six inches to better retain and slowly filter rainwater. The crew also partially sank into the ground two 1,800-gallon cisterns to capture all the rainwater that was now being channeled through the new roof gutters and downspouts. When needed, small pumps would recycle the cistern water to irrigate the plantings, including several new trees.

Lipkis couldn't wait to see how it all performed, but winter rains were months away. "So, I thought," he recalls, "hey, this is Hollywood. We can make a storm." On August 14 he invited the press and skeptical water bureaucrats to witness his first prototype's trial by fifteen tons of water. Lipkis handed out umbrellas to the assembled, two men ascended to the roof with fire hoses connected to a rented water truck, and the deluge began: Four thousand gallons of "rainwater" gushed from the hoses in a ten-minute mock thunderstorm. "It was quite a media event," recalls Lipkis. "A helicopter full of TV cameras documented that none of that water ran off into the street. It was a life-changing experience for a number of city and county officials, including Carl Blum, deputy director for L.A. County Public Works."

The next morning Blum, who had spearheaded the concreting of the Los Angeles River, telephoned Lipkis, who recalled: "This was such a profound moment for me. Here is the authority, the icon on the other side.... Blum said, 'I'm calling to say I'm sorry. I did not understand before. We think you've cracked it. We need to take this single-house model and see if it will work on a larger scale as fast as we can.'" A planned $42 million installation of conventional storm drains in the flood-prone Sun Valley community in northwest Los Angeles was put on hold, Blum reconceived the project as a Second

Nature retrofit, and an initial demonstration project on Elmer Avenue gained approval in June of 2004. "Essentially what happened," explained Lipkis, "is that we are now putting in an urban forest instead of a storm drain."

Not long after the bungalow "rainstorm," the Open Charter Elementary School offered its campus as Second Nature's pilot school. Again the crews rolled in and retrofitted the heavily paved landscape to become a "forest" of 150 trees, with gardens and grassy play areas and a 110,000-gallon underground cistern to capture rainwater from roofs and hardscapes. "This project transformed our school into a beautiful oasis," said principal Robert Burke in 2004. "And the kids love tending the trees and plants and watching things grow."

TreePeople, meanwhile, continued to host each year thousands of schoolchildren at its Coldwater Canyon Park headquarters (now with its own Second Nature community forestry building) to train citizen foresters and to plant trees. When the new mayor had announced Million Trees LA in mid-2005, Lipkis and his staff had been skeptical, knowing how gargantuan a task this would be. Despite this, TreePeople had also signed on as a partner to plant in the parks, only to find there was no detailed plan or significant committed money.

In 2006, after a year of discussion and general disarray, the mayor's people contacted Greg McPherson and his USFS team, who had worked with the city's schools earlier to show how trees could cool and shade school property. The mayor's office realized it had to determine such basic facts as the size and nature of the existing canopy, whether there was even room for another million trees in the five-hundred-square-mile city, and, if so, where were the best places to plant them. McPherson's study established that the Los Angeles urban forest had about eleven million trees, with tree canopy covering 21 percent of the city, and identified 1.3 million spots to "realistically" plant, many of which were in yards of private homes.

City officials had also directed McPherson to create a map charting the canopy cover in each of Los Angeles's fifteen City Council districts. The result was a revelation: Trees aligned with wealth. Rich neighborhoods like Bel Air had a luxuriant 53 percent canopy cover while gritty South Central had a meager 7 percent. "When we went around with this map," said one official, "people who didn't care about trees started to care. Council members in east and south L.A. wanted to know why they didn't have the same level of trees as wealthier neighborhoods." Unwittingly McPherson and his team had just developed another influential way to deploy tree-canopy data—as a marker of social inequity. If trees were so important to human well-being, shouldn't every urban community have a healthy canopy?

In the wake of the report, tree-planting efforts were stepped up in poor residential neighborhoods, where residents preferred lemon, lime, and orange trees, whose fruit was expensive to buy at grocery stores. The emphasis of Million Trees LA had shifted. "We all knew there were places with fewer trees, but with the map you can really *see* it," said executive director Lisa Sarno. "It's become a matter of social and environmental justice." Politicians were learning that planting a million trees was not just about numbers, as Lipkis had warned, but also about where those trees were most needed and how they would survive over time.

Farther north in Sacramento, California, the Sacramento Tree Foundation had continued its partnership with the local utility, planting as many as 20,000 young saplings a year in private yards, proud partner in "the largest shade tree program in the nation." On National Arbor Day, April 28, 2001, Ray Tretheway and his nonprofit reached their 1989 goal of planting a million trees over a decade, the final tree a native valley oak across from City Hall. And yet . . . "Despite the noble beginning and recent efforts," reported the Sacramento Tree Foundation in its 2000 *A State of Trees Report, A Call to Action for the Sacramento Region*, "the current management of Sacramento's urban forest remains a piecemeal effort." The report warned of "significant threats" to the overall park and street tree canopy of the City of Trees: "In the downtown area, thousands of aged plane trees, mature oaks and elderly elms are losing their battle against time and neglect. Seedlings planted a century ago now form a geriatric forest whose venerable members are dying. Each death leaves a gap." Underfunded public agencies were failing to provide care "worthy of Sacramento's street and park trees."

In 1996 McPherson had delivered an Urban Forest Ecosystem Study of the whole of Sacramento County, which allowed Tretheway to note that this larger urban forest delivered $50 million in environmental benefits. Yet "the vast majority of trees in the Sacramento region remain at the mercy of a management philosophy that can at best be charitably described as laissez faire, at worst as haphazard and neglectful." Tretheway challenged the entire Sacramento region "to develop a shared vision and regional tree management plan," with a thirty-year goal of planting three to five million new trees, while establishing "consistent tree planting and tree care policies." And so was born the Greenprint Initiative, which by 2005 would bring together elected officials from twenty municipalities to commit to this vision. The state and the U.S. EPA endorsed the goal of these future millions of trees cleaning the air with a $730,000 grant. "This was another huge moment in the history of the Tree Foundation," said Tretheway.

In Washington, Casey Trees had launched in the summer of 2002 by spending $1 million to orchestrate the city's first street tree inventory in decades. "We needed to hit the ground running," says board president Barbara Shea, "and make ourselves known." With McPherson's STRATUM not yet available for eastern cities, Shea remembers being thrilled to discover that the National Capital Planning Commission had just completed a three-year aerial GIS mapping project, perfect for plotting out inventory work grids, and it was happy to share its findings.

Still, she notes, "We had to develop the [inventory] software ourselves. Handheld computers similar to Palm Pilots were the collection devices. The city forester and his staff helped with the thirty data fields to be collected on size, species, condition, and site, as they would be using the results." By June, Casey had recruited and trained thirty-eight paid college team leaders, as well as almost five hundred volunteers. "We had thirty teams on the street six days a week," says Shea. "As you can imagine, the logistics were a challenge (even a nightmare). Each evening the interns returned to the office to download their data and recharge the units. Fortunately, one of our employees had previously managed a Congressional campaign."

On August 14 the Casey project announced the results of the inventory at a party at the National Building Museum. Its board was proud of having succeeded in this huge endeavor, and the data confirmed American Forests' grim 1999 report: One-third of Washington's 136,000 street trees were dead or dying. There were 32,000 empty pits. D.C. had lost 30,000 of its 38,000 American elms. Two-thirds of its much-diminished urban forest consisted of maples or oaks. Shea and her board were determined to keep the momentum going, and an embryonic staff created D.C.'s version of a citizen forester program to engage and further train the many enthusiastic inventory volunteers. That fall Casey Trees employees and volunteers began marking future tree pits and planting street trees.

"Before we started Casey Trees," recalls Shea, "the [D.C. Tree Management Administration] was moribund, with three employees, a budget of $700,000, and one or two arborists. But Mayor Williams was all on board with trees." In the wake of the CITYgreen report and just as Casey Trees was fully birthed, Williams hired Mark Buscaino to head the D.C. Tree Management Administration (soon to be renamed the Urban Forestry Administration and placed under the Department of Transportation). Buscaino had begun to revive and expand forestry when, not a year into the job, he accepted an offer to run the USFS's National Urban and Community Forestry Program. One of his major legacies as D.C.'s chief forester was the Urban Forest Preservation Act of 2002, which, he says, "creates disincentives to take down healthy trees more than 18 inches in diameter on private lands."

Mrs. Casey, meanwhile, had had occasion to meet Roger Holloway of Riveredge Farms, who soon converted her to replenishing the capital's dwindling population of American elms with his "Princetons." "We have been committed to the return of the American elm since our inception," says Shea. "The city was not planting any elms when we started, so it was a natural niche for us to fill in on historic elm streets. In 2003 we planted 78 elm trees in the snow on Barracks Row, one of the mayor's new 'great streets.'"

During these formative years, Casey Trees concentrated on refining its citizen forester program, planting street trees, and gathering further data on the parlous state of D.C.'s urban forest. Working with Forest Service scientist David Nowak and the National Park Service, Casey gathered street tree data for a UFORE study while citizen foresters conducted tree inventories at some of the city's public schools in preparation for greening up those spaces, as well as an inventory of all the remaining American elms on city streets. By March 2004 Shea had signed on five new board members, including Mark Buscaino, eighteen months into his tenure as head of urban forestry at the U.S. Forest Service. In June of 2006 the Casey board hired Buscaino to be the nonprofit's executive director.

This did not mean leaving behind i-Tree, Buscaino's major endeavor at the Forest Service. Casey Trees now joined as an i-Tree partner with the Forest Service, Davey Tree, the National Arbor Day Foundation, and the Society of Municipal Arborists. Two months after Buscaino took the helm at Casey, i-Tree was officially launched in Minneapolis at the International Society of Arboriculture's annual conference, and this far easier access to tree science was explained and celebrated in industry journals like *Arbor Age*. Certainly Shea of Casey Trees was impressed: "I think it's a miracle. It cost us $1 million in 2002 to do our tree inventory. Now you could just go get i-Tree and do it for almost nothing. It's fantastic—almost a perfect partnership of government, private, and nonprofit."

Once i-Tree 1.0 was launched, Maco and his tiny support staff found no shortage of feedback via e-mail: "We learned that we had to redesign [one] part because the box is too small, or [deal with questions like] 'Where is the data field for invasive species?'" As Maco and Davey Tree upgraded the i-Tree software, rewriting code to remove the bugs, they observed with fascination their emerging user base: A third was municipal arborists, another third was either professors or their students, and the rest were divided among urban-forestry consultants, nonprofits, and citizen groups. In 2007 more than a thousand people downloaded i-Tree, one hundred of whom lived in countries outside the United States.

One spin-off project, developed in partnership with Casey Trees, was a simple, personalized Tree Benefits Calculator, which could charm homeowners by allowing them to determine the value of their own trees. My twenty-five-year-old backyard pin oak, for example, was tallied as providing the following benefits: It intercepted and absorbed 7,669 gallons of rainwater per year ($75.92), raised my property value ($75.67), saved 229 kilowatts of electricity per year ($17.36), and improved air quality and stored carbon ($17.58). Of course, my family also enjoys the pin oak's beauty, the squirrels frolicking about its branches, and the many cardinals and other birds it attracts. It buffers us from a nearby busy street, abates noise, and once held a rope swing for my daughter.

In certain forward-thinking cities—Los Angeles, Philadelphia, Seattle, and New York—officials were now viewing trees as an essential part of green infrastructure, incorporating them into plans to cool down urban heat islands, clean polluted air, and mitigate the scourge of polluted storm water while providing the bonus of making city streets more beautiful, more healthful, and friendlier to humans. More than a century ago, the first tree evangelists made similar arguments about the importance of trees. But now hard science was making those claims credible to the skeptical.

CHAPTER TWENTY

"Help Restore a Lost Piece of American History": Return of the Elm

Tom Zetterstrom of Elm Watch poses with the Baldwin Elm in the Berkshires, Massachusetts. *(Photograph by Lisa Vollmer.)*

For the multitudes yearning to restore the American elm to the cityscape, the twenty-first century promised a wondrous return to the cathedrals of shade. "For the first time in more than 40 years," *USA Today* reported in May of 2007, "the American elm tree is being sold in large numbers to homeowners and other retail customers. Home Depot, the giant retailer, is trying to sell 12,000 disease-resistant elm trees in 400 stores in the eastern USA. The American elm's return to the mass market is a watershed event in the comeback of a tree that once dominated the nation's landscape, only to be wiped out by Dutch elm disease."

Roger W. Holloway of Riveredge Farms had proven to be a master promoter, persuading Home Depot to buy 12,000 of his "Princetons" to help restore to American streets these "beloved but stigmatized trees." (To put this comeback and gamble into perspective, Home Depot would that year sell 100,000 to 150,000 red maples and pin oaks.) For a generation not so familiar with *Ulmus americana,* Home Depot tagged each young Princeton with a special label reading, "Return of the American Elm," extolling its heartwarming revival and urging: "Plant this tree and help restore a lost piece of American history."

With widespread loss of the ash tree, the elm's resurrection could prove an ironic godsend to beleaguered city arborists. "It's come full circle over the last 50 years," said Marc Teffeau of the American Landscape and Nursery Association. "Elm trees are being planted to replace ash trees that were planted to replace elm trees."

In 2005 Roger Holloway had scored an even greater coup: the planting of eighty-eight of his Princeton elms in front of the White House, creating an allée lining a new pedestrian promenade in front of 1600 Pennsylvania Avenue. "These trees," wrote the *Washington Post,* "symbolize the rebirth of America's Main Street after a decade of security-driven restrictions and barriers. But they also mark the return of a great American tree . . . the quintessential civic tree." During the previous year workers had removed concrete barricades and redesigned the street to serve as a pedestrian boulevard. The existing willow oaks—favorite tree of Thomas Jefferson—would be removed to make way for this old-new American icon.

Holloway's fourscore and eight White House Princetons were twice the size of your average new street tree, their sturdy trunks four to five inches across, their root balls so hefty it took a crane to lift each one off the flatbed truck and position it in its new tree well. Once planted, these five-year-old elms stood twenty feet tall. For the first time in decades, a new generation would in time witness a young allée of American elms (sure to thrive under the most pampering of federal care) achieve a statuesque maturity, their arching branches intermingling in sinuous beauty above all who strolled under their boughs to behold the nation's most famous address.

Holloway, this elm's "fiercest champion," was thrilled to promote its rehabilitation as part of the urban landscape and not at all surprised by its fervent reception. "There's no other tree in our country that seems to stir up so much passion," he said. So visible a public relations coup as the Princetons being planted on Pennsylvania Avenue just drove more sales for Riveredge Farms, which was, for the moment, the only nursery shipping these elms in any quantity. Holloway was busy ramping up, intending to have 34,000 trees ready to ship in 2008 and more than double that by 2009.

The Pennsylvania Avenue Princetons were not, in fact, the first of Holloway's elms to find a home in the nation's capital. Casey Trees, reassured by Townsend's 2001 paper, had begun buying Princetons from Holloway that year, and over a four-year span enthusiastic volunteers had planted 750 of them throughout the capital, which still had 8,500 mature elms that had escaped or survived Dutch elm disease. In 2007 Mark Buscaino, executive director of Casey Trees, reported, "We have new data on trees that we've planted, and we're finding that of the 30 to 40 species we've planted, elms have the highest survival rate."

What the *Washington Post* article "The American Elm Stages a Comeback" failed to mention was that the nation's capital was among those rare American cities where one could still easily luxuriate in wondrous American elm canopies—most notably on and around the National Mall. In 1902 the McMillan Commission, which designed the Mall, specified that four rows of American elms would frame its two-mile length from the Capitol down to the Lincoln Memorial. In an epic case of bad timing, the National Park Service finally began planting those six hundred American elms in 1933, just as Dutch elm disease was confirmed in the United States. Undeterred, the commission planted yet another two thousand American elms to unify the overall look of the 1,100-acre Monumental Core, beautifying the Ellipse, the Lincoln and Jefferson memorials, West Potomac Park, and a number of major streets with these stately trees. Once Dutch elm disease arrived in 1947, the National Park Service—despite acting swiftly to save diseased trees, lost American elms at a rate of only 1 percent or 2 percent—and always replaced each lost tree with another American elm. Even today one can stroll under the Mall's still-soaring American elms in almost any season or weather and agree with Jim Sherald that "it is the majestic American elm that is the soul of the landscape."

In the same month in 2005 that the eighty-eight Princetons took up permanent residence in Washington, Denny Townsend released his final paper: "Evaluation of 19 American Elm Clones for Tolerance to Dutch Elm Disease." Townsend, who was about to retire, could cap his career by stating in the conclusion to this paper: "This large-scale study enabled us to identify the best American elm cultivars and selections for tolerance to DED. This information can be used for choosing specific trees for nursery production, landscaping, and tree breeding; for possible naming and release of the best clones to the nursery industry and for ultimately increasing the genetic diversity of American elms planted in the future." For many in the nursery trade this was the reassurance they needed to again take a chance on raising and selling these new American elm cultivars.

In a field trial conducted in Glenn Dale, Maryland, Townsend had expanded his number of test cultivars, including for the first time "Jefferson" (known only as Selection N3487), which had been raised and planted on the Mall by the National Park Service since the late 1970s. While the same cultivars from his previous trial, including Princeton, prevailed as best able to survive DED, Jefferson showed "promising levels of DED tolerance," and Townsend pronounced it worthy of both further testing and cultivation.

Jim Sherald of the National Park Service, who had long had charge of more than a quarter of the city's American elms, had been impressed by the performance of the original and very handsome seventy-five-year-old Jefferson elm growing on the Mall across from the Freer Gallery, as well as the tree's numerous offspring planted on the Mall to replace DED casualties. In February of 2005 the National Park Service and the USDA officially released Jefferson to the nursery trade.

The following year Jefferson vaulted to quiet stardom when President George W. Bush and First Lady Laura Bush selected it to be one of a few dozen elite White House trees personally planted by presidents. On October 2, 2006, Sherald found himself on the White House front lawn, watching as the presidential couple ceremoniously shoveled dirt onto a Park Service–grown Jefferson that was replacing a hundred-year-old American elm uprooted the previous spring during a violent rainstorm. That elm had begun life in 1902 or 1903, when Teddy Roosevelt romped the grounds with his children, and had achieved immortality as part of the White House scene depicted on the twenty-dollar bill.

Among those who were very pleased to have Townsend's reports on DED tolerance was Tom Zetterstrom of Canaan, Connecticut, a fine-arts photographer known for his tree portraits. "I remember feeling just euphoric when I first heard about Denny Townsend's work in 2001," he says, "and 2005 was further affirmation." In 1999 Zetterstrom, son of an arborist, had founded the nonprofit Elm Watch to document and save his region's remaining magnificent old elms. In 2001 he had expanded that mission to include restoring to the landscape for posterity the trees that had so enchanted him, and so had also begun planting Princetons bought from Holloway, as well as Townsend's "Valley Forge" and "New Harmony."

Like so many elm activists, Zetterstrom had witnessed firsthand the carnage wrought by DED. "Canaan, where I was born and have always lived, had hundreds of elms," he says. "From the time I was five years old, they were taking them down." In high school and when home from Colorado College, where he earned a BA in sculpture and photography, Zetterstrom worked at his father's firm, Ollie Zetterstrom's Tree Experts. After college Zetterstrom avoided the Vietnam War draft by teaching photography in inner-city Washington, D.C., schools from 1968 to 1970. Over the next couple of decades, operating out of Canaan, Zetterstrom traveled the world as a photographer. Locally he was an environmental activist engaged in land-trust work.

Elm Watch had been born when Zetterstrom was driving back and forth along scenic Route 7 from his house to his girlfriend's place in Great Barrington, Massachusetts. "I just could not stand to see yet another beautiful old tree succumb to Dutch elm," he says. "We had Tom Prosser of Rainbow Science Tree Care come in from Minneapolis and show thirty-four local arborists how to do macroinjections of Arbotect properly. The first tree we did in 1999 was the Baldwin Elm, a pasture elm such as would not be seen again, and the second was the Great Barrington Elm, just a beautiful fountainlike specimen." The Baldwin Elm became both the symbol of Elm Watch and the subject of Zetterstrom's best-known photograph, one of a number in the collections of the Metropolitan Museum of Art in New York, the Getty in Los Angeles, and the Corcoran Gallery in Washington, D.C.

By 2005 Zetterstrom had designated one hundred or so heritage elms growing near Route 7 in the Berkshire Taconic region as part of the "Majestic Elm Trail." Through his "Adopt-an-Elm" program he had persuaded homeowners, elite prep schools, country clubs, town councils, and shopkeepers to make contributions to pay for local arborists to inject these trees against DED every third year. Zetterstrom urged their preservation as part of understanding "a world that existed 150 years ago. It is an aesthetic experience so different from what we know today." He also appealed to local pride, casting the preservation of these trees as a way to save New England from the "homogenization" engulfing the rest of America.

Zetterstrom emerged as a self-appointed ambassador of the elms, a zealous apostle who made it his mission to cultivate other elm activists, whether academics, arborists, nurserymen, or average tree huggers. At age sixty he was energetic and lean from years of tree work, his Nikon camera at the ready, slung around his neck. When talking elm structure and pruning, he liked to wield a pen laser to highlight bad branching or limbs that had to be removed. By the early 2000s he had developed an elm network and become one of the few to actively go forth and proselytize, spreading the gospel of planting the New and Improved American Elm while preserving the old "heritage" specimens.

In 2003 advocates like Zetterstrom and Holloway gained an academic ally in Thomas J. Campanella, whose *Republic of Shade: New England and the American Elm* served as a lyrical and detailed history of what Americans had largely lost. "To me," wrote Campanella, "as a child of New York City, elm trees were magical emissaries from the countryside. My parents would point them out on summer road trips. . . . Peering out of the Rambler's rear window, I learned to watch for their telltale form, a gushing fountain of limbs and leaves." He told of a time before Dutch elm disease when, "Collectively, America's elms formed the most expansive urban forest ever planted, a verdant parasol soaring above the quotidian, casting it in a dappled and flattering light."

For serious American elm activists, the biggest news of 2005 was the launching of a National Elm Trial, the logical follow-up to the work of Denny Townsend, Eugene Smalley, and George Ware. For years, says Dr. William Jacobi, a tree pathologist at Colorado State University's Department of Bioagricultural Sciences and Pest Management in Fort Collins, "a bunch of us like-minded insect and tree disease people would get together, including Gene Smalley, and say, 'Jeez, we ought to bring back the elm,' but we knew we needed more trials. I'm a cheerleader, that's what I call myself, and especially once we had EAB and were going to lose the ash, we needed the elm back. So I decided to help do these trials. And while Gene died in 2002, he really was the inspiration."

By 2005 Jacobi had signed on academic scientists and extension specialists at seventeen land-grant universities to cooperate on a decade-long "volunteer effort to evaluate and promote the use of Dutch elm disease–resistant American and hybrid elms." There was still so little real-world experience with these new cultivars, believed Jacobi, that "much of the public is hesitant to purchase and plant any elm tree." Each Elm Trial "cooperator"—in states ranging from Vermont to Florida to North Dakota to Utah to California—would plant seventeen elm cultivars, "each cultivar represented by one tree in each of five blocks." The goal was to document how different trees fared in the harsh winters of northern states or the near-drought conditions of the West and each year assess specific cultivars for "height, diameter, crown characteristics, and fall color." Keith Warren at J. Frank Schmidt & Son Company donated more than a thousand young bare-root trees of all the elm cultivars the company was growing—whether pure American elms or Ware's Asian hybrid elms, while Princeton Nursery donated Princetons, and Lee Nursery in North Dakota provided "Prairie Expeditions." In 2006 Jefferson was added, but those trees turned out to be Princetons. Professor Jacobi also contacted the Elm Research Institute so he could include American Liberty cultivars. He says, "I could not get them to give me any trees."

In the spring of 2016 Jason J. Griffin, a PhD at the John C. Pair Horticultural Center at Kansas State University, and Jacobi completed the study, publishing the results in a scientific paper titled "Ten-Year Performance of Elms in the National Elm Trial." The compiled data from the seventeen sites on survival, growth rate, branching patterns, form, and insect and disease damage showed that "the preferred cultivars of American elm were 'New Harmony' and 'Princeton,' and the preferred cultivars of Asian elm were the Morton Arboretum introductions and 'New Horizon.' These findings will help green-industry professionals determine what elm cultivars will perform the best in different regions."

While Tom Zetterstrom with his Elm Watch was the most high profile of the self-appointed grassroots-activist American elm lovers, landscaper/groundskeeper Bruce Carley of Acton, Massachusetts, also played an important role. In 1990, while pursuing a certificate in native plant studies, Carley took the required native tree course at the Arnold Arboretum. Alerted by a handout about American Liberty elms, in 1992 he ordered, received, and planted his first few Liberty whips from the Elm Research Institute. In subsequent years Carley obtained about forty more, which he planted largely on local "conservation lands, where they always will be safe from indifferent landowners."

In December of 1997 Carley created what would become the Internet's most informative Web site dedicated to all aspects of "Saving the American Elm," www.elmpost.org, hoping his own project would inspire others. Gracefully written and thorough, Carley's postings on the site waxed poetic about the tree's early history and downfall, and then guided his readers through the pros and cons of the new DED-tolerant elms, which he was now raising and planting—mainly Valley Forge, Princeton, and a few New Harmony, using photos to illustrate his points.

In 2000, to advance the cause, he began selling American elms himself and offered neophytes a variety of cultivation tips, including his home remedies for fending off browsing deer, whose nibbling of tender saplings could kill them. Advised Carley: "A beaten egg mixed with a half gallon of water . . . should be reapplied at regular two-week intervals or immediately after rain, and used sparingly when applied to newly emerging shoots. As an alternative to such regular attention, a piece of soap tied in a nylon rag and hung from the sapling at browse level appears from my experience to be highly effective." Like a patient uncle, he guided the ranks of prospective and new DIY tree growers through the thicket of horticultural pitfalls. "Elm saplings should be pruned each year while they are young, with the goal of developing a sturdy trunk before the main limbs become prominent. . . . Removing these low branches up to about half the height of the tree, not to exceed the lowest main scaffold crotch, is a useful guideline."

On June 11, 2006, Carley and his brother went to photograph his favorite American elm, the lone Valley Forge elm at Meeting House Hill in Acton Center. "On that particular day, the young elm which I had planted on this site [in 2000] seemed healthy and vigorous, showing no hint of any problems," wrote Carley, "and I remember thinking that it was perhaps the best-looking Valley Forge elm with which I ever had worked. . . . [F]or an American elm of its age, it was a first-rate specimen . . . and I had no reason to imagine that something dramatic was about to begin happening to it."

A week later, though, Carley returned and noticed leaves and branches starting to wilt. He began monitoring the tree. "I thought it noteworthy that the affected regions of the tree were almost entirely lateral branches at various heights along the main trunk. Indeed, most of the tree looked healthy enough, but the conspicuous wilting in certain sections of the tree increasingly bothered me as the days went by." Determined to confirm his suspicion that his Valley Forge had been stricken with Dutch elm disease, he submitted samples to the University of New Hampshire's Plant Diagnostic Laboratory. On July 7 the lab confirmed an aggressive form of *Ophiostoma novo-ulmi*. "Though I told few people at the time," wrote Carley, "I was acutely aware that the outcome of this one tree's battle with DED could have profound implications for my entire project with elm trees." He kept a close eye on the tree, and though some wilting continued, that fall it was still pretty healthy-looking as it went dormant.

In early June of 2007, Carley began filing regular dispatches from his own DED battle zone: "The Valley Forge elm at Meeting House Hill was engaged in an all-out war against the fungus, and its victory, should it happen, would not be an easy one." On June 6 he "noticed for the first time that the bark was splitting in a six-foot, vertical crack along the height of the main trunk, exposing the inner wood along most of this length. . . . I also noticed that the inner bark of the tree already appeared to be closing in on the wound vigorously from each side, as though trying to heal the split. On June 18, the entire crown was showing signs of wilt, although the leaves were mostly green and intact toward the tips of the branches, whereas the wilted leaves were much more prevalent toward their bases."

By midsummer the tree appeared to have stabilized for the season; once again, although it no longer looked attractive, it showed no further wilting or other symptoms during the remainder of the year. By October the tree had survived a severe case of Dutch elm disease for two full growing seasons. In addition, the long split in the trunk was almost completely healed, with most of the previously exposed wood now covered with new bark. "The overall appearance of the tree at this point left much to be desired," Carley acknowledged, "but once again I considered its behavior to be remarkable and anticipated that it would survive another winter and would begin to recover from its weakened condition in the spring."

For the many elm lovers following this drama, Carley reported the following year that the Valley Forge had "filled out well and developed no trace of wilt during the crucial months of May, June, and July. The importance of this fact cannot be overstated; the tree's behavior was in complete contrast to that of a typical American elm infected with *Ophiostoma novo-ulmi*, and there was no other way of looking at it without a considerable stretch of imagination. Through whatever mechanism, the tree had recovered fully from a naturally induced, severe infection with aggressive Dutch elm disease and had done so entirely on its own." In real-world conditions, this cultivar of Denny Townsend's had not only survived but defied death and defeated Dutch elm disease.

Carley did not have as favorable a report when it came to one of his American Liberty elms after it too was stricken with DED. "Half the tree was leafless and trying to resprout from its lower skeletal remains; the other half was holding on only because the infection had more recently reached the tree's lowest limb. I have not cut down any of the Liberty elms which I had planted in the past, nor do I suggest cutting down any healthy trees, but I have come to regret placing as much faith in the Liberty elms as I once did." While he was grateful to John Hansel's Elm Research Institute for inspiring what had become a passion, he and others who had lost Liberty elms to DED felt the institute should acknowledge that some of the six different clones that confusingly made up Liberty were better at surviving DED than others.

Despite the experience of Bruce Carley and other elm activists, however, plenty of Liberty elms have achieved a comely adulthood, and they do have ardent fans. One such is Joseph R. Pickens, who, when chief scientist at the Martin Marietta Corporation in 1994, persuaded its then-CEO, active with the Boy Scouts, to buy five hundred or so trees from the Elm Research Institute. As the tiny Liberty whips arrived at the corporate campus in Catonsville, Maryland, says Pickens, "I organized planting parties and established hundreds in a little nursery. As they grew bigger, I gave them to my neighbors."

Pickens is a New Yorker who grew up in Brooklyn. "I used to walk with my father on Ocean Parkway, which was lined with rows of American elms that had this wonderful cathedral effect." Hoping to recapture some of that arboreal enchantment, between 1995 and 1998 Pickens planted sixty of his Liberty elms along the small country road that passed his own driveway and along the front of his and his neighbors' properties. By 2015 the branches of his fifty-foot-tall allée of Liberty elms were just beginning to mingle above the country road. John Hansel's dream was visibly manifest in this Maryland suburb.

In 2004 Pickens, like Carley, saw one of his Liberty elms become afflicted with Dutch elm disease. "I immediately went to it," Pickens recalls. "I got a big ladder and cut off the branches that looked infected. I cut back to where I saw no black rings and threw the diseased stuff away. It bounced back beautifully, and it's been doing magnificently ever since. I call it 'The Survivor.'" He had two other Liberty elms show flagging, and after conferring with the Elm Research Institute, he injected them successfully with fungicide the institute dispatched.

For his part, John Hansel had now been engaged for more than forty years in his crusade to restore to the nation's towns and cities the American elm. By 2007 the institute's Web site reported that "approximately 350,000 disease-resistant Liberty Elms have been planted across the U.S.

with less than 1% loss to DED." And should one of the institute's trees succumb to Dutch elm disease, its lifetime warranty guaranteed that it would be replaced at no charge.

While Hansel had much to be proud of, he was still brooding about the perfidy of the USDA and the way Townsend had panned his Liberty clones. "One of the regrets I have," Hansel says, "is that we worked with a midwestern professor [Smalley], because he didn't have the same view of the elm as I did, that Northeast tradition of that gorgeous, tall structure." The Elm Research Institute had always prided itself on tracking its trees once they were living in the real world beyond field trials. Who, wondered Hansel, was doing the same for all the other new American elms that Townsend had pronounced superior? The institute's Web site noted: "At ERI we register every tree and attach a permanent brass tag for identification. In this way we can certify losses." However well intentioned that plan, it was impractical, for most of the institute's elms were sent out to new homes as tiny whips—easily done in by neglect, deer, and lawn mowers long before they achieved any appreciable size. It also depended on fallible, busy humans to take the time to attach the brass tags, let alone call in to report the trees' failures.

Hansel, had, however, posed a very good question: Who *was* tracking the real-world fates of other new American elms—the ones Townsend had pronounced superior to the Liberty? How well were *they* faring out in the streets and parks and yards of America? Certainly Bruce Carley and Tom Zetterstrom in New England had been scrupulously documenting their experiences over the years. In around 2008 Zetterstrom had organized an online Listserv network to connect elm enthusiasts, scientists, nurserymen, and arborists so they could share their knowledge and expertise as the elm trials got launched and encourage and warn one another about any setbacks.

If it was any consolation to Hansel, Carley and Zetterstrom were coming to find serious flaws in elms Townsend had rated above Liberty. By late 2015 Carley had planted a total of forty of the new cultivars around Acton, Massachusetts. Of the fourteen Liberty elms he had planted, he had lost five to DED. He found the survivors to be "shapely and attractive, but of course I know they are not immune to DED." His two Jeffersons were under ten years old and doing well.

While Carley admired Valley Forge's fighting spirit and had planted twenty-six in Acton and then raised and sold many more, twenty years into his elm-planting campaign, he wrote, "I and many others have been more than slightly distressed by the bad structural habits of the Valley Forge elm. While I have lost only a few VF elms to DED, I have watched many of these trees split down the middle of a central crotch, lose limbs, or otherwise become badly disfigured. This is the major reason I feel we need additional and better cultivars to work with, using VF and other notable elms as parents."

Nor was the Princeton from Riveredge Farm proving to be much better. The U.S. Park Service in Washington, D.C., told Zetterstrom that after 2009 it would no longer plant the cultivar, due to "the problems with the crown as we get it from the Nursery," which it found "hard to correct—and PA [Pennsylvania Avenue] is a perfect example. We pruned these pretty hard the second year, were criticized for removing too much and leaving large wounds, and they are still not where they need to be as far as the spacing of the scaffold branches are concerned."

Zetterstrom was so distressed by the state of the 2005 White House allée after seven years that he condemned the whole enterprise as a botch, "88 ill-trained elms" made worse by "inadequate understanding of proper pruning, and layers of bureaucracy." Here, he declared, was "the most highly visible and significant elm restoration project in America, and one would argue, the world," and its elms were not looking like classic American elms at all. "I have suggested a number of solutions," wrote Zetterstrom, "none of which alone is ideal but is guided with the hundred-year perspective in mind." He regretted that tree lovers possessed "more enthusiasm than understanding about the limitation and habits of cultivars like Princeton." However, the Jacobi Elm Trial had shown that Princeton elms raised by Princeton Nursery grew very nicely, driving home the importance of properly raising young trees.

In 2010 a little-known scientist began promoting an exciting new direction for restoring the American elm. With Denny Townsend called back east and then retired, Jim Slavicek, a USFS geneticist, had found himself taking over the American elm work in Delaware, Ohio. In the summer of 2008 Slavicek walked into a young woodlands in the Mohican-Memorial State Forest near Cleveland and noticed that tree seedlings were shooting up among the thirty American elms he had planted five years earlier. "It was absolutely a eureka moment," says Slavicek. Eighty years after Dutch elm disease first began wiping out America's most iconic tree, Slavicek was thrilled to witness the first growth of these particular young American elms emerging from seed in the wild.

For these were not just any elm seeds or trees. The Mohican-Memorial State Forest seedlings represented the select progeny of the decades-long quest to bring back the American elm. In this forest clearing, once a helicopter landing spot, Slavicek had planted six each of five DED-tolerant elms. Those hardy five-foot-tall cloned elms—including New Harmony and Valley Forge—had grown, bloomed, released their pollen to the wind, and then fertilized the seeds of Slavicek's other elms. For good measure Slavicek had planted them near a huge old American elm, a survivor of the DED waves, hoping its DNA might further strengthen the DNA of his USDA elms. Later tests would tell.

The little elm trees growing from seed in that Ohio woodlands—each genetically unique in contrast to its five parental cloned elms—were guaranteeing the future of the American elm, a feat once thought impossible. Says Slavicek, "American elms are going to be back in the landscape where they can naturally regenerate, spread, coevolve with the Dutch elm disease fungal pathogen, and survive." By the fall of 2010 Slavicek and groups like Tom Zetterstrom's Elm Watch and the Nature Conservancy had planted hundreds more DED-tolerant elms in natural sites and watched with excitement as yet hundreds more elm seedlings shot up nearby. "These restoration sites can serve in the future," Slavicek explains, "as the source of a new American elm genotype and a resource for nurseries. . . . You need to create genetically individual trees grown from elm seed so the trees can evolve along with the DED fungus."

CHAPTER TWENTY-ONE

"Oh, My God! They're Really Here": Further Conquests of the Asian Beetles

Before and after removing trees infested with Asian long-horned beetles on Granville Avenue in Worcester, Massachusetts, February 2009. *(Courtesy of Massachusetts staff, USDA APHIS.)*

The spring of 2008, year thirteen of the beetle wars, brought much-needed excellent news from Chicago. A decade after Barry Albach had first spotted the strange black-and-white beetle waving its long antennae on his truck's rearview mirror, Mayor Daley stood before the press on April 17 to declare victory over the insect invaders. His Chicago troops had—as he had vowed they would—successfully expelled the beetle from the city, scoring a rare triumph over an invasive species. No live Asian long-horned beetle had been sighted in Chicago in three years. All told, 1,553 trees had been lost, almost all of them in Ravenswood, but the beetle had been defeated. The city's jubilation would, however, be short-lived.

Two months later, on June 20, 2008, the *Chicago Tribune* reported: "The dreaded emerald ash borer has made its first appearance within the city limits of Chicago." Mayor Daley left the announcement of this "long-expected event" to Streets and Sanitation commissioner Mike Picardi. Two years earlier Illinois had become the sixth state invaded by the borer. Edith Makra, Mayor Daley's former "tree lady" and then his liaison to the Chicago Climate Project, had joined the Morton Arboretum in 2002 as its community tree advocate. One of her duties was to convene a coalition of federal, state, and local officials to formulate a state EAB readiness plan. "I pulled the Climate Plan report off my shelves," she says, "and opened to the pages that had data showing that 20 percent of Chicago's street trees were ash. We were able to use that in all our communications: One in five street trees threatened—that was very motivating." A news story noted that the Morton Arboretum had one of the largest ash collections in the nation: 432 trees representing 56 kinds of ash.

In the two years it had taken EAB to advance into Chicago from its first Illinois stronghold in Wilmette, it had also invaded Pennsylvania, West Virginia, and Missouri. Nonetheless, Commissioner Picardi gamely declared, "We want to contain it [EAB], to make sure that it doesn't spread to other communities." Makra, perhaps more realistically, had focused on how to repurpose some portion of the millions of doomed ash trees and had secured a $100,000 grant to work with students at the Illinois Institute of Technology on ways to "harvest and market urban timber." By late August of 2008 the arboretum was also hosting "Rising from Ashes," a traveling exhibition showcasing designer ash-wood furniture—twenty-nine tables, chairs, and dressers priced at one thousand to seven thousand dollars. Presented as "a way of softening the emotional blow of losing ash trees," Makra viewed the furniture as a way to keep the tree alive "as a piece of furniture you can use. It tells a great story."

Chicago owners of large ash trees soon found arborists offering to save their imperiled trees with insecticides. Many were using what had been working against ALB—imidacloprid—though it had no particular track record against the emerald ash borer. In April the state had approved the use of a new insecticide, emamectin benzoate, known as TREE-äge. Some homeowners whose yards and homes were heavily shaded by mature ash were eager to try anything, even as state officials cautioned that no one knew how well these treatments worked over the long term. Moreover, they asserted the state's "right to remove an infected tree whether it's been treated with insecticide or not."

The tragic fact was the emerald ash borer had arrived, and in the coming decade Chicago was likely to lose 95,000 of its street trees to this pest.

On Saturday morning, August 2, 2008, the scenario that had long been feared regarding Asian long-horned beetles—namely, that they would make their way to the sugar maples of New England's woods and forests—began to play out. Responding to a cell-phone photo from a Worcester, Massachusetts, resident, the USDA's plant health director for New England, Patty M. Douglass, and a colleague walked into the small backyard of 41 Whitmarsh Avenue, a cream-colored four-family clapboard at the foot of the city's Greendale neighborhood. The night before, Douglass had examined photos taken by Donna Massie, and she had arranged to meet that morning here at her home. After greeting Massie, Douglass walked over to the maple trees bordering the back chain-link fence, where an Asian long-horned beetle obligingly stepped out and waved its black-and-white-banded antennae. "Oh, God!" exclaimed Douglass, grabbing Massie's arm. "They're really here." They were fourteen miles south of the New England Hardwood Forest.

A few days earlier Mrs. Massie, a slender woman with an ash-blond pageboy, had turned into her driveway during a warm summer dusk. Her husband, Kevin, and a friend were examining the door of Kevin's gold Hyundai Sonata, where a large, black, almost iridescent beetle with white spots waved its polka-dotted antennae before it whirred off into the fading light. Two evenings later the Massies were hosting a cookout in their small backyard when she noticed more of the insects, which were highly visible as they crawled up the shiny gray trunks of the big maples. One flew onto her husband's trouser leg and took some pulling to remove.

Getting up to inspect them more closely, Mrs. Massie noticed that one of these beautiful bugs was emerging from a sizable hole at the base of the tree, dappled with sawdust. "Truth be told," she said later, "the first thing that came to my mind was, can it hurt my grandchildren?" The next morning she was searching the Internet when she came across a Web page from the State of Florida warning about the Asian long-horned beetle. The picture looked very much like her insects, but she hesitated and over several days continued to do more research, wondering if she was being needlessly concerned.

On Friday, August 1, Massie decided to take action and telephoned Jennifer Forman, an invasive species ecologist with the Massachusetts Department of Agricultural Resources, who in turn requested that Massie send cell-phone photos of the insect, as false alarms were the norm. (Not only are there 350,000 known beetles species, but there are 26,000 species of long-horned beetles alone.) That evening around 7:00 p.m. in her office in Wallingford, Patty Douglass finally had a chance to study Massie's images. "Within five seconds," she later recalled, "I knew life in Worcester was going to change." She visited the city the following morning and by Monday she had the official positive ID.

In 2001 Worcester journalist Evelyn Herwitz had published *Trees at Risk: Reclaiming Our Urban Forest* (subtitled *A Case History of Worcester, Massachusetts*), hoping to mobilize her fellow citizens to restore their much-diminished canopy by detailing the rich story of her city's arboreal living legacy. Now a semifaded Gilded Age town, Worcester had long since seen its glory days, but it was still New England's second-largest city, a railroad nexus with mixed manufacturing, and home to Clark University and College of the Holy Cross. It had, like many American cities, once been shaded by a beautiful canopy of mature maples, elms, and American chestnuts. By 1923 the chief forester had reported, "Blight has taken the last of our chestnut trees. Over four hundred were removed during the year." Next came the great New England hurricane of 1938, smashing into Worcester and destroying a third of the city's 50,000 public trees. In 1953 a freak killer tornado shredded another 1,000 trees. One had to wonder about Worcester and its tree karma. By 1960 Worcester was down to 39,000 public trees, and 80 percent of those were Norway maples.

In the ensuing decades, hard times and a waning appreciation for trees took a further toll. By 1996 the city had only seventeen thousand public trees—still mainly Norway maples. As Evelyn Herwitz began penning her dirge for the city's diminished urban forest, she noted that "of those [seventeen thousand], the city's chief forester estimates that nearly a third are dead, damaged or diseased." She warned in her book that the Asian longhorned beetle found in Brooklyn posed "a serious potential threat to maple-dense cities" like hers. Having predicted just such a fate, she recalls, "I was just really sad to be proven correct."

Worcester, unfortunately, seemed fairly emblematic of the state of the American urban forest as the nation entered the twenty-first century. Despite a promising start for the urban forestry movement, few American cities had rallied to the cause as they lost population and manufacturing jobs. No president after George H. W. Bush showed any great passion for the environment, much less trees.

Ken Gooch, a forester with the Massachusetts Department of Conservation and Recreation, arrived in Worcester shortly after the discovery of the infestation. A veteran of combating beetles in Chicago and New York, on his first morning in town he scouted the public swimming pool just down the hill from the Massie home. In the lawn area he inspected several Norway and sugar maples, the most commonly planted trees in Worcester after the loss of most of the city's American elms. "They were big," he said of the trees, "and made nice shade on a hot August day. People would hang out under them." As his eyes adjusted to the shadows, though, he was astonished to see that "those trees were just raining down beetles."

Not far from the pool stood the city's major longtime employer: Norton Abrasives, a subsidiary of the French company Saint-Gobain since 1990, its offices and factory buildings sprawling across a dowdy twenty-acre industrial campus. At one end a sixty-foot-tall sugar maple spread its cooling canopy near Monolithic Refractory Plant No. 5. Standing beneath this soaring maple, Gooch was even more stunned: *Never* had he seen so many bullet-sized exit holes on a single tree. When he glanced over at another part of the campus, he noticed a mountain of discarded wooden shipping pallets. Later one federal ALB official estimated, based on old Google Earth images, that at one point over twenty thousand pallets had been piled there.

When Gooch and his fellow ALB warriors returned to officially inspect the tree, they would count 350 beetle holes and 1,500 egg sites on this one sugar maple. Dubbed the Mother Tree, for Gooch it spoke volumes about the unprecedented scope and scale of the Worcester invasion. If the beetle armies did finally make the push northward into the great New England hardwood forest and established themselves in those millions of arboreal acres, they would wreak untold economic and ecological destruction on the timber, tourism, and maple-syrup industries. In Maine alone, where timber contributed $6.5 billion to the local economy, half of the state's 17.7 million acres of forest were vulnerable. "This insect scares us to death," said Tom McCrumm, the coordinator of the Massachusetts Maple Producers Association.

On August 6, 2008, the Worcester city manager held a press conference to announce the discovery of the Asian long-horned beetle in the city. Residents puzzled over the implications of the imposition of a martial law of sorts, the creation of a designated "regulated zone" that encompassed their homes and yards, and the novel sight of people inspecting trees wearing jackets with ALB INCIDENT COMMAND TEAM emblazoned on the back. Two weeks later, at a standing-room-only town meeting, citizens were informed of the severe fines that would be levied for taking so much as a stick of wood outside the now sixteen-square-mile "regulated zone" and of widespread tree removals. Such was Norton's local economic importance and political sway that no official ever alluded publicly to the Mother Tree or the heap of pallets.

Gooch, serving now as the state's ALB eradication program director, found his troops scooping up beetles by the bucketful. He created a new code to categorize the disaster-level infestations. The antibeetle forces spray-painted *D* on trees with more than one hundred exit holes, *C* on those with ten to one hundred exit holes, and *B* for fewer than ten exit holes, while a spray-painted *A* meant there were eggs or oviposits on the bark. How long, Gooch wondered, had the beetle armies been silently conquering new territory beyond that Mother Tree? Ford's Hometown Pest Control Service provided an unsettling answer. In 1997 the exterminating firm had added a dramatic but unidentified bug—an Asian long-horned beetle, as it was now realized—to the insect display it took around to schools and clubs. The beetles had been proliferating beyond Norton Abrasives for eleven years, eleven years during which people had thought nothing of hauling firewood out to their cabins up in the woods.

In October of 2008 Clint McFarland, thirty-four, an APHIS veteran who had served under Gittleman for eight years in New York, where he had charge of Brooklyn, arrived to assume command in Worcester. An entomologist, McFarland had long brown hair pulled into a ponytail that reached halfway down his back. He wore a tan polo with USDA on it, his silver APHIS badge, green khaki cargo pants, laced hiking boots, and a USDA jacket. Tall and sturdy, he had blue eyes, a square jaw, and dimples. McFarland had deliberately chosen to rent a house with his wife that had no trees in its yard.

Like Gooch, when McFarland began to look around, he was stunned. "It was even more devastating than I had imagined. I felt almost sick to my stomach. I didn't see anything on that level in my years in Brooklyn." He was staggered at how thoroughly the Chinese beetles (which he noted were larger than those in New York) had occupied parts of the city and how long they had gone undetected. Especially puzzling was the failure of a Davey Tree team to spot the beetles four years earlier while conducting a tree inventory for the city. Within two weeks McFarland's antibeetle forces had condemned 1,500 infested trees.

Working initially out of the National Guard Armory in Worcester, McFarland and his lieutenants could see from a large aerial map that if Norton Abrasives was ground zero, the beetles had been riding the prevailing easterly winds up the hill into residential neighborhoods. (The Massies lived at the foot of that hill, across four-lane Lincoln Street from Norton's campus.) Within a month of McFarland's taking on leadership, the "regulated" or quarantine zone had quadrupled in size to sixty-four square miles, containing some 600,000 host trees. Such was the level of alarm in Worcester that the federal and state governments sent in seasoned ALB troops, spending $40 million to combat the beetles in just that first year. The western smoke jumpers soon followed, and of the 10,000 trees they inspected, a third were found to be infested. Within weeks the number of affected trees needing removal had reached 6,000.

On the night of December 11, a freezing rain fell. That morning McFarland stepped out of his front door to a world coated in a quarter inch of ice. "I thought as I was going to work at six thirty, *What a winter wonderland!* But then I started hearing all this cracking." The weight and cold of the ice caused trees and tree branches to snap and fall everywhere, landing on cars, houses, and yards, clogging many streets. "Contractors up and down the East Coast, from as far south as Florida, began arriving in the city in pursuit of debris-removal work," wrote reporter Peter Alsop, "many of them unaware of the ordinance against removing wood from a quarantined area." A week later an exhausted McFarland told Alsop, "We know that wood has been moved out of the city. That's our paramount concern right now. It can't happen again."

As year fourteen of the beetle war had begun in 2009, Clint McFarland had been racing against time in Worcester, trying to destroy all the beetle strongholds he could before the warm weather set off their spring campaign. As his antibeetle forces identified and spray-painted thousands of doomed trees, McFarland realized he would be almost clear-cutting entire neighborhoods in Worcester's eastern hills.

As the tree-cutting crews had roared through that winter, removing seventeen thousand public and private trees, residents felt helpless fury. "What once looked like Worcester," said one man, "now looks like Arizona, and people are angry." One father said, "It's devastating. It feels like we were raped. You come home, and your trees are gone." In especially hard-hit neighborhoods, residents felt as traumatized as civilian victims of a whirlwind war—returning home to streets and homes so ravaged by the disappearance of their trees that they almost did not recognize them.

"We've had nothing but sheer living hell in this neighborhood from the ALB situation," wrote one resident to the mayor. Without the trees, the wind suddenly seemed far more ferocious, and when it rained, the water—no longer absorbed by all those trunks and leaves and roots—began to buckle the streets. Realtors warned those who considered moving that their houses had just dropped a third in value.

All spring and summer McFarland and his forces found more and more infested trees, and they did not dare err on the side of sparing a potential host. By the end of 2009, they had inspected 198,707 trees for beetles, larvae, and eggs. By year's end USDA tree crews had cut down another 6,000 trees. The only good news was that after the first year of gathering grown beetles by the bucketful, in 2009 they found only thirty live ones.

As year fifteen of the beetle wars opened in 2010, McFarland's budget was doubled. For the first time, chemical warfare was used, and a total of 62,320 trees were injected with imidacloprid. In Worcester more than 500,000 trees were inspected, which caused the regulated zone to be expanded from 74 to 104 square miles, and only 120 live beetles were found. Of course, crews discovered larvae and eggs, but tree cutters took down only 1,057 trees, a fraction of the number that had been felled in the prior year. Faced with removing so many huge trees in Worcester, many from fenced backyards, McFarland and APHIS had pioneered a lightning-fast system: On a breezy morning that September, McFarland arrived at 18 Park Villa Street, a sweet mint green clapboard Cape Cod with white shutters, a white picket fence, a towering dark pine, and a front-yard wishing well and deer statue. He watched as an arborist in a hard hat climbed astride a crane's dangling heavy ball, rode the ball up into the air, and was then slowly lowered into the upper branches of the home's sixty-foot-tall backyard maple. From the side driveway, where the hundred-foot-tall crane sat on a flatbed truck, McFarland could see the arborist swiftly attaching the crane's rope to various tree limbs before climbing to the ground, where he grabbed a chainsaw. The arborist roared the machine into action, in moments slicing through the maple's thick trunk at its base.

The entire maple tree, severed from the ground, rose straight up into the air, twirling as it sailed above the roof of the house, a dreamy and beautiful sight—like some fanciful balloon tree departing on an arboreal adventure. It briefly snagged an old TV antenna atop the roof, then twirled balletically above the street, as if contemplating where to head next. In a children's tale the maple would now sail forth on lovely travels—perhaps to visit other far-off trees in Brazilian rain forests or west to see the giant redwoods.

In real life, here in Worcester on September 14, 2010, the maple was instead laid down in front of the house, where tree guys in hard hats began hacking off its huge branches and tossing them into the maw of a powerful chipper, which spit out a wide river of mulch into a waiting covered truck. Only fifteen minutes elapsed from the time the arborist rode the ball into the tree to its lifeless trunk's being flung onto the truck. Nineteen such cranes were operating throughout Worcester.

The homeowner of 18 Park Villa Street, Helen McLaughlin, a slender eighty-two-year-old widow, stepped out on her side steps. "I went from window to window watching," she said, clutching a Kleenex in her hand. "It was so fascinating. My son called, and I was crying. I didn't think I'd care, but the kids used to play in a hammock that hung between that tree and another one that's gone now. I didn't think I'd care, but I was really crying." She wiped her eyes again.

McFarland recalls the far more painful removal on a nearby block of two huge and venerable silver maples. Set on a front lawn, they had long served as the cool front porch for a pair of elderly sisters, who often sat beneath them enjoying the to-and-fro of their neighbors, reading the papers, taking naps. Where would they go now on hot summer days and evenings? They too had wept.

"Is this a winnable war?" McFarland asks. "I am an optimist. This is the second year that we saw very few beetles. The Asian long-horned beetle is lackadaisical. It's really a homebody by nature and it doesn't like to move if it doesn't have to. I would never say that about the emerald ash borer or the gypsy moth. In New York you got one generation a year, but I think up here I believe we're seeing one generation every two years." On another occasion he remarked, "Look, we can do this. I've been studying this beetle for years and I think eradication is really possible, and that's hard to say about most insects. And we don't have a choice, do we?"

In an otherwise encouraging year, 2010 did deliver one unnerving scare to McFarland. "On Friday, July 2," he recalls, "I got a phone call from Faulkner Hospital in Boston saying they thought they might have beetles.

They are right across the road from the Arnold Arboretum. I couldn't sleep that night. I didn't say anything to my wife, Amanda, because we had plans to spend Fourth of July weekend together. So I suggested an outing to the arboretum for Saturday. As we got into the car, she wondered why I had binoculars in the backseat.

"To my wife's credit," he continues, "once I explained, she spent eleven hours with me at Faulkner as we looked at all the possible host trees, and she was the one who found the two maples in the parking lot. There was so much frass [sawdust created by boring beetles], just handfuls of it at the base of the tree, that most people would have mistaken it for mulch."

The following day McFarland returned with eight APHIS foresters and scientists. "We spent a really trying day looking at the host trees for six hours with binoculars to detect the movement of antennae. We found forty adults and sixty more pupae. By 4:00 a.m. Tuesday morning we had the tree-removal people taking out six maples."

Stephen W. Schneider, the Arnold Arboretum's director of horticulture, recalls being at home an hour north of Boston that day. "I made the mistake of turning on my BlackBerry and saw many, many messages freaking out about the Asian long-horned beetle. The Arnold Arboretum has the most important collection of *Acer*, maple trees, in the world. It's a historic collection and extremely comprehensive, 482 *Acers* representing 137 taxa. The first two paperbark maples brought to North America are there. Wilson brought those back from his first trip for the Arnold as seedlings. All we could think was if you have infested trees, there is no question they will be removed. There was a lot of anxiety."

By the time of the press conference Tuesday morning in front of Faulkner Hospital, all of the hospital's six infected trees had been reduced to sawdust. The presence of Mayor Thomas Menino only underlined the crisis, and in the *Boston Globe* the long-dreaded appearance of the Asian long-horned beetle was above-the-fold front-page news.

Schneider's anxiety was not assuaged when he met Clint McFarland and the APHIS crew after the news conference. "They said there was a particular maple tree they were worried about. It was our *Acer mono*, and it was the largest specimen in North America. That tree came to us in 1902 as a seed from the Imperial Botanic Garden in Tokyo, Japan, and it is now roughly forty feet tall with a sixty-foot spread. They wanted to survey it first, even though it was far away from ground zero, as was our whole maple collection. But they told us that *Acer mono* is like caviar to the Asian long-horned beetles, and in China they use it as a lure tree."

Clint McFarland recalls walking over with Ken Gooch, Schneider, and the new Arnold Arboretum director, Dr. Ned Friedman. "We wondered," says McFarland, "whether the beetles would have flown across the street to this particular tree. I know that to them it was horrifying that this could be happening to a tree with such lineage. So we all walked over and looked with our binoculars and saw nothing. So there was momentary relief. But then in following weeks, they had their staff out looking and they thought everything they saw was a sign of infestation. So we went out to work in tandem with their staff to calm their fears. We took our climbers again to look at those maples, explaining that this was not the Asian long-horned beetle signs, but just a squirrel scratch, or that hole was from a different borer. We then worked hand in hand not only to educate their staff but because our own individuals had to certify the inspections."

The key question was, as ever, where had these particular beetles come from? McFarland was hugely relieved when he learned "from the DNA profile of the beetles that they *did* come from Worcester. So we speculate that someone came to visit the hospital and parked near the trees. Some people use firewood to weigh down the back of pickup trucks, so we think someone visiting the hospital had wood in the back of their truck, and a few beetles emerged."

Elsewhere in the invasive beetle wars, by 2009 emerald ash borers had made their way to such far-flung new territories as Wisconsin, New York, Kentucky, and northern Illinois. In September 2009 Dru Sabatello, the village forester of Arlington Heights, a northwest Chicago suburb bisected by the old Union Pacific rail line, was checking the ten purple-panel EAB traps provided by the Illinois Department of Agriculture when he found a beetle. "Obviously, we had heard EAB was coming," he says, "but we had some hopefulness back then that it could be stopped or we wouldn't be hit that hard." On October 7, 2009, the invasion was officially confirmed.

The Village of Arlington Heights, population 76,000, 16.6 miles square, 25 miles by rail from downtown Chicago, had 36,315 street trees and, as Sabatello well understood, developers of its subdivisions had planted far too many green or white ash. About thirteen thousand—or more than a third of the street (or parkway) trees—were ash. "Twenty years ago," recalls Sabatello, "I said, 'If anything ever happens to ash trees, it will be like losing the elms. It's just too much of a monoculture.'" Knowing that the EAB were coming, he journeyed to Fort Wayne, Indiana, which had already endured the battle with the borers, "and their forester told us of the nightmares of getting all those trees down." Sabatello learned about the "death curve, the crest, and the wave" and plotted out a removal plan for Arlington Heights and a budget, envisioning removing all the dead ash trees by 2017.

In November Sabatello's crew had just begun marking beetle-infested ash with green dots on Embers Lane when, as he recalls, "this one woman came out of her house in her wheelchair going crazy, crying, and yelling at us, 'What are you doing to my tree?' It was Patty Mora. She seemed to think we were already there with the chain saws and chippers, and there's nothing like chain saws and chippers to get people crazy." Mora was shocked and furious to learn that these sixty-foot-tall trees were going to vanish in coming weeks. Not every tree on her street was an ash, but most were. Sabatello, a genial man who had lived in Arlington Heights since the eighth grade, was immediately conciliatory. "I told Patty Mora, 'We'll stop.' I said we'll reevaluate the trees; we'd see in the spring whether the trees leafed out." He and his crews got in the trucks and left. Mora went inside and called the media. Laurie Taylor, who lived in a nearby subdivision, was watching the television news that evening. She recalls: "CBS had a reporter in Arlington Heights interviewing this lady who was all upset because they were going to take down all the ash trees in her subdivision. That's when I realized our trees were at risk."

Sabatello now wondered how best to alert residents to two crucial facts: (1) that the emerald ash borer had arrived, and (2) that in coming years that voracious insect army would kill every parkway ash on their streets. "We started by holding about a dozen outreach meetings at the library or with neighborhood associations," says Sabatello.

In early winter of 2011 he and Ashley Karr, a young arborist hired largely to help with EAB work, realized how few village residents had any idea which parkway trees were ash and decided to identify each of the doomed trees: "All that winter," he says, "we tied ribbons around every single ash tree in the whole village." This was no small endeavor in a suburb with 160 miles or so of streets. "The goal," says Sabatello, "was to get people to know they had an ash tree and to generate phone calls where we could then talk to people. Probably about two hundred people did call."

As Sabatello connected with the community in early 2012, he also learned that a new ash treatment was gaining favor. Deborah G. McCullough, whose work at Michigan State University had been completely focused on EAB research since 2002, hailed the emergence of the systemic pesticide TREE-äge (emamectin benzoate) as the best news on the EAB battlefront in years. TREE-äge, injected at the base of the trunk in the spring as the sap began to rise, killed the beetles for as long as two years, maybe three. Sabatello told a number of Arlington Heights' residents about the treatment, cautioning that it was new and thus still experimental.

Laurie Taylor was president of the Northgate Civic Association, representing eight hundred homeowners in an Arlington Heights subdivision of pleasant winding roads, trim front lawns, and magnificent trees. She now learned that two thirds of their parkway trees—about six hundred—were mature ash. "We had a homeowners meeting not long after," she recalls, "and we asked, 'Do we want to get involved in potentially saving the trees?' And everyone there said, 'Yes!' That was amazing to me." One of Northgate's board members was a chemist who investigated the science and economics of possible treatments and found TREE-äge the most plausible alternative.

Now that treating the trees seemed a real option, Sabatello worked to mobilize other subdivisions to "save their ash." Taylor's Northgate Civic Association contracted with a commercially licensed pesticide company to treat their trees for $135 per Northgate homeowner. As the civic association explained on its Web site: "If we do not take action, all of our ashes will almost surely die and be removed over the next few years. We will see a combination of adverse effects on the beauty and environment of our neighborhood and the values of our homes. With a pretty modest expenditure, we can do a lot to help ourselves." On May 30, 2012, the Village Board announced a "cost-sharing plan" of up to $50 per tree as an incentive to stave off clear-cutting a third of the parkway trees, bringing the Northgate cost down to $85 per household. A few other subdivisions signed on.

With TREE-äge as an option, some municipalities were treating *all* their large ash street trees, with the plan of gradually (over the course of two decades) replacing them with diverse species. You could certainly argue that at $500 a tree for removal and another $500 to plant a new tree, treatment and gradual replacement made financial sense. The city of Milwaukee, Wisconsin, with 28,000 ash trees, was firmly in the "treat with TREE-äge" camp. Argued David Sivyer, Milwaukee's forestry services manager: "The treatment is so effective and so much cheaper than removal and replacement that I can't get a single elected official to weigh in on the side of removing healthy trees because we don't have to, and that is never popular with the public. The injections allow *us* to decide what happens to those trees, not the beetle."

Still, other municipal arborists and local governments preferred to condemn and remove their doomed ash as quickly as possible. This group, which included Minneapolis, was skeptical about committing to years of pesticide treatments whose long-term efficacy was unknown. As Ralph Sievert, director of forestry for the Minneapolis Park and Recreation Board, observed, "Just the word pesticide has garnered a negative reaction." These arborists also tended to view the huge trees as inevitable casualties. Minneapolis had removed one hundred ash that looked just fine along Penn Avenue South. "We describe them as not-yet-infested trees," explained Sievert. "We don't call them healthy trees."

As the weather warmed in May of 2012, Northgate residents could see Sunrise Tree Care employees and trucks working their way along the neighborhood's streets, slowly injecting TREE-äge into 454 of the subdivision's 600 ash trees at a cost of about $60,000. Elsewhere in Arlington Heights, the roar of chain saws and chippers signaled the start of wholesale removal of infested ash trees. In Northgate, the trees leafed out as usual, and birds perched and nested and sang from their boughs. Laurie Taylor was relieved and elated to see how vibrant the treated trees looked.

Despite all of Sabatello's and Ashley Karr's months of village outreach, when the forestry crews arrived to cut down infested trees in a subdivision that was *not* treating their ash, a resident or two would often emerge—surprised, distressed, sometimes crying—when they heard the machines roaring. Why were their big trees coming down? Laurie Taylor heard similar laments: "There was a lot of regret. People would say, 'Why didn't you tell us?' or 'Why didn't the Village organize something?' One woman told me after losing all the ash in her subdivision, her electric bill the next summer was $400 higher. I can believe that." And yet, of the 13,315 parkway ash alive in 2011 in Arlington Heights, residents had been willing to pay to treat only 3,000 trees in a handful of subdivisions. The emerald ash borer invasion was rapid and the occupation complete. Every one of 1,119 village parkway ash trees removed in 2013 was infested, as were the 4,726 ash taken down in 2014.

Arlington Heights was but one suburban village confronting the most destructive forest insect to ever invade the United States. Since 2009 the beetle army's ever-expanding ranks had swarmed into Iowa, Tennessee, Connecticut, Massachusetts, Kansas, New Hampshire, North Carolina, Georgia, Colorado, New Jersey, and Arkansas. "The potential economic and ecological impacts of EAB are staggering," wrote Deborah McCullough. "Hundreds of millions of mature urban ash trees are growing on municipal and private land in the U.S. A 2010 analysis in *Ecological Economics* examined the potential costs of either treating or removing 50 percent of landscape ash trees in urban areas affected by EAB. Projected costs would exceed $10.5 billion by 2019. If suburban ash trees are included, costs nearly double."

The overall scale of the loss from the relentless advance of the emerald ash borer was just hard to fathom. In coming decades scientists anticipated the eventual near-extirpation of the entire twenty-two species of native ash. While local governments and arborists wrestled with the finances and logistics of this ecodisaster, for researchers working to calculate the value of trees to humans, especially in and around cities, the rapid disappearance of the ash presented a unique scientific opportunity.

In spring of 2013, R. J. Laverne, manager of education and training at the Davey Tree Expert Company headquarters in Kent, Ohio, could not fail to notice, while driving through Arlington Heights, the miles of parkway ash tied round with green ribbons sporting large ID tags. "Some of these streets, like Canterbury Drive," he says, "were just the poster child of what a tree canopy should be." Laverne was seeking an EAB-afflicted community for his PhD fieldwork at Cleveland State University.

A native of Detroit, as a child he too had watched the American elms disappear from his street. After graduating from Michigan Technological University with a degree in forestry, he worked for Champion International and then Mead Paper Company on the paper-pulp side of the industry before joining a forestry consulting firm. In 1989 he began to hear about the new field of urban forestry: "Rowan Rowntree . . . was very instrumental in opening my eyes to the wide range of benefits of urban trees. So it was a fork in the road. Instead of tangible products like paper pulp provided by trees, there would be trees absorbing storm water, cleaner air, and beneficial shade." Laverne earned a master's degree in remote sensing and began working as a consultant on urban tree inventories, generating data to quantify the collective economic, environmental, and social benefits of a city's trees.

By 1997 he had joined Davey Tree, where he spent five years specializing in tree inventories before becoming head of education and training, a position that required a PhD. In his doctoral research Laverne wanted to build on the work of Rachel and Stephen Kaplan of the University of Michigan, investigating whether "restful time in outdoor environments" did, as the Kaplans had argued, not only support human "functioning" but also "permit tired individuals to regain effective functioning." In short, "Does the environment around us affect our ability to concentrate?" The mass die-off of the ash trees provided an unparalleled research opportunity to examine questions like "Will the loss of tree cover affect the quantity and quality of soundscape?" and "Will changes in the soundscape affect our directed attention?"

Laverne contacted Sabatello, who told him that in the next two years there would be wholesale clear-cutting of ash on some of these very same lovely streets. This was perfect for Laverne, who needed just such dire "before" and "after" scenarios for his research. "I still can't believe," says Laverne, "I stumbled onto Arlington Heights at a time when they had labels on the ash trees and maps on their web site showing the location of all ash trees."

In July 2013 he made the first of sixteen trips to Arlington Heights, driving the 420 miles from Kent, Ohio, in his Ford pickup truck to set up his digital recording equipment on the grass of a parkway and on certain streets, during each of the four seasons. He focused most of his effort on six sites, first recording the sounds of the street during each season while all the towering ash trees still provided living, mature canopies. He then returned to the same spots when they had been removed and recorded what the environment sounded like with all the trees gone.

Over the next two years Laverne recorded 376 hours of sound, which he classified as either biophony (sounds from insects, frogs, and birds that find habitat in and around ash trees); anthrophony type 1 (mechanical sounds like traffic or lawn mowers); anthrophony type 2 (sounds of human voices or music); or geophony (sounds from nonliving elements of nature like wind, rain, and storms). He then spent hundreds of hours editing and analyzing his chosen recorded sounds and then randomly selected human volunteers to listen to the recordings made before the trees died and others to listen to sound recordings made after the trees were removed. "Then we subject the volunteers to a collection of simple tasks designed to measure the strength of their directed attention." "The driving purpose of this research," Laverne explained, "is to determine if we can build better cities, better neighborhoods, indeed better places to function as humans by understanding the benefits of access to nature, including the soundscape."

Another creative study of the EAB disaster data was undertaken by Geoffrey H. Donovan, an economist working as a research forester in the Portland office of the USFS's Pacific Northwest Research Station. With a PhD in forest economics from Colorado State University, he had started his career publishing papers on such pragmatic matters as "Estimating the Impact of Proximity of Houses on Wildfire Suppression Costs in Oregon and Washington" and "Sources of Product Information Used by Consumers When Purchasing Kitchen Cabinets." Donovan continued to investigate issues surrounding the ever-worsening western wildfires, but he also had turned his attention to urban forestry.

His first foray had yielded a 2010 paper titled "Trees in the City: Valuing Street Trees in Portland, Oregon," in *Landscape and Urban Planning.* "There was remarkably little stuff on the impact of trees on house prices," says Donovan, "so this was low-hanging fruit, so to speak." Starting in the summer of 2007, he and his colleagues looked at 2,608 single-family homes in Portland, Oregon, visiting each one to gather data on "the number of street trees that fronted the property . . . diameter and height of each tree . . . the type of the tree . . . [as well as] data about the house: the number of blocks from a busy street (a street designed for through travel) . . . and a subjective judgment of the house's condition (poor, average, or good)."

They used a "hedonic pricing model"—one that aims to break down the total value of, say, a house into the "contributory value" of each part, for example, the addition of a third bathroom. When they had factored in all the possible variables and crunched all their numbers, wrote Donovan, they "found that the number of street trees fronting the property and crown area within 100 ft (30.5m) of a house positively influences sales price. Combined they, on average, added $8870 to the sales price of a house." This additional value would not only benefit the seller of a home but also generate higher property taxes for the City of Portland.

"We came up with high values," says Donovan, "and that's great, but after a while the numbers started to bother me a bit. You should be able to 'de-compose' that value of $8,870 that's getting added on to the median Portland house price. Why was this figure so big? You could glibly say aesthetics, but that's no explanation at all. That's a lot of pretty for $8,870. What could feasibly generate values of that magnitude? I thought it would have to be some sort of health and well-being. So I started thinking about health."

In 2011 he published "Urban Trees and the Risk of Poor Birth Outcomes" in the journal *Health & Place*. "Pregnancy is a good thing to study," Donovan explained. "A pregnancy only lasts nine months and so you can look at the exposure to trees during pregnancy and look at the outcome for the baby." For this analysis Donovan and his collaborators drew on the data for 5,696 babies born in 2006 and 2007 to mothers in single-family homes in Portland, "geocoded" the addresses, and linked them to aerial imagery to "calculate tree-canopy cover in 50, 100, and 200m buffers around each home in our sample."

They compiled more detailed data on "maternal characteristics and additional neighborhood variables" from birth certificates and tax records. The outcome: "We found that a 10% increase in tree-canopy cover within 50m of a house reduced the number of small for gestational age (SGA) births by 1.42 per 1000 births. . . . Results suggest that the natural environment may affect pregnancy outcomes and should be evaluated in future research." Donovan and his colleagues posited that "stress reduction is a plausible biological mechanism linking trees" to healthier babies. Scientists in Spain, France, and California conducting similar tree/birth studies confirmed his findings. "Birth outcomes can set you on a trajectory through life," says Donovan, "so it's a profound thing if a tree can improve that. . . . And we tried to control for the nice tree/nice neighborhood effect, but that question was a tickle at the back of my brain."

In 2011, as this paper was published, Donovan, while sitting bored at a conference, "began thinking about this issue of trees and health and the hedonic value of trees and houses. I knew about the emerald ash borer and the loss of tens of millions of trees. Suddenly I thought, *Why not flip the question on its head? If trees are good for you, then if there are no trees, or many fewer trees, then that's bad for you.*. One of the great things about EAB: It doesn't care about demographics; it kills trees everywhere and it does it quickly. The spread of EAB is pure gold from the statistical point of view."

In June of 2013 Donovan was the lead author of "The Relationship Between Trees and Human Health: Evidence from the Spread of the Emerald Ash Borer" in the *American Journal of Preventive Medicine*. As Donovan and his coauthors noted in their introduction, "The spread of the borer is a unique natural experiment allowing the evaluation of the effect of changes in the natural environment on public health." The loss of 100 million ash trees across 1,296 counties in fifteen states would enable the investigators to examine whether "the spread of the ash borer is associated with increased mortality related to cardiovascular and lower-respiratory tract illnesses," the first- and third-most-common causes of death in the United States.

The sheer scale of the historic loss of ash from EAB seemed to have confounded most observers. How to think about the looming demise of what had long been the default urban tree in the Midwest? Donovan had come to believe a basic hypothesis: Trees improve human health. And he was after data. So Donovan's ingenious observational inquiry asked very simply: What did it mean for human health when 100 million trees died and were visibly removed from the living landscape? It appeared that more people died.

As his paper concluded: "Across the 15 states in the study area, the borer was associated with an additional 6,113 deaths related to illness of the lower respiratory system, and 15,080 cardiovascular-related deaths. . . . This finding adds to the growing evidence that the natural environment provides major public health benefits." When Donovan saw the results of this research, he says, "I was surprised as anybody. And I found a bigger effect in wealthier counties, where there were bigger, well-maintained trees, so there was bigger health impact when trees are killed. When you get results like this, you don't tell anyone, and then you sit in your office and try and poke holes. So I looked at accidental deaths and there was no relationship there." Donovan has come to believe, as he says, "Trees, quite literally, can be a matter of life and death for people."

CHAPTER TWENTY-TWO

"A Tree Is Shaped by Its Experiences": The Survivor Trees

Above: Bobby Zappala poses with the Survivor Tree at the NYC Department of Parks and Recreation Van Cortlandt Park Nursery. *(© Michael Browne, RLA.)* *Below:* The Survivor Tree at Ground Zero in Manhattan in 2016. *(Photograph by Jin S. Lee, 9/11 Memorial.)*

Over the course of almost a decade, the Ground Zero Callery pear (*Pyrus calleryana* "Cleveland Select") steadily recovered, regaining its thirty-foot stature. Each April, at the Arthur Ross Nursery in the north Bronx, the tree's dainty green white flowers bloomed, a diaphanous affirmation of life. Police and firemen still came on personal pilgrimages to commune with the still visibly battered tree, seeking solace and hope from its vigorous regrowth despite its many wounds. Richard Cabo, who took over as nursery manager when Bobby Zappala retired in 2007, said, "No one had any idea of how it would become so symbolic." For first responders the ornamental pear became simply known as the "Survivor Tree."

Ronaldo Vega, the 9/11 Memorial's director of design and construction, had been one of those first responders who labored for months after the attacks, clearing the hellish ruins of Ground Zero. In early 2005, as planning for the memorial had advanced, Vega had bonded with a delegation of Oklahoma City bombing survivors and family members. That April he flew there to view their memorial, which he found "so poetic, with its gates framing the time of the attack and empty chairs symbolizing lives lost." And then he approached their "living memorial," the old American elm Survivor Tree, ensconced in its circular promontory. "It was here," he would later write, "that I learned to see trees with a more discerning eye. To assign to them human qualities. To believe in their healing powers."

Oklahoma's memorial elm had assumed that role of wise if unruly spirit of that city's sacred ground. Many people assumed its bowing northward, away from what had been Fifth Street, was due to the force of the blast, but the tree's trunk had assumed this stance long before the bomb exploded—as revealed in a black-and-white photo from the 1940s, when it was only a few decades old and growing in a front yard.

"A tree is shaped by its experiences," wrote biologist Bernd Heinrich.

"Like us, the longer it lives, the more different and independent it becomes from other trees. A record of what has happened to it becomes inscribed on its body." Oklahoma urban forester Mark Bays feels there is no one explanation for "how the Survivor Tree got its unique shape you see today. . . . Trees have a remarkable ability to respond and adapt to many environmental and cultural changes through time. The Survivor Tree has been through so much throughout its entire life and it continues to hold its ground and stand strong."

New York City's National 9/11 Memorial would feature a pair of somber waterfall pools in the approximate footprints of the two downed towers, both to be surrounded by a formal "engineered forest" of four hundred memorial swamp white oaks, living pillars of strength and longevity, their leaves and canopy lending calm to this place of "contemplation and remembrance." The Manhattan site would become the most famous of the U.S. Forest Service's many 9/11 "Living Memorials." After experiencing the power of Oklahoma City's elm, Vega, who had heard rumors about New York's own Survivor Tree, thought that it too should be part of the memorial. He asked around the office to try to learn its location. Only when he e-mailed other agencies did Rebecca J. Clough, the Callery pear's original savior and now an assistant commissioner at the Department of Design and Construction, respond: "I know where that tree is."

On April 30, 2009, Vega drove up to Van Cortlandt Park to officially "tag" the Callery pear for planting at Ground Zero. Bobby Zappala returned for the occasion. He was proud of his robust protégée, her white spring blossoms mingling with new tender green leaves. When Vega stepped forward to meet the tree, "his strong hands patted the furrowed bark with almost tender affection. 'You're one of us now, girl,'" he said as he "slipped a metal 'WTCMEM' identifying tag (No. 350) around one of the biggest branches." As he told *New York Times* reporter David Dunlap, "This is the first time I've seen it flowering, which is pretty awesome." After a pause he added, "It's made it through its hardest times."

On December 21, 2010, Vega again journeyed to the Arthur Ross Nursery in the Bronx, this time with a crew from Bartlett Trees and a big flatbed truck to transport the Callery pear, its roots contained by a huge, square-angled wooden box, back down to the National 9/11 Memorial. Richard Cabo, now the tree's primary caretaker, had developed a deep affinity for the tree, feeling it had completed his own healing from the lingering trauma of the violent gunshot wounds. Back in March, Cabo had ministered to the tree after a second, lesser disaster—a severe windstorm that had whipped through the nursery, half wrenching the pear from the ground. Cabo had raced over when he heard what it had suffered, partially lifted the tree up, and then cosseted its roots with compost and mulch. After a few weeks, he had brought in a boom truck to gently right the tree, which bloomed in April as ever.

A little more than nine years after Rebecca Clough and Michael Browne had first extracted the charred pear from the smoking ruins of the World Trade Center towers, the Survivor Tree was welcomed back to Ground Zero. A crowd that included Mayor Michael Bloomberg and many reporters had assembled to witness the tree's homecoming. With its wintry limbs completely bare, its disfigurement was still evident—but so was the vigor of the new branches that reached toward the sky. Vega had not been willing to allow the pruning of one particular low-hanging branch, insisting "It took ten years for her to grow that branch, so there is no way we would lop it off."

The mayor invoked the tree as "the symbol of our unshakable belief in a brighter future." Cabo watched with a smile as the pear was gently lowered into the earth, as all around cameras whirred. Keating Crown, a member of the memorial's board, was one of the survivors of the attack, having escaped his office on the seventy-eighth floor of the World Trade Center's south tower via the last usable stairwell. Crown looked at this once-ordinary tree, now back where it had grown unheralded for twenty-five years, and like others, he admired its innate will to live, saying, "The fact that this tree survived such devastation reminds us all of the capacity the human spirit has to endure."

Michael Browne was excited to learn that the tree had been formally incorporated into the memorial at Ground Zero, at that stage still a major construction site with tightly controlled access. "I made a point to travel through the area to take a glimpse of the tree from outside the fence between the fabric," he said. "It stood in the middle of a very sparse area, still very much a disturbed construction site, but the tree stood out. Although leafless, it was very robust, upright and dwarfing the new [oak] trees nearby."

There was great irony, of course, in the fact that the Survivor Tree was a Callery pear, given its unpopularity with committed tree lovers. After city foresters, developers, and homeowners planted it by the millions, as the trees aged their fatal flaws emerged: Their fast-growing limbs turned out to be weak and highly prone to failure. A passing storm of no great force could split off major branches or whole sections of the tree.

Even John Creech, the director of the National Arboretum who had introduced the Bradford pear, had expressed serious misgivings. He reportedly confessed to an audience of arborists, landscapers, and nursery growers that "of all the thousands of plants he introduced by one way or another, the Bradford pear was the one he regretted the most. . . . He never dreamed it would be used to line streets, replace long-lived trees in urban environments and become the most over-used tree in recent memory." Apparently he had originally envisioned it serving only as a "back-yard ornamental specimen. . . . When the tree collapsed, it could be replaced by another one."

Arborists and landscapers felt otherwise for many years, and when the Bradford's dismal track record could no longer be ignored, they championed instead a less fragile Callery pear with a longer and better performance—one "discovered" on the streets of Cleveland, Ohio, and sold in the nursery trade since 1965 as "Chanticleer" or "Cleveland Select." In 2005 the arborist trade magazine *City Trees* voted this Callery "Urban Tree of the Year." Forester Steve Nix explained why Chanticleer had now become the default urban tree: It was, he said, a "beautiful tree" blessed with a "unique combination of good traits including great form and resistance to blight and limb breakage" and, probably more important to the public, "white flowering magnificence in the spring" and "plum fall color tinged with claret." While it could also grow in many different soils and tolerate drought, heat, cold, and pollution, Nix did acknowledge: "Odor is a concern. Where Callery pears are in full bloom an undesirable odor can be noticed."

The same year that Chanticleer was crowned Urban Tree of the Year, the proudly profane online Urban Dictionary (self-described as a "veritable cornucopia of streetwise lingo, posted and defined by its readers") bestowed its own hip moniker on the Callery pear: the "semen tree," saying this was "another name for the Bradford pear, an ornamental pear tree. Characterized by greenish-white flowers which smell like a cross between old semen, dirty vagina, and rotting fried shrimp." In 2010 blogger Susannah Breslin picked up on this, writing, "Spring has sprung, and so, here I am again, haunted by the tree that smells like semen. Yes, you heard what I said. *A tree that smells like semen.* I didn't make that up, it's not my imagination." She identified the guilty tree as Chanticleer and referenced the Urban Dictionary entry.

She did not say where she had encountered this tree that "smells like semen," but so complete was the Callery pear's arboreal dominance of American cities and suburbs, it could have been anywhere. Clearly, many endowed with lesser senses of smell had not even noticed the "odor." But Breslin was by no means alone. A Pittsburgh woman complained to NPR's *All Things Considered* in spring of 2015, "They told us they were putting up nice trees with flowers on them. . . . Everywhere you see one of these trees with the white on them, they smell like dead fish."

Perhaps its faults were what made the Survivor Tree an especially fitting choice as a symbol of New York City's spirit of survival at Ground Zero. The Callery pear had arrived as an immigrant and been recruited to serve in the grittiest of city environments. As Michael Dirr says, "We thought, gee, this is a panacea. You can stick it into any planting space in an urban situation—in concrete, heavy soils, clay soils, limestone-y soils, acid soils, and it's going to grow." As Callery pears colonized urban and suburban spaces, and its flaws emerged, so did its critics. But this was clearly a tree with not only moxie but also a huge number of fans, fans who kept planting it despite the scorn of the best-informed experts. Its unattractive scent only added to its street cred as a tree with attitude, really a perfect tree to survive the Big Apple's worst disaster and emerge stronger and better, rebounding beyond all expectations. And in a New York fairy tale way, the Survivor Tree was about to become an all-out world-famous celebrity, entering that select pantheon of trees that become destinations for travelers.

The first and most prominent public pilgrim to visit the Callery was President Barack Obama. On May 5, 2011, after the death of Osama bin Laden, the president made his first visit to Ground Zero. It was a clear blue day, and construction at the memorial had been halted for this ceremony of mourning and, now, justice done. Obama gave no speech but simply laid a red, white, and blue flowered wreath at the foot of the Callery pear, whose glossy leaves rustled in the strong breeze. The president stood silently, his head bowed, his hands clasped before him. This was the most public stop of his four-hour visit to the city; he also met privately with police, firemen, rescue workers, and relatives of the survivors. When his motorcade pulled away from Ground Zero, work began again, for the first section of the sixteen-acre site was set to open officially that September on the tenth anniversary of the attack. The New York Survivor Tree had now begun its official duties as an emblem of the emotional strength of ordinary living beings.

Jane Goodall, famous for her work with gorillas and her environmental activism, had been in Manhattan during the 9/11 attacks. On a clear spring morning in 2012, she made her own pilgrimage to the Survivor Tree. "As we walked toward this special tree," she wrote, "on that special April day, I felt as much in awe as if I were going to meet one of the great spiritual leaders or shamans. She is not spectacular to look at—not until you realize what she has been through and understand the miracle of her survival. We stood together outside the protective railing. We reached out to gently touch the ends of her branches. Many of us, perhaps all, had tears in our eyes—my own eyes were too blurred to be sure."

She had heard from Ron Vega that some people carped about having the tree on the plaza, feeling it "spoiled" the intended symmetry of the landscaping, being a mundane and different species. "Indeed," wrote Goodall, "she is different. For one thing, she is much older, and wiser. And on the tenth anniversary of 9/11, when the memorial site was opened to survivors and family members, very many of them tied blue ribbons onto Survivor's branches.

"One last memory," wrote Goodall, "Survivor should have been in full bloom in April when I meet her. But like so many trees in this time of climate change, she had flowered about two weeks early. Just before we left, as I walked around this brave tree one last time, I suddenly saw a tiny cluster of white flowers. Just three of them. Somehow it was like a message symbolizing, for me, the power of the life force that had enabled Survivor to be brought back into the world."

In 2013, on a perfect late morning in August, crowds of people were converging on the National 9/11 Memorial. The four hundred perfect swamp white oaks, sturdy twenty-five-foot-tall trees, presided over the austere plaza, their canopies as regularly spaced as sentries. Near the south Memorial Pool stood the Survivor Tree, the commonplace Callery pear that had lived through the attacks and, despite its permanent scars and deformities, was flourishing. People gravitated toward it—couples, families, teenagers—creating eddies of life all around its low square railing. Four guide wires anchored the tree securely to the ground.

A mother knelt in front of it, holding her young son close as her husband framed a photo. Others waited their turn to admire the tree, looking up into its boughs, touching its trunk or branches, caressing a glossy leaf, needing to commune with this authentic witness and survivor of the attacks. A brown-haired ten-year-old boy in an "I Love New York" T-shirt posed next to the tree with two older ladies in floppy summer hats as friends and family snapped away with their phones. They lingered and then moved on, having absorbed something of the tree's palpable strength and good karma. This one tree embodies a spirit that no built structure or imported white oak can now match.

On the anniversary of 9/11 in 2013, the National 9/11 Memorial announced its Survivor Tree seedling program, a partnership with Bartlett Tree Company, which had planted all the swamp oaks at Ground Zero, as well as the Survivor Tree itself. Each year the memorial would present three communities, each afflicted by calamity but buoyed by resilience, with a six-foot-tall Callery pear offspring of the Survivor Tree. These were grown by students at the John Bowne High School in Flushing, Queens, working with Bartlett arborists. The initial recipients were Boston, still reeling from the marathon bombing; Prescott, Arizona, where nineteen firefighters had died battling a June wildfire; and Far Rockaway in New York City, which had been devastated by Superstorm Sandy.

In New York City neither Michael Browne nor Rebecca Clough would actually see the Survivor Tree in the plaza until spring of 2014. While Browne had been invited to an April preview of the 9/11 Museum, he says, "I took a solo visit in March so I could be prepared. I was unsure how I would react emotionally. Seeing the tree up close was amazing, and a bit overwhelming, because it came with a flood of flashbacks. I could see where the new growth had taken off from Bobby Z's strategic pruning. I remembered the deep piles of dust and granulated glass in the crotches. Now, no longer there." Alone in the plaza, Browne peered closely into the canopy, recalling "where the mourning dove took nest and hatched new life," and inspected the bark and new growth. "It looked great and could last another 30 to 40 years with proper care," he says. "During my investigation I was startled when a docent appeared out of nowhere to explain to me that this was the 'Survivor Tree.' I politely waited and listened to his story. And as usual, it had a few things missing. I was happy to give him my angle, and he was genuinely interested and engaged." A small crowd soon pressed in to listen. For Browne "the biggest takeaway was that the tree was the most uplifting thing in the entire place."

As for Rebecca Clough, thirteen years had passed since she had watched the scorched tree depart Ground Zero and the smoking rubble. Unlike Michael Browne, she had not sought the tree out. In May of 2014, when the 9/11 Museum opened and honored the first responders, she was finally reunited with the now-very-famous Callery pear. Like so many others seeking solace and communion, she approached slowly, in her case looking to see how the shattered branches had regrown. Standing under the tree, she gazed up into its canopy, the light filtering through, and marveled at the pear's robust recovery. "It was a very strange feeling," she says. "I felt very connected to it. I put my hand on it for a few minutes. It was a strange sensation to look up into that tree after all that time. A lot of firemen and policemen were also stopping at the tree, conveying its significance. A lot of people were taking pictures. The tree has definitely become a piece of Americana."

AFTERWORD

"The Answer Is Urban Forests"

MillionTreesNYC plants trees with the NFL at Randall's Island in spring 2013. *(Photograph by Anne Tan. Courtesy of the New York Restoration Project.)*

On a late August afternoon, I enter the grounds of what remains of the Philadelphia estate of William Hamilton, the ardent collector of trees whose letter to Thomas Jefferson opened this book. My guide is David Hewitt, a Harvard-trained PhD biologist who has fallen in love with The Woodlands historic site, now a 54-acre cemetery where aged, giant trees shade the crumbling gravestones, mausoleums, and funerary angels crowded amid well-tended green lawns. Not unlike Hamilton, Hewitt is convivial, charming, and eager to show off his arboreal treasures.

Adjacent to the University of Pennsylvania campus (which largely occupies the other 540 acres of Hamilton's former estate), The Woodlands Cemetery is an oasis of green, a lush necropolis where the crescendo of late-summer cicadas soon blots out traffic and other city noises. We wander past weathered nineteenth- and twentieth-century graves and tombstones (31,000 people are buried here), heading toward Hamilton's fieldstone mansion with its creamy two-story columned front, its grandeur faded but still impressive.

We are in search of the famous Woodlands ginkgoes, but Hewitt is intent on first showing me a rare Caucasian zelkova, a gargantuan tree that appears to have materialized out of a Russian fairy tale. Its multiple trunks have long since fused together into a massive foundation for hundreds upon hundreds of thick branches that spring straight up, also braiding here and there, as Hewitt describes them, "like a mazed tangle of upright hair, knotting together, and then unknotting again" toward the sky. French plant explorer André Michaux discovered this particular specimen of zelkova in Persia, and soon Hamilton had to have his own.

We peer up into this tree's dense, magically entwined canopy, feeling certain that some species of large owl must be up there napping. As far as anyone knows, this is the only such specimen in all of Philadelphia, and maybe even the state. Hewitt had determined that despite the tree's immensity, it was a century-old offspring that had emerged from one of the half dozen such trees that long ago lined Hamilton's front driveway, suitable sentinels for colonial-era visitors in search of botanical wonders.

And what of Hamilton's ginkgoes? These "living fossils" can flourish for hundreds, even a thousand, years. Alas, the original Woodlands ginkgoes are gone. But David Hewitt is quite certain that one of the current trees, which looks to be thirty to forty years old, had sprouted from Hamilton's late-eighteenth-century imports. (Rumor has it that in the early 1980s a groundskeeper took a chainsaw to the venerable lady ginkgo after his dog became ill from eating ginkgo drupes.)

And indeed, there to the left rear of the mansion stands a tall, luxuriant ginkgo, grown far past its adolescent gawkiness. Of course, Hamilton's original ailanthus are long dead, but their descendants have efficiently occupied this and every other North American city, as confirmed by David Nowak's 2007 UFORE analysis of Philadelphia's urban forest. The tree of heaven triumphs as the third most common species among the city's 2.1 million trees.

In the Philadelphia of Hamilton's day, which had a population of about fifty thousand, local inhabitants took for granted an abundance of all kinds of trees, many of them large and ancient, as well as open fields, thick forests, and an easy enjoyment of the bounties of nature. More than a century of fossil-fueled industrialization, urbanization, and the domination of the automobile have long since changed that landscape, not just in Philadelphia but in most of the urban United States. With our modern cities becoming ever more paved over, Americans can no longer take for granted even the small daily pleasures of walking under towering canopies, strolling through fields or parks, hearing birds, or simply savoring the ordinary sights and sounds of the natural world.

No one gave our denatured cityscapes much thought until 1984, when Texas A&M University landscape architecture professor Roger Ulrich published a paper in *Science* titled "View Through a Window May Influence Recovery from Surgery." He reported that those patients convalescing from gallbladder surgery who looked out on a small stand of trees recovered faster and required less pain medication than a control group whose view was a brick wall. This landmark study caused a sensation, Ulrich recalled, garnering worldwide media coverage. The idea for it came from his own experience: As a teenager suffering from kidney disease, he himself endured "long periods spent at home in bed feeling quite bad, looking out the window at a big pine tree. I think seeing that tree helped my emotional state."

As urban construction and concrete continue to subsume the trees, woods, and other remaining fragments of the natural world, new science and technology have begun to illuminate how essential nature actually is for our well-being. "When we're in a setting with a great deal of stimulation, like a city," writes journalist Eric Jaffe, "we expend a great deal of direct attention on tasks like avoiding traffic and fellow pedestrians. When we're interacting with nature, however, we use an indirect form of attention that essentially gives our brain a chance to refresh, much like sleep." Trees and the urban forest—the most visible manifestation of nature in most cities—and green spaces like parks demonstrably improve human health.

We know this from the work of University of Michigan professors Rachel and Stephen Kaplan and their disciples, and from University of Illinois computer engineer William C. Sullivan and psychologist Frances E. Kuo, who pioneered research on the restorative power of nature for weary city dwellers. In one of the most famous of these studies, Sullivan and Kuo compared residents of two identical Chicago public housing projects—one landscaped with trees and greenery, the other bleak and barren. Those living in the project with views of trees and nature exhibited better concentration, discipline, and lower aggression. "Those data were astounding," said Kaplan. "That's a miserable environment, and for [nature] to make a difference in it, that was awesome."

In 2011, Kuo, by then director of the Landscape and Human Health Laboratory at the University of Illinois at Urbana-Champaign, could sum up years of research: "People with less access to nature are more prone to stress and anxiety, as reflected not only in individuals' self-report but also in measures of pulse rate, blood pressure, and stress-related patterns of nervous system and endocrine system anxiety."

In 2012, another of the Kaplans' University of Michigan graduate students, Marc Berman, a psychologist who describes his work as environmental neuroscience, directed one set of study participants to walk around the school's 123-acre Nichols Arboretum, while a second group perambulated the nearby city streets. Those in the arboretum group had a 20 percent increase in cognitive control. "To consider the availability of nature as merely an amenity," concluded Berman, "fails to recognize the vital importance of nature in cognitive functioning." Eric Jaffe reported on the study and then mused, "Be great to know just how many trees it takes to give a city brain the break it needs from crossing traffic, or navigating subway platforms, or sitting in our cubicles."

In 2015, Berman, now at the University of Chicago, responded to that challenge. He and his colleagues selected Toronto as the target city because of Canada's universal health care system. They analyzed two large data sets—one detailing the 530,000 trees on the city's public land, the second featuring 39,000 respondents to a city health survey. "We find," wrote Berman and his colleagues, "that having 10 more trees in a city block, on average, improves health perception in ways comparable to an increase in annual personal income of $10,000 . . . or being seven years younger."

Why, wondered Elizabeth Preston, reporting on this work for *Discover Magazine* online, might urban trees lead to better health? Cleaner air? More beautiful streetscapes? More motivation to be outside and exercise? "My guess," said Professor Berman, "is that a few different mechanisms are at play." While these are still early days in determining specifically how and why trees are good for people, scientists are showing with data what people have long known intuitively—that time spent among trees refreshes our tired psyches, promoting happiness and health.

What we far better understand, thanks to a decade of improving the science and delivery of i-Tree in its various guises, are all the *practical* ways the urban forest serves cities. Since i-Tree's launch in 2006, "It has fundamentally changed the way we view the urban canopy," says Scott Maco, i-Tree's director of research and development at Davey Tree. "The fact that i-Tree can quantify benefits—that didn't exist years ago. You don't manage an urban forest anywhere now without knowing its structure and benefits. This thing just took off and usership has increased significantly every year." In 2008, the i-Tree website had almost 9,000 unique visitors with almost 19,000 page loads; in 2013, it had 125,000 unique visitors with almost 367,000 page loads.

One surprise? From the start, foreigners were a quarter to a third of those using i-Tree's expanding menu of public domain programs. By 2015 i-Tree customers spanned 120 countries, ranging from 744 in Canada to 199 in China to 16 in Peru to 8 in Ethiopia. The previous year Craig Alexander, senior vice president and chief economist for TD Bank Group in Toronto, had been able to advocate confidently for the fiscal value of increasing the urban tree canopy. "There are 10 million trees in Toronto, covering close to a third of the city," he said to an interviewer. "Take the impact of trees in terms of absorbing air pollutants. We estimate the trees are removing 19 thousand metric tons. To put this in context, that's equal to the pollution emitted from over 1 million automobiles each year. . . . Trees are returning between $1.35 and $3.20 (a 220% return) in benefits, and we aren't including things we can't measure like the intangibles of being able to go to a park and enjoy the trees."

One of the enduring charms of tree advocacy is how its institutions can be as long-lived as sturdy oaks. Harvard's Arnold Arboretum, the New York Botanical Garden, American Forests, the Morton Arboretum, the National Arbor Day Foundation—all are still with us, their missions rejuvenated to confront twenty-first-century phenomena like urbanization, social inequity, climate change, and globalization.

When Charles Sprague Sargent established Harvard's Arnold Arboretum, he described it as a "Living Tree Museum," whose "first and real duty is to increase knowledge." Professor William (Ned) Friedman, an evolutionary biologist appointed the arboretum's eighth director in 2011, has been keeping that vision alive, adding cutting-edge botanical science to the mission with the opening of the 44,000-square-foot Weld Hill Research Building. Here, Friedman hopes to "unlock the secrets of evolution, tell stories about the history of science and humankind," and tackle "biodiversity and climate change." Toward those ends the arboretum is systematically upgrading its herbarium, adding pizzazz to public programs, and updating its historic "living collections."

Plant exploration, meanwhile, is enjoying a comeback. Where Sargent sought specimens to enrich America's horticultural landscape, Friedman's goals reflect our more parlous times: "One of the really important things we're after are species on the brink of extinction," he says. "Once a species is gone, it's gone forever unless we can get it into a botanical garden." With an eye to climate change, the new plant explorers are also seeking specimens able to thrive in a warmer Boston. Perhaps a new dawn redwood awaits discovery.

The New York Botanical Garden has also been reimagining itself for the twenty-first century, vigorously promoting its mission as "an advocate for the plant kingdom" on multiple fronts. As a public garden, it dazzles and engages vast throngs by showcasing its thirty thousand mature trees, imaginative gardens, conservatories, exhibitions, classes, concerts, tours, and children's programs. It trains scientists, horticulturalists, and science teachers and develops curricula for all grade levels. The LuEsther T. Mertz Library, founded in 1899, houses the nation's largest and most comprehensive botanical collection. Ambitious digital initiatives have made these riches increasingly available through the Internet, as well as those of the garden's more than seven million herbarium specimens.

One suspects that "The Naturalist," aka William Alphonso Murrill, would approve of all these initiatives, especially the garden's Pfizer Plant Research Laboratory, opened in 2006, with its heavy focus on plant genomics, or how genes function in plant development. Its scientists have been working to establish a DNA library of the world's flora (some rare or now extinct), stored on-site in twenty freezers. NYBG scientist Damon Little has served as the leader in the national tree Barcode of Life project, which is creating a DNA catalog of the world's 100,000 tree species by sequencing "a standard region of DNA as a tool for species identification." When giving a tour of the twenty-seven lab processes from dried tree leaf to barcoded tree specimen, he mentions that New York homicide detectives had shown up with leaves retrieved from a suspect's car. They hoped the evidence's DNA would place the suspect where his wife's body had been found in the woods. (It did not.)

One can also imagine Murrill's ghost hovering about on April 18, 2012, when the New York Botanical Garden welcomed professors William Powell and Charles Maynard of the SUNY College of Environmental Science and Forestry. Working under the aegis of the American Chestnut Foundation, the two men had been invited to lecture on their twenty-year quest to genetically engineer a blight-resistant American chestnut. The breakthrough proved to be "the gene we borrowed from wheat," which neutralized the deadly acid of the blight. Their work marked a further major advance in demonstrating how to dramatically speed up tree research "by testing the leaves of chestnut trees that are only a few months old [for disease resistance], so we no longer have to wait three years to see if our experiments are working."

When Murrill had prophesied that a century might pass before American chestnuts could again thrive, he had been uncharacteristically too optimistic. A hundred years later, the blight was as virulent an enemy as ever. In 2004 environmental activist Bart Chezar of Brooklyn had planted nine young American chestnut trees (grown from nuts ordered from the

American Chestnut Foundation) in Prospect Park and other borough locales. By the time the two professors came to speak at the botanical garden in the Bronx, several of Chezar's original young trees were struggling with blight, but trees planted in subsequent years were looking healthier, and would by 2015 bear a handful of genuine American chestnuts. (Chezar conceded in a 2016 report, "Trying to restore the American chestnut in Brooklyn can be a humbling experience.")

After professors Powell and Maynard delivered their lecture, the two scientists and other chestnut lovers gathered on the grounds for a ceremony to plant ten of their tiny (one-to-two-feet-tall) transgenic trees, marking the official return of the American chestnut to the botanical garden. Biologist and author Bernd Heinrich was less sanguine about this approach: "Yes, I would love to see the American chestnuts restored. . . . But do we need to alter the chestnut's genome—the code of life that has evolved over millenniums? I don't think so. . . . How will these trees evolve over time with their altered genome. . . . The consequences of genetic engineering can be unpredictable." Professor Powell defended his wheat-gene trees as benign, and intends—if they prosper as he hopes—to seek less-restricted government approval for further plantings. Under current permits for these transgenic trees, when they mature and bloom, the botanical garden will have to remove all their flowers to prevent pollen release.

What to do about these recurring tree-killing plagues? "Forest pests are the only threat that can decimate an entire tree species within just decades," writes Dr. Gary Lovett, a forest ecologist at the Cary Institute of Ecosystem Studies, and a team of sixteen scientists in a recent paper in *Ecological Applications*," . . . [and] the loss of trees has heightened significance today as communities turn to trees as 'green infrastructure' to help mitigate and adapt to climate change." This new alliance of scientists asks: Why should we spend huge *public* sums fighting (and losing to) foreign tree pests and pathogens when shippers and merchants, who cause the problem, could also largely solve it?

Under what they are branding as "tree-SMART-trade," these scientists propose a new policy that will end *all* use of solid wood packing and pallets, a measure that a global shipper like IKEA has already taken. The second major policy change would be to restrict or ban the further import into the United States of woody plants. Is a beguiling new foreign shrub imported to garden centers really worth the possible extirpation of yet another tree species?

Outside of Chicago, the Morton Arboretum has launched a new Center for Tree Science, its response to ever-shrinking government funding for USDA and university tree scientists solving practical problems in arboriculture. Seeing "an urgent need to invigorate tree-focused research and training," Morton proposed to gather together scientists and funding in configurations small and large to "collaborate, generate new knowledge, and deploy that in the real world."

Locally Morton, newly branded as the "Champion of Trees," has been rallying the tree troops post-EAB into a ten-partner 2014 Chicago Region Trees Initiative, with the "lofty" goal of making "a significant, measurable improvement to the regional forest and the lives of its inhabitants by the year 2040." The first step was an i-Tree census of the city and its surrounding suburban counties, which revealed about 157 million trees and a 21 percent tree canopy. With this data in hand, the Morton Arboretum has established working groups.

One can only wish them well, as few American cities have succeeded in increasing and restoring their urban forests. In 2010, when USFS i-Tree team leader David Nowak analyzed canopy cover using aerial photographs of twenty American cities, from New York to Los Angeles, the news was

Planting a street tree in Baltimore, Maryland. *(Photograph by Peggy Fox.)*

not good. With one exception, Nowak reported that "tree cover tends to be on the decline in U.S. cities while impervious cover [paved-over space] is on the increase." Since then many cities have launched concerted tree-planting campaigns, so future surveys may show progress.

New York City's MillionTreesNYC campaign wrapped up two years early, a rare shining example of the tree gods aligning: Mayor Michael Bloomberg, a powerful leader persuaded by i-Tree data; ample funding; talented city foresters able to orchestrate high-volume, high-quality tree plantings; and the involvement of Bette Midler and her savvy nonprofit, which served as a partner handling marketing, mobilizing volunteers for vast tree-planting days, and raising major financial contributions. On November 20, 2015, Mayor Bill de Blasio planted an American linden in Joyce Kilmer Park on the Grand Concourse in the Bronx, right up from Yankee Stadium. "There are now one million more reasons why New York is the greatest city in the world," observed Midler. New York's approach has always been strategic: USFS scientist Greg McPherson's work enabled city foresters to target neighborhoods that lacked trees—almost always home to poor minorities.

And how have our nation's two historic Cities of Trees fared since concerned citizens stepped forward to restore their canopies? Mark Buscaino of Washington D.C.'s Casey Trees says that old aerial images show that the city had a 50 percent tree canopy in 1950. By the year 2000, that was down to 35 percent. "When we started, there were 90,000 city street trees," he says. "Now that is up to 135,000. And we and City Forestry have gone from planting 3,400 trees a year to 13,000, three times what it was. The canopy number is still about 35 percent, which means we have preserved our overall tree canopy, and that is no small matter at a time when almost every other city's canopy has declined."

For Buscaino, one of the most important means of saving mature trees has been D.C.'s Urban Forestry Preservation Act of 2002. In any city most trees are on private land, and this local ordinance, on track to be further strengthened, obliges private property owners to pay a $1,500 fine (or more, depending on trunk size) to remove a tree. Take down a big tree without a permit, and the fine is triple that sum. "Restoring the canopy is a marathon, not a sprint," says Buscaino. "You need advocacy, legislative action, you need to train the grass roots, and you have to keep planting."

In Sacramento, where the canopy has grown from 13 percent to almost 24 percent since 1998, executive director Ray Tretheway looks back and says, "At first, it was all about planting trees, then it was about stewardship and tree survival. With McPherson and Nowak, it became about all the benefits—straightforward air quality and storm water. But now we're moving into how trees affect climate change and public health." He remains as bullish as ever about the power of science-fueled advocacy, for he has seen its power to mobilize policy, officials, and money.

When California launched the nation's first carbon trading markets in 2012 to reduce greenhouse gases, those "carbon offsets" emerged as a rich new source of dollars for ever-penurious local urban forestry programs. Trees store carbon! As California State urban forester John Melvin explained, "If your company's emissions exceed their limit, you have to buy credits; not only can we tell you how much carbon trees are storing, we can say how many emissions have been saved by their shade." In 2014, the California Department of Forestry and Fire Protection (CAL FIRE) awarded the first state cap and trade funding, $15.6 million for twenty-nine urban forestry projects. To put this in perspective, the U.S. Forest Service's annual budget line for urban forestry has not budged much above $30 million for twenty years.

And so, for the longtime tree groups, these are dynamic times. American Forests, an early activist in urban forestry and climate change, was hard hit by the 2009 economic collapse. After regrouping, by 2013 it was able to launch a new Community ReLeaf program with its longtime partner, the U.S. Forest Service, aimed at supporting grassroots groups and environmental justice in inner cities. In Atlanta, writes CEO Scott Steen, they planted "hundreds of trees around underserved schools to improve student health and lower stress levels," while in Detroit they "turn abandoned lots into public greenspaces." The National Big Tree program is more popular than ever, crowning 750 national champions for its seventy-fifth anniversary, with the stars featured on YouTube. American Forests has also embraced a timely new role as urban forest ambassadors, working to spread the gospel of the trees to those who design, build, and run our cities.

The National Arbor Day Foundation is going strong and estimates it has dispatched some 250 million tiny trees over the past four decades to its million loyal members. The foundation has also become far more involved in urban forestry. Municipal arborists still love Tree City USA, and now colleges can strive to become a Tree Campus USA. In 2007 the Arbor Day Foundation became "the lucky and humble hosts" of a reconfigured annual urban forestry conference. "We have brought together all the groups that work on trees," says President Dan Lambe, who succeeded John Rosenow in mid-2014, "municipal arborists, the utilities, the nonprofits. We need to be together." It also now runs the Alliance for Community Trees, a network of grassroots groups.

The threats to survival for a city tree have not changed substantially since Andrew Jackson Downing, the original tree evangelist, wrote in 1846: Trees "are placed in a hole quite too small for them, a little rubbish is thrown about their roots, the bricks are laid carefully as tight and near the root as possible . . . nobody ever thinks of watering it, it dwindles and dies. The wonder is that any city tree ever survives the treatment." It's amazing how many city trees still suffer this fate.

Lambe acknowledges that when it comes to trees and cities, "it is a constant battle, teaching foresters and homeowners about planting the right kind of tree, pruning properly. But we're also smarter and more connected as a community. We *are* making progress."

With more Americans—especially the young—opting for urban life (95 percent of Californians, for example, live in cities), and with climate change upon us, the time has come for politicians and city managers to get serious about creating the lushest tree canopies we can nurture. After years of proselytizing, it seems this message is finally reaching opinion leaders. As one environmentalist wrote recently, "Quick, name a climate solution for cities which helps lower carbon emissions, protects vulnerable people who live there, and even helps students get better grades? Give up? The answer is urban forests, and you're not alone if you didn't come up with the answer. After all, most of us see trees as woven into our city streets and parks as just a pretty, cinematic backdrop for urban life."

The more climate scientists study how best to stave off further warming of our planet, the more they understand the importance of trees and forests. "Every time I hear about a government program that is going to spend billions of dollars on some carbon capture and storage program, I just laugh and think, what is wrong with a tree?" said Nigel Sizer, president of the Rainforest Alliance. "All you have to do is look out the window, and the answer is there."

Acknowledgments

During the past eight years, many wonderful people and institutions have aided the research and reporting that shaped this book, and I would like to express my deep appreciation.

In Washington, D.C., I thank the Woodrow Wilson International Center for Scholars for a productive period of writing in the fall of 2011, with special acknowledgment of Steve Lagerfeld. Much gratitude to Mark Buscaino and Barbara Shea of the nonprofit Casey Trees; Jim Sherald and Bob DeFeo of the National Park Service; Irving Williams, longtime (now retired) White House head gardener; Melanie Choukas-Bradley, author of *City of Trees;* Susan Fugate and Diane Wunch at the USDA's National Agricultural Library; Boss Shepherd scholar John Richardson; Diana Parnell, biographer of Eliza Scidmore; and Robin Murphy of the Conservation Fund. Nancy Korber and Brooke Lemaire of Fairchild Tropical Gardens supplied USDA-related research and photos.

In New York City, for the reporting on MillionTreesNYC, I am much indebted to Jennifer Greenfeld of the NYC Forestry Division, Department of Parks and Recreation, along with her colleagues Fiona Watt, Andrew Newman, Matthew Stephens, Bram Gunther, Jason Stein, and a variety of folks at the New York Restoration Project, including Leah Silver, who provided valuable photography help, as did Bridget Miles at Sesame Shop. At the New York Botanical Garden, scientist Damon Little was very helpful, as was librarian Esther Jackson with the illustrations. It was a pleasure to take a tour with former parks commissioner Henry Stern.

For the story and photographs of the New York's Ground Zero Survivor Tree, deepest thanks to Michael Browne and Rebecca Clough, and also

Richard Cabo. In Oklahoma City, arborist Mark Bays was a heartfelt guide during my visit to the Survivor Tree at the memorial, while Kari Watkins, MaryAnn Eckstein, and Gary Marrs kindly shared their knowledge.

At the U.S. Forest Service, the urban forestry pioneers always went above and beyond with time and information over the years: Rowan Rowntree (with extra thanks for photography duty!), Greg McPherson, David Nowak, Paula Peper (including kind hospitality while in Sacramento), Geoffrey Donovan, Matthew Arnn, and Alice Ewen. For Dayton history, my thanks to Fred Bartenstein; and for Chicago tree history, Edith Makra. At U.S. APHIS, my gratitude to Joe Gittleman, Clint MacFarland, and Rhonda Santos for the numerous Asian long-horned beetle tours in New York and Worcester. And for explaining their parts of the ALB experience, thanks to Ingram Carner, author Evelyn Herwitz, and the Arnold Arboretum's Stephen Schneider. In Chicago, forester Joe McCarthy was a terrific guide.

For the story of the emerald ash borer, grateful thanks to Professors Deborah McCullough and Daniel Herms, as well as Nate Siegert, Dave Roberts, David Smitley, Craig Kellogg of APHIS, and Tom Holden of Davey Tree. In Arlington Heights, arborists Dru Sabatello and Ashley Karr shared their frontline experience, while Laurie Taylor supplied invaluable detail on resident response.

I appreciate Davey Tree's cooperation on a variety of topics. Thank you, Greg Ina and Scott Maco of i-Tree; and Janis Hittle, Matt Fredmonsky, and Jennifer Lennox for historical matters; and for the research perspective, R. J. Laverne.

To research the story of the American elm and its comeback, I am much indebted to John Hansel and Yvonne Spalthoss of the Elm Research Institute; the USDA's Denny Townsend (now retired) and Jim Slavicek in Delaware, Ohio; retired professors Ray Guries and William Jacobi; Tom Zetterstrom for numerous elm tours; Joe Pickens, Bruce Carley, and Richard Olsen of the National Arboretum; Jason Griffin; and Keith Warren and Nancy Buley of J. Frank Schmidt & Son Company. At the Wethersfield Historical Society, thanks to Rachel Quish, Kayla Pittman, and Elizabeth Thompson.

For the story of the ongoing revival of the American chestnut, I thank Ruth Gregory at the American Chestnut Foundation for information and photos, along with Bart Chezar in Brooklyn and Sara Fern Fitzsimmons.

A number of nonprofit tree groups and their leaders and staff have been more than generous with their time: At American Forests, former executive director Deborah Gangloff, nurseryman Jeffrey Meyer, and current vice president, Lea Sloan. At the National Arbor Day Foundation, I thank John Rosenow and Dan Lambe for numerous favors and help from the start. The Arbor Day Community & Urban Forestry conferences have been a tremendous forum for learning about urban forestry and meeting its devoted practitioners. At the Sacramento Tree Foundation, thanks to Ray Tretheway and Anne Fenkner for education and tours, and Misha Sarcovich at SMUD for much data. At TreePeople in L.A., thanks to Andy Lipkis, Mashani Allen, Jessica Jewell, Jim Hardie, Rachel Malarich, and Danny Carmichael. And to Lisa Sarno for her help on L.A. tree policy. At Berkeley, thanks to Trish Roque and David K. Smith at the University of California Museum of Paleontology for their efficient help with Chaney Expedition photos.

At Harvard's Arnold Arboretum, Peter Del Tredici has been very kind over the years with tree tours, lunches, and answering so many questions. I also owe much gratitude to the arboretum's excellent librarians Sheila Connor (now retired), Lisa Pearson, and Larissa Glasser. At the Morton Arboretum, thanks to Gerard Donnelly, Rita Hastert, and Patricia MacMillan, as well as Joy Morton biographer Jim Ballowe. The arboretum's 2011 Urban Tree Growth symposium was a deeper view into tree science.

Thanks to Jessica Baumert of the Woodlands Cemetery for her hospitality on William Hamilton's back porch, and to David Hewitt for his arboreal tour of the wondrous cemetery grounds.

In Cincinnati, Ohio, city foresters Jim Burkhardt and Dave Gamstetter kindly showed me Eden Park and its various historic groves.

Also, many friends in other cities made my Tree Travels most enjoyable. A big thank-you to Judith Weinstein and Matt MacCumber and Lew and Sue Erenberg in Chicago for much hospitality and fun over the years. Dear friend Jennifer Brown in Ohio put me up and drove me here and there, entertaining me with tales of her woodchucks. Ann Boulton and Hank Young were outstanding hosts in Oklahoma City. In New York City, Peggy and Bob Sarlin were once again gracious pals as I came and went over the years. In L.A., it was a pleasure to stay with the ever-lovely Amy Gelfand and David Misch. In Portland, Oregon, old friend Peggy Noto and her family were fine company. And I truly want to thank Nancy Buley for a memorable day at J. Frank Schmidt & Son Company in Oregon learning about the 100-plus steps of raising a young tree and for her help in various realms. In Boston, Kathy Jacob, the swashbuckling archivist and longtime friend, often put me up and drove me about. In Tokyo, my dear friend Mutsuko Murakami and her family were such kind hosts for ginkgo viewing and then the cherry blossom festival, as were my brother Denis and his family in Fukuoka.

Here in Baltimore, I have treasured getting to know the local tree huggers, launching the Baltimore Tree Trust with a great group of idealists, as well as serving on the city's Forestry Board, all of us working to make this a city where everyone lives on a tree-lined street.

And much love to my own family—husband, daughter, brothers and their families, in-laws, cousins, and all—and many friends who have been very good-natured about my transformation into a tree hugger and encouraging through all the long years it seemed to take to get this book done. While I regret that my wonderful parents are not alive to see this book, I am indebted to them for so much, including my own love of nature and the outdoors.

And during this project's slow gestation, I was happy once again to be working with my longtime agent, the excellent Eric Simonoff at William Morris Entertainment, and my lovely three-time editor, Rick Kot of Viking Press, who even took the photograph of the New York Survivor Tree. Diego Nunez, Viking Penguin editorial assistant, has been an invaluable ally as we wrangled images, photos, and many details. I look forward to working with Lindsay Prevette, director of publicity, Kristin E. Matzen, Emma Mohney, Lydia Hirt, and Caitlin Kleinschmidt.

Notes

INTRODUCTION: "We Appreciate the Symmetry of Human and Sylvan Life"

xiv **"stainless steel version":** Blake Gopnik, "Art Review: Roxy Paine's 'Graft' at the National Gallery Sculpture Garden," *Washington Post,* November 1, 2009.

xv **"an unspeakable degree":** Margaret Bayard Smith, *Forty Years of Washington Society* (New York: Charles Scribner's Sons, 1906), pp. 10–12.

xvi **natural air conditioners:** Colin Tudge, *The Tree: A Natural History of What Trees Are, How They Live, and Why They Matter* (New York: Three Rivers Press, 2005), p. 255.

xvii **"foster republican sentiments":** Michael Pollan, "Look Who's Saving Elm," *New York Times Sunday Magazine,* October 31, 1993.

xvii **"our own lives":** Thomas J. Campanella, *Republic of Shade: New England and the American Elm* (New Haven, CT: Yale University Press, 2003), pp. 4–5.

xvii **"out of traffic":** Nalini M. Nadkarni, *Between Earth and Sky: Our Intimate Connections to Trees* (Berkeley, CA: University of California Press, 2008), p. 11.

xix **"ozone and smog":** Greg McPherson, interview with the author, March 20, 2010.

xix **"do for us":** Rowan Rowntree, interview with the author, July 20, 2010.

xix **"names: like people":** "Tree Museum: A public art project by Katie Holten," pamphlet from Bronx Museum of the Arts, Fall 2009.

xx **"the Bronx sky":** Ian Frazier, "On the Street Dial-a-Tree," *The New Yorker,* July 20, 2009.

xx **"difficult and unrewarding":** David Allen Sibley, *The Sibley Guide to Trees* (New York: Knopf, 2009), p. xi.

xx **"it's a shrub":** Ibid.

CHAPTER ONE: "So Great a Botanical Curiosity" and "The Celestial Tree": Introducing the Ginkgo and Ailanthus

2 **"to your garden":** William Hamilton to Thomas Jefferson, July 7, 1806, Thomas Jefferson Papers Series online, General Correspondence.

2 **"seen in England":** Karen Madsen, "To Make His Country Smile: William Hamilton's Woodlands," *Arnoldia* 49, no. 2 (Spring 1989): 14.

2 **their "balmy odours":** Oliver Oldschod, "The Woodlands," *The Portfolio* 2, no. 3 (December 1809): 506.

2 **"had not procured":** Rev. Dr. Manasseh Cutler, letter dated November 22, 1803, in *Pennsylvania Magazine of History and Biography* v.8 (1884), p. 109.

3 **"it with regret":** Oldschod, "The Woodlands," 507.

3 **“that precious blossom”:** Sarah P. Stetson, “William Hamilton and His Woodlands,” *Pennsylvania Magazine of History and Biography* 73, no. 1 (January 1949): 32.

3 **“collection of plants”:** Ibid.

3 **“and his Madeira”:** James A. Jacobs, “William Hamilton and the Woodlands: A Construction of Refinement in Philadelphia,” *Pennsylvania Magazine of History and Biography* 30, no. 2 (April 2006): 182.

3 **“good eatable nut”:** William Hamilton to Thomas Jefferson, July 7, 1806, Thomas Jefferson Papers Series online, General Correspondence.

3 **“tree, magnolia,” etc.:** John W. Harshberger, “The Old Gardens of Pennsylvania: The Woodlands,” *Garden Magazine,* April 1921, p. 120.

3 **“the Dogs too”:** Benjamin H. Smith and William Hamilton, “Some Letters from William Hamilton, of the Woodlands, to His Private Secretary,” *Pennsylvania Magazine of History and Biography* 29, no. 1 (1905): 71.

3 **specifically his ginkgo:** Peter Crane, *Gingkgo* (New Haven, CT: Yale University Press, 2013), p. 221.

4 **“disposition of them”:** Thomas Jefferson to William Hamilton, March 22, 1807, Thomas Jefferson Papers Series online, General Correspondence.

4 **“of your country”:** Thomas Jefferson to William Hamilton, March 1, 1808, Thomas Jefferson Papers Series online, General Correspondence.

6 **“four in circumference”:** Peter Del Tredici, “The Ginkgo in America,” *Arnoldia* 41, no. 4 (July/August 1981): 153.

6 **“of this circle”:** David Schuyler, *Apostle of Taste, Andrew Jackson Downing, 1815–1852* (Baltimore: Johns Hopkins University Press, 1996), p. 86.

7 **“the pleasure ground”:** Andrew Jackson Downing, *A Treatise on the Theory and Practice of Landscape Gardening,* (New York: Wiley and Putnam, 1841), pp. 262–63.

8 **“massy and irregular”:** Andrew Jackson Downing, *A treatise on the theory and practice of landscape gardening,* 2nd ed. (New York: Wiley and Putnam, 1844), p. 206.

8 **“when frost commences”:** Ibid.

8 **“New Yorkers’ dwellings”:** “The Ailanthus,” *New York Daily Times,* December 27, 1852, p. 3.

9 **“her elegant furniture:”** Ibid.

9 **“become completely orientalized”:** Ibid.

9 ***“planted with trees”*****:** Andrew Jackson Downing, “Article 1—No Title,” *Horticulturalist and Journal of Rural Art and Rural Taste* 1, no. 9 (March 1847): 393.

10 **“greater metropolitan favorite”:** Ibid.

10 **“the Great Metropolis”:** Andrew Jackson Downing, “The London Parks,” *Rural Essays,* September 1850.

10 **“the lecture room”:** Robert Twombly, ed. *Andrew Jackson Downing: Essential Texts* (New York: W. W. Norton, 2012); p. 224. From Downing, “A Talk About Public Parks and Gardens” (1848).

11 **“‘like a rose’”:** George B. Tatum and Elizabeth B. MacDougall, eds. *Prophet with Honor: The Career of Andrew Jackson Downing* (Philadelphia: Athenaeum of Philadelphia, 1989), p. 302.

11 **“piece of ground”:** “Shade Trees in Cities,” *Horticulturalist and Journal of Rural Art and Rural Taste* 8 (August 1, 1852): 345.

12 **“regal as Zenobia”:** Ibid.

12 **“before they open”:** “The Ailanthus or ‘Tree of Heaven,’” *New York Daily Times,* December 27, 1852, p. 3.

12 ***“or measuring worms”*****:** “Julia Pleads for the Ailanthus,” *New York Daily Times,* July 4, 1855, p. 2.

13 **"it will flourish":** Oliver E. Allen, "A Tree Grows in America," *American Heritage* 35, no. 3 (April/May 1984).
13 **"the same direction":** Albert de Broglie, ed. *Memoirs of the Prince of Talleyrand,* vol. 1 (New York: G. P. Putnam's Sons, 1891), p. 176.
13 **"for his grandchildren":** George B. Emerson, *A Report on the Trees and Shrubs Growing Naturally in the Forests of Massachusetts* (1846), p. 2.
14 **"rains of autumn"** Ibid., p. 5
14 **"away dusty cobwebs":** Downing, "Article 1—No Title," 393.
14 **"an excellent work":** "Rare Trees and Pleasure Grounds of Pennsylvania," *Horticulturalist and Journal of Rural Art and Rural Taste* 6 (1851): 128.
14 **"of absolute necessity":** E. Gregory McPherson and Nina Luttinger, "From Nature to Nurture: The History of Sacramento's Urban Forest," *Journal of Arboriculture* 24, no. 2 (March 1998): 74.
15 **"sting, and bite":** Elbert Hubbard, *A Brother to the Trees* (East Aurora, NY: Roycrofters, 1909), p. 13.
15 **"friend and ally":** Francis Parkman, "The Forests and the Census," *Atlantic Monthly* 55, no.6 (June 1885), p. 83.
15 **"all other trees":** Emerson, *A Report on the Trees and Shrubs Growing Naturally in the Forests of Massachusetts,* p. 288.
15 **"the American army":** George B. Emerson, "A Report on the Trees and Shrubs Growing Naturally in the Forests of Massachusetts," 4th edition (Boston: Little, Brown & Company, 1887), p. 333.
16 **"decrepit and condemned":** Downing, "Article 1—No Title," 393.
16 **"planter of trees":** Ibid.

CHAPTER TWO: "No Man Does Anything More Visibly Useful to Posterity Than He Who Plants a Tree": Inventing Arbor Day and Cities of Trees

18 **"heritage of trees":** Nathaniel Hillyer Egleston, *Arbor Day: Its History and Observance* (Washington, D.C.: Government Printing Office, 1896), p. 13.
19 **"to American authors":** John B. Peaslee, *Thoughts and Experiences In and Out of School* (Cincinnati: Curts & Jennings, 1900), p. 110.
19 **"of my boyhood":** Ibid., p. 272.
20 **"the immortal Washington":** "Tree Culture," *Cincinnati Commercial Tribune,* April 27, 1882, p. 4.
20 **"on the continent":** Peaslee, *Thoughts and Experiences In and Out of School,* p. 118.
21 **"homes of the people":** Ibid., p. 142.
21 **"shade to the horizon":** J. Sterling Morton, "Arbor Day," *The Youth's Companion,* April 30, 1896, p. 227.
22 **"and malignant administration":** James Ballowe, *A Man of Salt and Tears: The Life of Joy Morton* (De Kalb: Northern Illinois University Press, 2009), p. 19.
22 **"and bleak prairies":** "The Origin of Arbor Day," *Nebraska City News,* April 25, 1885, p. 1.
22 **"a fever patient":** Morton, "Arbor Day," p. 227.
22 **"this dollaring age":** Eric Rutkow, *American Canopy* (New York: Scribner, 2012), pp. 132–33.
22 **"of the future":** Morton, "Arbor Day," p. 227.
23 **"on every side":** E. Gregory McPherson and Nina Luttinger, "From Nature to Nurture: The History of Sacramento's Urban Forest," *Journal of Arboriculture* 24, no. 2 (March 1998): 74.
23 **"with an axe":** "Vandalism," *Sacramento Transcript* 1, no. 21 (May 18, 1850).
23 **"of the summer":** Ibid., p. 75.

23 **beautifying the landscape:** "State Arbor Day," *Sacramento Daily Union* 56, no. 19 (September 13, 1886).
24 **"or fruit-bearing trees":** Ibid.
24 **"the best results":** Ibid.
24 **"generation will play":** Egleston, *Arbor Day*, p. 33.
25 **"gay deciduous neighbors":** John B. Peaslee, *Trees and Tree-Planting with Exercises and Directions for the Celebration of Arbor Day* (Cincinnati: Ohio State Forestry Association, 1884), p. 46.
25 **"of our times":** Egleston, *Arbor Day*, p. 13.
25 **"green his memory":** Ibid.
25 **"plants a tree":** Ibid., p. 65.
26 **"(objection to Sycamores)":** Andrea Wulf, *Founding Gardeners* (New York: Knopf, 2011), p. 14.
26 **"Tulip-Poplar rose conspicuous":** Margaret Bayard Smith, *The First Forty Years of Washington Society* (New York: Scribner's, 1906), pp. 10–12.
27 **"in their bareness":** Erle Kaufman, *Trees of Washington: The Man, the City* (Washington, D.C.: The Outdoor Press, 1932) p. 56.
27 **"leisure by people":** Ibid.
28 **"of the poor":** Smith, *The First Forty Years of Washington Society*, pp. 10–12.
28 **"for their preservation":** Margaret Bayard Smith, *A Winter in Washington* (New York: E. Bliss & E. White, 1824), p. 41.
28 **"of our valleys":** Ibid., p. 43.
28 **"that great thoroughfare":** Pamela Scott, "The City of Living Green: An Introduction to Washington's Street Trees," *Washington History* 18, no. 1–2 (2006): 22.
29 **"of this country":** "Alexander R. Shepherd, The Man Who Redeemed and Beautified Washington," *The Evening Star*, September 24, 1887.
29 **"to the nose":** Ibid.
29 **"to see it":** Ibid.
29 **"wanted to do":** Wilhelmus Bogart Bryan, *A History of the National Capital* (New York: Macmillan Company, 1916) vol. 2, p. 596.
30 **"and half-paved sidewalks":** Ibid., p. 607.
30 **"is damned room":** Thanks to John Richardson, Shepherd biographer, for this quote from a family descendant.
30 **"of the plan":** Bryan, *A History of the National Capital*, p. 601.
30 **"diversion of forms":** *Report of the Board of Public Works of the District of Columbia* (Washington, D.C.: Chronicle, 1973), p. 89.
31 **"of dust-laden siroccos":** Scott, "The City of Living Green," *Washington History*, p. 27.
32 **shape proper structure:** Ibid., pp. 89–90.
32 **"health and comfort":** "The Trees in Washington," *New York Times*, May 31, 1885.
32 **"in a forest":** Peter Henderson, "Street Trees of Washington," *Harper's New Monthly Magazine* 77 (June 1888), p. 285.
33 **"ornamentation of public streets":** Ibid.
33 **"a poor man":** Eugene L. Meyer, "Boss," *Washington Post Magazine*, April 22, 1991, p. 45.
33 **"two Army ambulances":** Ibid.

CHAPTER THREE: "A Demi-God of Trees" and "The Tree Doctor": Charles Sprague Sargent and John Davey

36 **"very sticky molasses":** Donald Worster, *A Passion for Nature: The Life of John Muir* (New York: Oxford University Press, 2008), p. 334.

36 **"with exquisite taste":** S. B. Sutton, *Charles Sprague Sargent and the Arnold Arboretum* (Cambridge, MA: Harvard University Press, 1970), p. 158.

36 **"formal and informal":** Ibid.

37 **"and marked influence":** Henry Wintrop Sargent, supplement in *A Treatise on the Theory and Practice of Landscape Gardening,* 6th ed. (New York: Wiley and Putnam, 1859).

37 **"of my correspondents":** Charles Sargent to John Muir, October 28, 1895, Arnold Arboretum Archives.

38 **"dollars a year":** C.S. Sargent, "The First Fifty Years of the Arnold Arboretum," *Journal of the Arnold Arboretum* 3, no. 3 (January 1922), p. 130.

38 **"to the imagination":** Sutton, *Charles Sprague Sargent,* p. 73.

38 **"and aesthetic value":** W. T. Councilman, "Charles Sprague Sargent," *Later Years of the Saturday Club, 1870–1920,* ed. M. A. DeWolfe Howe (Boston: Houghton Mifflin, 1927), p. 289.

39 **"after four o'clock":** Sutton, *Charles Sprague Sargent,* p. 179.

39 **or anywhere else:** E. H. Wilson, "Charles Sprague Sargent," *Harvard Graduate Magazine,* June 1917, p. 610.

39 **"a Demi God":** Char Miller, *Gifford Pinchot and the Making of Modern Environmentalism* (Washington, D.C.: Island Press, 2001), p. 131.

39 **"made everything light":** John Muir, "Sargent's *Silva,*" *Atlantic Monthly* 92, 9 (July 1903), p. 22.

40 **"as a nation":** "American Trees for America," *Garden and Forest,* December 29, 1897, p. 509.

40 **"in the shade":** Elbert Hubbard, *A Brother to the Trees* (East Aurora, NY: Roycrofters, 1909), p. 25.

40 **"in the leaf":** John Davey, *The Tree Doctor* (New York: Saalfield, 1907), p. 14.

40 ***The Tree Doctor:*** Ibid., p. 32.

41 **"practical curative process":** Robert E. Pflegger, *Green Leaves: A History of the Davey Tree Expert Company,* (Kent, OH: Davey Tree Expert Company, 2007), p. 8.

41 **"local 'tree butchers'":** Ibid.

42 **"boy or girl":** Davey, *The Tree Doctor,* p. 32.

43 **"for large trees":** Ibid., pp. 50–52.

43 **customer for life:** Pfleger, *Green Leaves,* p. 27.

44 **"of his children":** "Washington Trees Associated with Memories of Famous Men," *Washington Post,* July 28, 1907, p. R-7.

44 **"interest in forests":** Worster, *Passion for Nature,* p. 368.

44 **"North American trees":** Stephen Fox, *John Muir and His Legacy: The American Conservation Movement* (Boston: Little, Brown, 1981), p. 115.

44 **"John Quincy Adams":** "Washington Trees Associated with Memories of Famous Men," *Washington Post,* July 28, 1907, p. R-7.

46 **"must have trees":** "President Urges to Observe Arbor Day," *Los Angeles Times,* April 12, 1907, p. 12.

46 **died in 1957:** Irving Williams, retired White House gardener, interview with the author, April 2, 2015.

46 **"passed under it":** Melanie Choukas-Bradley, *City of Trees: The Complete Field Guide to the Trees of Washington, D.C.* (Charlottesville: University of Virginia Press, 2008), p. 33.

46 **"round, fat dumpling":** John Whitcomb and Claire Whitcomb, *Real Life at the White House* (New York: Routledge, 2000), p. 62.

CHAPTER FOUR: "This Fungus Is the Most Rapid and Destructive Known": A Plague Strikes the American Chestnut

49 **"all the trees":** Eric Rutkow, *American Canopy: Trees, Forests, and the Making of a Nation* (New York: Scribner, 2012), p. 214.

49 **"middle of it":** Michael Kimmelman, "Where Art Grows on Trees (and Everywhere Else)," *New York Times*, December 30, 2005.

49 **"an oak stump":** William A. Murrill, *Trees* (Gainesville, FL: self-published, 1945), p. 11.

50 **"the fungus grew":** Susan Freinkel, *American Chestnut: The Life, Death, and Rebirth of a Perfect Tree* (Berkeley: University of California Press, 2007), p. 34.

51 **"years at best":** William A. Murrill, "A New Chestnut Disease," *Torreya* 6, no. 9 (September 1906): 186–89.

51 **"the Zoological Park":** Hermann W. Merkel, "A Deadly Fungus on the American Chestnut," Tenth Annual Report of the New York Zoological Society, January 1906, p. 98.

51 **"is so great":** Murrill, "A New Chestnut Disease," pp. 186–89.

51 **"in the country":** "States Are to Act on Chestnut Blight," *New York Times*, February 18, 1912, p. 6.

51 **"cabmen and newsboys":** Henry David Thoreau, *Wild Fruits* (New York: W. W. Norton, 2000), pp. 210 and 213.

51 **"'a chest-nutting'":** Henry Ward Beecher, "This Darling Old Fellow," in *Mighty Giants, an American Chestnut Anthology*, ed. Chris Bolgiano (Bennington, VT: Images from the Past, 2007), p. 27.

51 **"of chestnut festival":** Bolgiano, ed., *Mighty Giants*, p. 45, from "Nutting Parties," in *The American Girls Handy Book* (New York: Charles Scribner's Sons, 1887).

52 **"his faithful dog":** William A. Murrill, *Autobiography* (Gainesville, FL: self-published, 1945), p. 71.

52 **"a promising future":** Ibid., p. 73.

52 **told a reporter:** "The Costly Blight of the Chestnut Canker," *New York Times*, May 31, 1908, p. 9.

52 **"distressed to lose":** "Chestnut Trees Face Destruction," *New York Times*, May 21, 1908, p. 4.

53 **"ease in climbing":** Beecher, "This Darling Old Fellow," in Bolgiano, ed., *Mighty Giants*, p. 26.

53 **them with sulfur:** "All Chestnut Trees Here Are Doomed," *New York Times*, July 30, 1911, p. 6.

53 **"fame and influence":** Murrill, *Autobiography*, p. 70.

54 **"beginning to wither":** "Chestnut Trees Face Destruction."

54 **"chestnut trees again":** "Mysterious Blight Kills Chestnuts by Thousands," *New York Times*, October 2, 1910, p. 2.

54 **"general construction purposes":** "The Costly Blight of the Chestnut Canker."

55 **"their pockets bulged":** "Caught Roosevelt Felling a Tree," *New York Times*, July 17, 1910, p. 5.

55 **"their distinctive beauties":** Theodore Roosevelt, *An Autobiography* (New York: Da Capo, 1985), p. 339.

56 **"was given them":** "The Costly Blight of the Chestnut Canker."

56 **"or fifteen years":** Freinkel, *American Chestnut*, p. 47.

56 **"remains elsewhere unchecked":** Ibid., p. 51.

56 **"it at night":** Murrill, *Autobiography*, p. 74.

57 **"native tree standing":** "States Are to Act on Chestnut Blight."
57 **"some concerted action":** "Conference on the Chestnut Tree Blight," *American Lumberman*, February 24, 1912, p. 73.
57 **"foredoomed to failure":** Murrill, *Autobiography*, p. 79.
58 **would be "un-American":** Freinkel, *American Chestnut*, p. 55.
58 **"come upon us":** Ibid., p. 58.

CHAPTER FIVE: "Washington Would One Day Be Famous for Its Flowering Cherry Trees": Eliza Scidmore and David Fairchild

60 **"two thousand years":** Eliza Ruhamah Scidmore, "The Cherry Blossoms of Japan," *Century Magazine* 79, no. 5 (March 1910): 643.
61 **"the blossoming trees":** Eliza R. Scidmore, *Jinrikisha Days in Japan* (New York: Harper & Bros., 1891), p. 73.
61 **"blossoms it celebrates":** Eliza R. Scidmore, "Japan Is Joyous in Cherry Time," *Chicago Daily Tribune*, June 6, 1904, p. 1 (her dateline is April 16).
62 **"Nothing happened":** Eliza R. Scidmore, "Capital's Cherry Blossoms Gift of Japanese Chemist," *Sunday Star* (Washington), April 11, 1926.
62 **"of interesting people":** "For and About Women," *Washington Post*, February 4, 1894, p. 13.
63 **"the whole lot":** Scidmore, "Capital's Cherry Blossoms Gift."
63 **"the firing line":** "Letter from Miss Scidmore," *Chicago Daily Tribune*, December 19, 1903, p. 4.
64 **"the sweet potato":** David Fairchild, *The World Was My Garden* (New York: Charles Scribner's Sons, 1941), p. 20.
64 **"of the world":** Ibid., p. 31.
65 **"the United States":** Marjory Stoneman Douglas, *Adventures in a Green World: The Story of David Fairchild and Barbour Lathrop* (Coconut Grove, FL: Field Research Projects, 1973), p. 12.
65 **"want to go":** Douglas, *Adventures in a Green World*, p. 15.
66 **"with Mr. Lathrop":** Fairchild, *World Was My Garden*, p. 84.
66 **"thing by correspondence":** Ibid., p. 117.
67 **"things to eat":** Ibid., p. 381.
67 **cherry (*Prunus sargentii*):** Anthony A. Aiello, "Japanese Ornamenal Cherries—a 100-Year-Long Love Affair," *Arnoldia*, April 2012, p. 3.
67 **"in his yard":** Fairchild, *World Was My Garden*, pp. 312–13.
68 **"in sheltered spots":** Ibid., pp. 317–18.
69 **trees were flowering:** David Fairchild, "How the United States and Japan Entered a League of Flowers," *Geographic News Bulletin*, undated, U.S. National Arboretum Collection, Cherry Tree Files, series I, box 1, folder 19, USDA Library (Special Collections), Beltsville, Maryland.
69 **"the river's bank":** Scidmore, "Capital's Cherry Blossoms Gift."
70 **"of the occasion":** Fairchild, *World Was My Garden*, p. 412.
70 **"flowering cherry trees":** Ibid.
70 **"in Potomac Park":** Scidmore, "Capital's Cherry Blossoms Gift."
70 **"them up then":** Ibid.
71 **"of floral appreciation":** David Fairchild to Spencer Cosby, April 4, 1909, U.S. National Arboretum Collection, Cherry Tree Files, series I, box 1, folder 19, USDA Library (Special Collections), Beltsville, Maryland.
71 **"make any show":** Scidmore, "Capital's Cherry Blossoms Gift."

71 **"of this Republic":** Col. Spencer Cosby to Hon. Yukio Ozaki, December 10, 1909, U.S. National Arboretum Collection, Cherry Tree Files, series I, box 1, folder 18, USDA Library (Special Collections), Beltsville, Maryland.

71 **"families in California":** Philip J. Pauly, *Fruits and Plains: The Horticultural Transformation of America* (Cambridge, MA: Harvard University Press, 2007), p. 148.

72 **"liberty, patriotism, union":** "Cherry Trees of Japan," *New York Times*, August 27, 1909, p. 6.

72 **"making them live":** Fairchild, *World Was My Garden*, p. 412.

73 **"to this country":** Roland M. Jefferson and Alan M. Fusonie, *The Japanese Flowering Cherry Trees of Washington, D.C.: A Living Symbol of Friendship* (Washington, D.C.: USDA, 1977), p. 10.

73 **Marlatt, was "imperative":** Charles Marlatt to the Secretary of Agriculture, January 19, 1910, U.S. National Arboretum Collection, Cherry Tree Files, series I, box 3, folder 44, USDA Library (Special Collections), Beltsville, Maryland.

73 **"the entire shipment":** Fairchild, *World Was My Garden*, p. 412.

74 **"of the Speedway":** Ibid., p. 413.

74 **"could possibly afford":** Helen Taft, *Recollections of Full Years* (New York: Dodd, Mead, 1914), p. 362.

74 **"growth next spring":** Scidmore, "Capital's Cherry Blossoms Gift."

74 **"any new infections":** "Wounding to Japanese Sensibilities," *New York Times*, January 31, 1910, p. 1.

75 **"the entire country":** $2,000,000,000 a Year to Feed Our Pests," *Washington Post*, February 6, 1910, p. MT 10.

75 **"of bugs develop":** "Gift Is Destroyed," *Washington Star*, January 29, 1910.

75 **"to this article":** Scidmore, "The Cherry Blossoms of Japan," p.643.

75 **White House aide:** Carl Sferrazza Anthony, *Nellie Taft* (New York: William Morrow, 2005), p. 302.

75 **"saplings were ready":** Ozaki Yukio, *Autobiography of Ozaki Yukio* (Princeton, NJ: Princeton University Press, 2001), p. 232.

76 **"or plant diseases":** Jefferson and Fusonie, *Japanese Flowering Cherry Trees*, p. 20.

76 **"the White House":** Col. Spencer Cosby to Mayor Yukio Ozaki, April 4, 1912, U.S. National Arboretum Collection, Cherry Tree Files, series I, box 1, folder 18, USDA Library (Special Collections), Beltsville, Maryland.

77 **"far into May":** Scidmore, "Capital's Cherry Blossoms Gift," p. 4.

77 **"and by dawn":** Ibid.

CHAPTER SIX: "I Knew That There Were No Roads in China": Plant Explorers Frank Meyer and E. H. Wilson

79 **told the journalist:** "All Chestnut Trees Here Are Doomed," *New York Times*, July 30, 1911, p. 6.

80 ***New York Times*:** "A Magnificent Chestnut Tree," *New York Times*, January 7, 1913.

80 **"with the blight":** David Fairchild, *The World Was My Garden* (New York: Charles Scribner's Sons, 1938), p. 405.

81 **"throughout the interior":** Ibid., p. 315.

81 **"that first interview":** Ibid.

82 **"green with mold":** Ibid., p. 316.

82 **"never forgot it":** Isabel Shipley Cunningham, *Frank N. Meyer: Plant Hunter in Asia* (Ames: Iowa State University Press, 1984), p. 147.

82 **"economic Chinese farmers":** Chris Bolgiano, ed. *Mighty Giants: An American Chestnut Anthology* (Bennington, VT: Images from the Past, 2007), pp. 165–66.

83 **"on Meyer's cultures":** Fairchild, *World Was My Garden*, p. 406.
83 **"eighties or nineties":** Ibid.
83 **"already too late":** Susan Freinkel, *American Chestnut: The Life, Death and Rebirth of an American Tree* (Berkeley: University of California Press, 2007), p. 68.
84 **March 31, 1908:** Cunningham, *Frank N. Meyer*, p. 277.
84 **"of the autocrats":** E. H. Wilson, "Charles Sprague Sargent," *Harvard Graduate Magazine*, June 1927, p. 614.
84 **"the Yang'tsi Valley":** David Fairchild, USDA, to E. H. Wilson, Arnold Arboretum, December 26, 1906, Archives of the Arnold Arboretum, Harvard University.
84 **"from his men":** S. B. Sutton, *Charles Sprague Sargent and the Arnold Arboretum* (Cambridge, MA: Harvard University Press, 1970), p. 233.
85 **"benignly upon me":** Daniel J. Foley, ed., *The Flowery World of "Chinese" Wilson* (London: Macmillan, 1969), p. 34.
86 **waiting egg cell:** Peter Crane, *Ginkgo* (New Haven, CT: Yale University Press, 2013), pp. 67–68.
86 **"the Lower Cretaceous":** Peter Del Tredici, "The Ginkgo in America," *Arnoldia* 41, no. 4 (July/August 1981): 152.
86 **"the male nucleus":** Fairchild, *World Was My Garden*, p. 255.
87 **"cilla wave energetically":** Crane, *Ginkgo*, pp. 67–68.
87 **"of sexual awareness":** Ibid., p. 66.
87 **"a wild state":** Charles Sprague Sargent, "Notes on Cultivated Conifers: I," *Garden and Forest* 502 (October 6, 1897): 390.
87 **beyond his report:** C. S. Sargent, "Ginkgo Biloba," *Arnold Arboretum Bulletin of Popular Information* , n.s., 2 (1916): 51–52.
88 **"shall never know":** E. H. Wilson, "The Romance of Our Trees—II. The Gingko," *Garden Magazine* 30, no. 4 (November 1919): 144.
88 **"fully appreciate them":** Fairchild, *World Was My Garden*, p. 415.
88 **had become diseased:** David Fairchild, "Plant Record" from In the Woods, National Arboretum Archives, National Agricultural Library, Beltsville, Maryland.
89 **"throughout the country":** Fairchild, *World Was My Garden*, p. 434.
90 **"from the Professor":** Ibid.
90 **"too many errors":** C. S. Sargent to David Fairchild, January 16, 1917, Archives of the Arnold Arboretum, Harvard University.
90 **"deeply to possess":** E. H. Wilson, "Japanese Cherries and Asiatic Crabapples," *Garden Magazine*, 27, no. 3 (March 19, 1916): 76.
90 **"color is ravishing":** E. H. Wilson, *Plant Hunting* (Boston: Stratford, 1927), p. 222.
91 **"sorts in Boston":** Fairchild, *World Was My Garden*, p. 424.
91 **"the Arakawan collection":** Anthony S. Aiello, "Japanese Flowering Cherries: A 100-Year-Long Love Affair," *Arnoldia* 69, no. 4 (April 2012): 6.
91 **"search for Meyer":** Papers of Frank N. Meyer, National Agricultural Library Archives, Beltsville, Maryland, p. 220.
91 **"up alive somewhere":** David Fairchild to C. S. Sargent, June 7, 1918, Archives of the Arnold Arboretum, Harvard University.
91 **"to depress me":** United States Bureau of Plant Industry, Foreign Seed and Plant Introduction, *South China Explorations, Supplementary Report*, September 21, 1918, p. 25. Frank N. Meyer Collection, Special Collections, National Agricultural Library.
92 **"work among humanity":** Ibid., p. 26.
92 **"of grey trousers":** Ibid., p. 31.

CHAPTER SEVEN: "A Poem Lovely as a Tree": Cherishing Memorial and Historic Trees

94 **"gloom about them":** Michelle Robbins, "Rooted in Memory," *American Forests*, 109, no. 1 (Spring 2003): 40.

94 **"ever had before":** "Trees: Memorials to Soldiers," *Washington Post*, March 22, 1919, p. 6.

94 **"and tree planting":** "Memorial Trees in 1920," *American Forestry* 25, no. 312 (December 1919): 1537.

95 **"also be gold":** "Living Memorials," *American Forestry* 26, no. 318 (June 1920): 348–49.

95 **"in the service":** Ibid.

95 **"of the world":** "Putting Towns on Dress Parade," *American Forestry* 26, no. 320 (August 1920): 484.

96 **"a brilliant promise":** Thomas Vinciguerra, "A Tree Grows and Grows," *New York Times Sunday Book Review*, August 18, 2013, p. 26.

96 **"make a tree":** "Urge Memorial Trees," *New York Times*, December 26, 1918, p. 14.

97 **"comrades to fall":** Robbins, "Rooted in Memory," p. 42.

97 **"back from France":** Ibid., p. 43; Ernest H. Wilson, *Aristocrats of the Trees* (Boston: Stratford, 1930), p. xvi.

97 **the Great War:** Erle Kauffman, *Trees of Washington: The Man, the City* (Washington, D.C.: Outdoor Press, 1932), p. 76.

98 **"tell the world":** "Fifty Years of Arbor Days," *American Forestry* 28, no. 341 (May 1922): 279–82.

98 **"tree is planted":** "Memorial Trees," *Arnold Arboretum Bulletin of Popular Information* 8, no. 16 (October 23, 1922): 61–63.

98 **"tomb of Confucius":** "Arbor Day Finds Old Trees Lusty," *New York Times*, April 25, 1926, p. 13.

99 **"with the builders":** Kauffman, *Trees of Washington*, p. 40.

99 **"trees from them":** "Historic Trees Here," *Washington Post*, May 25, 1902.

99 **"soldier and patriot":** Kauffman, *Trees of Washington*, p. 37.

100 **"stood so long":** Samuel Francis Batchelder, "The Washington Elm Tradition," *Cambridge Historical Society* 18 (October 1926): 46.

CHAPTER EIGHT: "The Two Great Essentials for an Arboretum, Soil and Money": Chicago, D.C., and Boston

102 **"not been misspent":** S. B. Sutton, *Charles Sprague Sargent and the Arnold Arboretum* (Cambridge, MA: Harvard University Press, 1970), p. 330.

102 **"in San Francisco":** Ibid., pp. 322–23.

103 **"of great age":** "Memorandum with Reference to Morton Arboretum Location," included in letter from C. S. Sargent to Joy Morton, November 26, 1921, Archives of the Morton Arboretum, Lisle, Illinois.

103 **"plenty of it":** James Ballowe, *A Man of Salt and Trees: The Life of Joy Morton* (DeKalb: Northern Illinois University Press, 2009), p. 209.

103 **"its coal smoke":** Joy Morton to C. S. Sargent, December 29, 1921, Archives of the Morton Arboretum, Lisle, Illinois.

103 **"from the lake":** C. S. Sargent to Joy Morton, January 19, 1921, Archives of the Morton Arboretum, Lisle, Illinois.

103 **"soil and money":** C. S. Sargent to Joy Morton, September 21, 1923, Archives of the Morton Arboretum, Lisle, Illinois.

104 **"fun doing it":** Ballowe, *Man of Salt and Trees*, p. 212.

105 **"generally unrecognized system":** C. S. Sargent, *Manual of the Trees of North America* (1922; repr., New York: Dover Books, 1965), vol. 1, p. iii.
105 **"streets of Washington":** David Fairchild, *The World Was My Garden* (New York: Charles Scribner's Sons, 1941), p. 408.
105 **"of the name":** David Fairchild, "What a National Botanic Garden Should Mean," *GSA Bulletin*, December 1920, p. 57.
105 **"range of specimens":** "U.S. National Arboretum Celebrates," *Arbor Friends*, Winter 2002, p. 4.
106 **"deal about plants":** Fairchild, *World Was My Garden*, p. 409.
106 **"through that channel":** Ibid.
106 **"of her hand":** Martha McKnight, "Draft Report re: Arboretum and Other DC Gardens/Parks," 1942, Garden Club of America Archives.
106 **"I gladly accepted":** Fairchild, *World Was My Garden*, p. 409.
106 **"socially there again":** Ibid., p. 410.
107 **"of plant collections":** Frank P. Cullinan, *Our National Arboretum*, p. 7, reprint from *The Herbarist*, 1968, National Agricultural Library, USDA, Beltsville, Maryland.
107 **"in every way":** Katherine Sloan to Mrs. Fairfax Harrison, December 26, 1924, Garden Club of America Archives.
107 **"for the Bill":** Janet Noyes to President Calvin Coolidge, February 17, 1925, Garden Club of America Archives.
108 **"Mrs. Noyes' living room":** Cullinan, *Our National Arboretum*, p. 3.
108 **"for a generation":** Frederick V. Colville to Mrs. Frank B. Noyes, January 20, 1926, Garden Club of American Archives.
108 **"and a half":** Charles A. Goodrum, "The History and Development of the National Arboretum," Library of Congress Research Paper, Washington D.C., 1950, National Agricultural Library, USDA, Beltsville, Maryland.
108 **"fields of Paradise":** Mary Sargent Potter, "Silhouettes: My Mother, My Father," *Spur* 42, no. 11 (December 1, 1928): 142.
109 **"or dead ones":** "Many Newspapers Voice Appreciation of John Davey's Work," *Davey Bulletin* 11, no. 23 (December 1, 1923): 16.
109 **"shade we enjoy":** D. Q. Grove, "Tributes of Love from Father John's Co-workers," *Davey Bulletin* 11, no. 23 (December 1, 1923): 4
109 **"be the answer":** "John Davey's Character as Revealed in His Own Writing," *Davey Bulletin* 11, no. 23 (December 1, 1923): 12.
110 **"years or more":** Ernest H. Wilson, *Aristrocrats of the Trees* (Boston: Stratford, 1930), p. xvii.
111 **"and barbarously slaughtered":** "The Use of Arbor Day," *Garden and Forest*, April 17, 1889, p. 181.
111 **"the poetry recited":** "Planting New Places," *Garden and Forest*, April 16, 1890, p. 185.
111 **"venerable old age":** "Arbor Day," *Garden and Forest*, April 13, 1892, p. 169.
111 **"for Arbor Day":** "Plans for Arbor Day," *Washington Post*, April 2, 1911, p. 4.
111 **"the flowering cherries":** Michael A. Dirr, *Dirr's Hardy Trees and Shrubs: An Illustrated Encyclopedia* (Portland, OR: Timber Press, 1997), p. 304.
112 **"throughout the country":** Fairchild, *World Was My Garden*, p. 414.
112 **"Merritt of Washington":** Ibid., p. 475.
112 **for "distinguished Americans":** "Miss E. Scidmore, 72, Dies in Geneva," *Washington Post*, November 4, 1928, p. M23.
112 **decked with flowers:** Ibid.
112 **"Miss Scidmore enjoyed":** http://www.agreatblooming.com/anniversary-eliza-scidmores-death/.

113 **"all our trees":** Charles F. Thurston, "Good-bye, Chestnuts," *American Forestry* 29, no. 360 (December 1923): 753.

113 **"the animal kingdom":** John Davey, "The Trees of the Lord," *Chautauquan* 41, no. 4 (June 1905): 319.

113 **up to ten dollars:** Oliver E. Allen, "A Tree Grows in America," *American Heritage* 353, no. 3 (April/May 1984).

113 **"other European cities":** Behula Shah, "The Checkered Career of *Ailanthus altimissima*," *Arnoldia* 57, no. 3 (Fall 1997): 26.

114 **"could destroy it":** Betty Smith, *A Tree Grows in Brooklyn* (New York: Harper & Brothers, 1943), pp. 442–43.

CHAPTER NINE: "Imagine the Wiping Out of the Beautiful Avenues of Elms": Battling to Save an American Icon

116 **"to microscopic examination":** Berton Roueché, "A Great Green Cloud," *New Yorker*, July 15, 1961, p. 48.

117 **"of war gases":** "An Imported Problem," *Morton Arboretum Quarterly* 2, no. 1 (Spring 1966): 11.

117 **were the "vectors":** Richard J. Campana, *Arboriculture: History and Development in North America* (East Lansing: Michigan State University Press, 1999), p. 209.

118 **"seems highly improbable":** Curtis May and G. F. Gravatt, "The Dutch Elm Disease," USDA Circular No. 170, May 1931, Washington, D.C., p. 2.

118 **"and aesthetic importance":** F. A. Bartlett, "The Dutch Elm Disease," *American Landscape Architect* 3:2 (August 1930): 40–41.

119 **"the American elm":** Roueché, "A Great Green Cloud," p. 49.

119 **"European bark beetle"** Curtis May, "The Dutch Elm Disease in New Jersey," *Plant Disease Survey/ Division of Mycology and Disease Survey* 17 no. 9 (August 1, 1933): 107.

119 **"felled at once":** Ibid., p. 109.

120 **"have included elm":** George H. Hepting, "The Threatened Elms: A Perspective on Tree Disease Control," *Journal of Forest History* 21, no. 2 (April 1977): 92.

120 **"of classical beauty":** Andrew Jackson Downing, *A Treatise on the Theory and Practice of Landscape Gardening*, p. 154.

120 **"the humblest farm-house":** C. S. Sargent, "The American Elm," *Garden and Forest*, June 11, 1890, p. 281.

120 **"in Arlington Cemetery":** Roueché, "A Great Green Cloud," p. 38.

121 **"beneath an elm":** Ibid., p. 39.

122 **"of supporting himself":** Susan Freinkel, *American Chestnut: The Life, Death, and Rebirth of a Perfect Tree* (Berkeley: University of California Press, 2007), p. 59.

122 **"like an Asian tree":** Ibid., p. 98.

122 **very early days:** Ibid., p. 100.

122 **"a clean sweep":** "Progress in the Dutch Elm Disease Fight," *New York Times*, July 29, 1934, sec. X, p. 8.

123 **"of these insects":** "Dutch Elm Disease Shows Alarming Gains," *American Forests* 40, no. 11 (September 1934): 420.

123 **"in the beginning":** Ibid.

123 **"our chestnut trees":** "Lehman Asks Fund to Save Elm Trees," *New York Times*, August 4, 1934, p. 13.

124 **"will be maintained":** "U.S. Destroyed Last of Infected Elms," *New York Times*, April 21, 1935, p. B20.

124 **"from New York City":** "Old Tree Felled as Town Mourns," *New York Times,* July 14, 1935, p. N7.

124 **"written about it":** "Poems and Trees," *Washington Post,* July 17, 1935, p. 6. Joyce Kilmer's mother-in-law, Ada Alden, widow of a former editor of *Harper's Magazine* and herself a Larchmont resident, called the *New York Times* even as the elm was being carted off to say that Mrs. Stevens's elm had not inspired "Trees." The poem had been written in 1913, and the poet had moved to Larchmont several years later.

124 **"to this country":** "Old Tree Felled as Town Mourns."

125 **"to the ground":** "New Poem Created as Old Elm Passes," *New York Times,* July 17, 1935, p. 17.

126 **"a billion dollars":** "The Elm Emergency," *American Forests* 43, no. 5 (May 1937): 209.

126 **"campaign in time":** J. H. Faull, "The Viewpoint of the Arnold Arboretum on the Dutch Elm Disease," *Arnold Arboretum Bulletin of Popular Information,* series 4, vol. 4, no. 3 (April 1, 1936): 15–20.

127 **"is adequately financed":** William P. Wharton, "Are the Elms Being Saved?" *American Forests* 44, no. 12 (December 1938): 545.

128 **"preserve our elms":** J. H. Faull, "The Dutch Elm Disease Situation in the United States at the Close of 1938," *Arnold Arboretum Bulletin of Popular Information,* series 4, vol. 6, no. 13 (October 14, 1938): 75.

128 **"of an eye":** Thomas J. Campanella, *Republic of Shade: New England and the American Elm* (New Haven, CT: Yale University Press, 2003), p. 154.

129 **"to the heart":** "Wethersfield Women Save Historic Elm," *Hartford Times,* December 5, 1925.

129 **"rum about her":** Donald Culross Peattie, *A Natural History of Trees of Eastern and Central North America,* 2nd ed. (Boston: Houghton Mifflin Co., 1966), p. 240.

130 **"thus be prolonged":** S. F. Willard Jr., "The Wethersfield Elm," *Cornell Countryman,* October 1908.

130 **"for two years":** "Wethersfield Women Save Historic Elm."

131 **"Connecticut to Vermont":** Campanella, *Republic of Shade,* p. 158.

131 **"taken into consideration":** Faull, "The Dutch Elm Disease Situation," p. 78.

131 **"institutions and homes":** Campana, *Arboriculture,* p. 218.

131 **"will be lost":** "The Case of the Elms," *American Forests* 47, no. 1 (January 1941): 31.

131 **"funds for eradication":** George H. Hepting, "The Threatened Elms: A Perspective on Tree Disease Control," *Journal of Forest History* 21, no. 2 (April 1977): 92.

CHAPTER TEN: "A Forest Giant Just on the Edge of Extinction!": Discovering the Dawn Redwood

133 **"edge of extinction":** "How to Fund Botanical Expeditions," *Arnoldia* 58, no. 4 (Winter 1998): 21.

133 **"practically no cost":** Ibid.

134 **"horticulture and botany":** Ibid.

134 **"lovable human being":** Richard A. Howard, "Elmer Drew Merrill," *Journal of the Arnold Arboretum* 37, no. 3 (July 1956): 200.

134 **"policies and accomplishments":** Ibid., p. 202.

135 **to the troops:** Ibid., p. 210.

135 **"shipments were confiscated":** Ibid.

136 **"joy and excitement":** Hsueh Chi-ju, "Reminiscences of Collecting the Type Specimens of *Metasequoia glyptostroboides,*" *Arnoldia* 45, no. 4 (Winter 1985): 10.

136 **"[saved] the tree":** Ibid., p. 18.

137 **"stems called peduncles":** William Gittlen, *Discovered Alive: The Story of the Chinese Redwood* (Berkeley, CA: Pierside Publications, 1999), p. 30.
138 **"States and elsewhere":** E. D. Merrill, "Metasequoia, Another 'Living Fossil,'" *Bulletin of Popular Information of the Arnold Arboretum* 8, no. 1 (March 5, 1948): 6.
139 **"back to San Francisco":** Milton Silverman, *Search for the Dawn Redwoods* (unpublished manuscript, 1990), Archives of the Arnold Arboretum, Harvard University, p. 28.
140 **"apart, and poorer":** Ibid., p. 58.
141 **"needles in winter":** Ibid., p. 66.
141 **"of this canyon":** Ibid., p. 84.
141 **"a living dinosaur":** Milton Silverman, "Science Makes a Spectacular Discovery," *San Francisco Chronicle*, March 26, 1948, p. 1.
142 **"me for years":** Silverman, *Search for the Dawn Redwoods*, p. 132.
143 **"the American garden":** "Dr. David Fairchild, Botanist, Is Dead," *New York Times*, August 7, 1954.

CHAPTER ELEVEN: "There Was No Question That People Wanted to Save This Tree": Crusading for a New American Elm

146 **"the tree's plight":** "Famous Wethersfield Elm Losing Battle to Disease," *Hartford Times*, July 28, 1948.
148 **the *Hartford Times*:** "Famed Wethersfield Elm Goes," *Hartford Times*, May 29, 1953.
148 **"an aircraft engine":** "Fights Elm Tree Disease," *New York Times*, October 24, 1946, p. 27.
149 **"seen one there":** Rachel Carson, *Silent Spring* (Boston: Houghton Mifflin, 1962), p. 106.
149 **"nine of them":** Ronald Rood, "Nature's Broken Vase," *Audubon*, May 1969, p. 9.
149 **"can't bear it":** Carson, *Silent Spring*, p. 112.
149 **"what is threatened":** Ibid., p. 105.
149 **"Dutch elm disease":** Ibid., p. 114.
150 **"over our heads":** Thomas J. Campanella, *Republic of Shade* (New Haven, CT: Yale University Press, 2003), p. 172.
150 **"save this tree":** Ibid.
150 **"bathed in shadows":** Phillip J. Hutchison, "When Elm Street Became Treeless: Journalistic Coverage of Dutch Elm Disease, 1930–80," *Journalism History*, Summer 2012, p. 107.
151 **"the [American] chestnut":** John Hansel, "Hope for the Future of Our Elms" (speech), March 1970, Archives of the Elm Research Institute, Keene, New Hampshire.
151 **"Elm Research Institute":** Ibid.
151 **"than an elm?":** "Hope for Elms," *Time*, December 12, 1969.
151 **"was too absurd":** Campanella, *Republic of Shade*, p. 175.
152 **"in Burlington, Vermont":** Rood, "Nature's Broken Vase," p. 10.
152 **"its small budget":** Ibid.
152 **"and disease-resistant varieties":** George H. Hepting, "The Threatened Elms," *Journal of Forest History*, April 1977, p. 96.
152 **"University of Wisconsin":** "$30,000 E.R.I. Grant Will Complete Dutch Elm Disease 'Pot' Project" (press release), July 1970, Elm Research Institute Archives, Keene, New Hampshire.
152 **"close the windows":** John Hansel, interview with the author, August 23, 2010.
153 **"we are working":** John Hansel to Elm Research Institute funders and board, July 7, 1971, Elm Research Institute Archives, Keene, New Hampshire.
153 **"cause serious illness":** Walter Sullivan, "Elm Blight," *New York Times*, April 4, 1965, p. E7.

153 **"to set up":** Campanella, *Republic of Shade*, p. 173.
154 **"It Yourself solution":** John Hansel, interview with the author, August 24, 2010.
154 **"red within hours":** Ibid.
154 **"least 12 months":** Mark A. Stennes, "Dutch Elm Disease Chemotherapy with Arbotect 20-S and Alamo," in *The Elms: Breeding, Conservation, and Disease Management*, ed. Christopher P. Dunn (Boston: Kluwer Academic, 2000), p. 176.
154 **"then be secure":** John Hansel, speech, March 1977, Elm Research Institute Archives, Keene, New Hampshire.
154 **"elm was sealed":** Keith Warren, interview with the author, December 15, 2010.
155 **"a control program":** "Dutch Elm Disease and You," *American Forests* 73 (June 1967): 45.
155 **"that was resistant":** John Hansel, interview with the author, August 24, 2010.
155 **"bark beetle infestation":** E. B. Smalley, R. P. Guries, and D. T. Lester, "American Liberty Elms and Beyond: Going from the Impossible to the Difficult," *in Dutch Elm Disease Research: Cellular and Molecular Approaches*, ed. M. Sticklan and J. Sherald (New York: Springer-Verlag, 1993), p. 27.
155 **"formed, vase-shaped tree":** Ibid.
156 **"Russia, China, Europe":** Hansel, interview with the author, August 24, 2010.
156 **"considered for release":** E. B. Smalley and R. P. Guries, "Breeding Elms for Resistance to Dutch Elm Disease," *Annual Review of Phytopathology* 31 (1993): 340–41.
156 **"'the American elms'":** Hansel, interview with the author, August 24, 2010.
157 **"improve on them":** Ray Guries, interview with the author, October 1, 2011.
157 **"elm at that":** Elm Research Institute, "Progress Report 1970," p. 7, Elm Research Institute Archives, Keene, New Hampshire.
158 **"plan and recommendations":** Minutes from November 12, 1975, Elm Research Institute board of directors meeting, Elm Research Institute Archives, Keene, New Hampshire.
158 **"parties working together":** John Hansel, interview with the author, August 24, 2010.
159 **"interested in trees":** Denny Townsend, interview with the author, July 27, 2010.
159 **"of moderate vigor":** E. B. Smalley, R. P. Guries, and D. T. Lester, "American Liberty Elms and Beyond: Going from the Impossible to the Difficult," in *Dutch Elm Disease Research: Cellular and Molecular Approaches*, ed. M. Sticklan and J. Sherald (New York: Springer-Verlag, 1993), p. 27.
160 **"elm bark beetle":** "New American Elms Restore Stately Trees," *USDA News & Events*, July 1996.
160 **"have escaped exposure":** Denny Townsend, interview with the author, July 27, 2010.
160 **"was the thinking":** Denny Townsend, interview with the author, October 6, 2010.
161 **"on that issue":** Guries, interview with the author, October 1, 2011.
161 **"of our progress":** John Hansel, "Hope for the Future of Our Elms" (speech), March 1970, p. 4, Elm Research Institute Archives, Keene, New Hampshire.
161 **"with other researchers":** Guries, interview with the author, October 1, 2011.

CHAPTER TWELVE: "Having Cities Work with Forces of Nature": The Rise of the New Urban Forestry

163 **"to look at":** Ronald Reagan, speech to the Western Wood Products Association, March 12, 1966, San Francisco.
164 **"own urban forests":** Dorothy Behlen, "Planting the President's Tree," *American Forests* 85 (August 1979): 21.

164 **and vacant lots:** Andy Kennery, "Erik Jorgensen: Canada's First Urban Forester," 2012 obituary written for the *Forestry Chronicle* and featured on https://treecanada.ca/en/programs/urban-forests/erik-jorgensen-canadas-first-urban-forester/.

164 **in removal costs:** Anne Whiston Spirn, *The Granite Garden: Urban Nature and Human Design* (New York: Basic Books, 1984), p. 171.

165 **"adjacent residences decline":** Ibid.

165 **"of high school":** Rowan Rowntree, interview with the author, February 5, 2014.

166 **"and industrial operations":** Fred Bartenstein, interview with the author, January 14, 2014.

167 **"trees to cities":** Ibid.

167 **"a pecuniary one":** Rowan Rowntree, interview with the author, July 21, 2010.

168 **"and surface runoff":** Rowan A. Rowntree et al., "Evaluating Urban Forest Structure for Modifying Microclimate: The Dayton Climate Project," *Proceedings of the Second National Urban Forestry Project*, October 10–14, 1982, Cincinnati, Ohio (Washington, D.C.: American Forestry Association, 1982), p. 136.

169 **"percent or more":** Ibid.

169 **"Central Business District":** Ibid., p. 139.

169 **"to work on":** Rowan Rowntree, interview with the author, July 21, 2010.

169 **"to the environment":** Jane Palmer, "A Career Takes Root," *Omaha World-Herald*, March 7, 2001, p. 41.

171 **"thinking behind it":** John Rosenow, interview with the author, May 21, 2014.

172 **"program increased dramatically":** John Rosenow, "Community Funding Sources," *Proceedings of the Second National Urban Forestry Project*, October 10–14, 1982, Cincinnati, Ohio (Washington, D.C.: American Forestry Association, 1982), p. 309.

172 **"ground that way":** Tim Jacob, interview with the author, February 20, 2014

172 **"organized and methodical":** Deb Derrick, "Embracing Uncertainty: Leadership in a Culture of Change," *Engineering Alumnae Magazine* (University of Nebraska at Lincoln), Spring 2002.

173 **"such immediate gratification":** Ann Linnea, *Keepers of the Trees: A Guide to Re-Greening North America* (New York: Skyhorse, 2010), pp. 12–13.

174 **"and private industry":** John R. McGuire, "Plant a Tree and Save a City," *Journal of Arboriculture* 1, no. 11 (November 1975): 208.

174 **"for next year":** Michael Seiler, "Andy vs. the Bureaucratic Deadwood," *Los Angeles Times*, April 23, 1973, p. F1.

175 **"TreePeople was born":** Linnea, *Keepers of the Trees*, p. 13.

175 **"create natural beauty":** Art Seidenbaum, "Andy's Conservation Ideas Take Root," *Los Angeles Times*, December 22, 1975, p. G1.

175 **"of a cityscape":** Art Seidenbaum, "Tree People in Full Flower," *Los Angeles Times*, March 15, 1978, p. OC-C1.

176 **"thinking and direction":** Andrew Lipkis, "One Million Trees for the 1984 Olympics in Los Angeles," *Proceedings of the Second National Urban Forestry Project*, October 10–14, 1982, Cincinnati, Ohio (Washington, D.C.: American Forestry Association, 1982), p. 98.

176 **"before the Olympics":** Ibid., p. 96.

176 **"of the cost":** Ibid.

177 **"pursuing their dreams":** Ibid., p. 97.

177 **"Dutch elm disease":** E. B. Smalley and R. P Guries, "Building Better Elms: The Wisconsin Breeding Program," *Proceedings of the Second National Urban Forestry Project*, October 10–14, 1982, Cincinnati, Ohio (Washington, D.C.: American Forestry Association, 1982), p. 361.

CHAPTER THIRTEEN: "Trees Are the Answer": John Hansel, Henry Stern, Deborah Gangloff, and George Bush

180 **"the American landscape":** Michael Pollan, "Look Who's Saving Elm," *New York Times*, October 31, 1993.

180 **"and USDA programs":** E. B. Smalley, R. P. Guries, and D. T. Lester, "American Liberty Elms and Beyond: Going from the Impossible to the Difficult," in *Dutch Elm Disease Research: Cellular & Molecular Approaches*, ed. M. Sticklan and J. Sherald (New York: Springer-Verlag, 1993), p. 26.

181 **"on my street":** Ellen Goodman, "Hope Takes Root with Elm Planting," *Chicago Tribune*, December 6, 1984, p. E1.

182 **"a comeback story":** Ibid.

182 **"American Forestry Association":** Robert Marshall Ricard, *Politics and Policy Processes in Federal Urban Forest Policy Formation and Change* (PhD diss., University of Massachusetts at Amherst, May 2009), p. 276.

183 **"this depressing trend":** Mark Johnston, "A Brief History of Urban Forestry in the United States," *Arboricultural Journal* 20 (1996): 273.

183 **"to individual porches":** Derrick Z. Jackson, "The Return of the Trees," *Boston Globe*, April 1, 2006, p. A-11.

184 **"on engineered sites":** John. R. McGuire, "Plant a Tree and Save a City," *Journal of Arboriculture* 1, no. 11 (November 1975): 210–11.

184 **along a parkway:** Gary Koller and Richard E. Weaver, "Replacing the American Elm: Twelve Stately Trees," *Arnoldia* 42, no. 2 (Spring 1982): 90–96.

185 **"of glowing embers":** Ibid., pp. 95–98.

185 **"is uniformly acidic":** Ann F. Rhoads, Paul W. Meyer, and Robert Sanfelippo, "Performance of Urban Street Trees Evaluated," *Journal of Arboriculture* 7, no. 5 (May 1981): 127–32.

185 **"of a pond":** Theresa M. Culley and Nicole A. Hardiman, "The Beginning of a New Invasive Plant: A History of the Ornamental Callery Pear in the United States," *Bioscience* 57, no. 11 (December 2007): 960.

186 **"'can save trees'":** Joe Klein, "Henry Stern Goes to the Park," *New York*, August 1, 1983, p. 43.

187 **"for personal convenience":** Deirdre Carmody, "East Siders Ask If Garage Is Worth Tree," *New York Times*, February 23, 1985.

187 **"the government responds":** Joe Mathews, "Force of Nature: Henry Stern," *Baltimore Sun*, July 19, 1998, Art Section front page.

188 **"not on purpose":** Deirdre Carmody, "New York, Again Tree City USA, Pays Homage to Its Finest Specimens," *New York Times*, April 27, 1985.

188 **"gardens and shrubberies":** City of New York Department of Parks and Recreation, *Great Tree Walk Guide*, 1990, Queens section.

189 **also constructed here:** Ibid., Manhattan section.

189 **"to Marie Antoinette":** Jeffrey G. Meyer, *America's Famous and Historic Trees* (Boston: Houghton Mifflin, 2001), p. 23.

189 **"was just chaos":** Jeffrey G. Meyer, interview with the author, March 30, 2011.

190 **"location and ownership":** Joseph L. Stearns, "Let's Find and Save the Biggest Trees," *American Forests*, September 1940, p. 413.

190 **"at 130 feet":** Corliss Knapp Engle, "John Adams, Farmer and Gardener," *Arnoldia* 61, no. 4 (Winter 2002): 12.

190 **"most media attention":** Deborah Gangloff, interview with the author, May 13, 2016.

190 **tree of heaven:** Erik Eckholm, "Great Trees: One Champ Is 16 Feet of Trouble," *New York Times*, April 20, 1986, p. 1.

191 **"so much easier":** Gangloff, interview with the author, May 13, 2016.

191 **"are left untended":** Robert Makla, "The Trees of New York City's Parks Deserve Professional Care" (letter), *New York Times*, June 18, 1988, p. 26.

191 **"people to Henry":** Former New York City official who prefers to remain anonymous, interview with the author.

191 **"trees a year":** David W. Dunlap, "Death Is Heeded: Sympathy, At Least, for City Trees," *New York Times*, July 24, 1988, in Week in Review section.

192 **"and everything changed":** Unpublished memo by USFS official who prefers to remain anonymous.

192 **"plant more trees":** Ibid.

192 **"he's the guy":** Ricard, *Politics and Policy Processes*, p. 313.

192 **"the White House":** Unpublished memo by USFS official who prefers to remain anonymous.

194 **"the same way":** "New Varieties of Elm Raise Hope of Rebirth for Devastated Tree," *New York Times*, December 5, 1989, p. C4.

194 **"classic American elm":** Ibid.

194 **"elm leaf beetles":** Ibid.

194 **"Smalley," recalls Townsend:** Denny Townsend, interview with the author, October 6, 2010.

194 **"have shot him":** John Hansel, interview with the author, August 24, 2010.

CHAPTER FOURTEEN: "Don't *Trees* Clean the Air?": Rowan Rowntree, Greg McPherson, and David Nowak

196 **"they won't thrive":** Debra Shore and Stephen Packard, "A Green Vision for Chicago," *Chicago Wilderness*, Spring 2000.

196 **"on my hands":** Edith Makra, interview with the author, October 21, 2014.

198 **"provide these answers":** R. A. Rowntree, "Ecology of the Urban Forest: Introduction to Part III," *Landscape and Urban Planning* 15 (1988): 6–8.

198 **"come to Chicago":** Rowan Rowntree, interview with the author, July 20–21, 2010.

198 **"getting us $900,000":** Rowan Rowntree, interview with the author, May 19, 2009.

198 **"work in Chicago":** Rowntree, interview with the author, July 20–21, 2010.

199 **"than the cost":** Gregory McPherson, interview with the author, September 15, 2014.

199 **288 kilowatts of energy:** E. Gregory McPherson, "Accounting for Benefits and Costs of Urban Greenspace," *Landscape and Urban Planning* 22 (1992): 41–51.

199 **"building on Tucson":** Gregory McPherson, interview with the author, March 16, 2010.

200 **"in urban forestry":** Rowntree, interview with the author, July 20–21, 2010.

201 **"an important area":** David Nowak, interview with the author, July 29, 2010.

201 **"its trunk size":** Ibid.

201 **"of a tree":** Rowntree, interview with the author, July 20–21, 2010.

201 **"the Civil War":** McPherson, interview with the author, September 15, 2014.

202 **"transplanting large trees":** "A Pearl Anniversary for a Gem of a Place," *Naturalist: Quarterly Nature Newsletter of the Chicago Park District*, October/November/December 2009, p. 1.

202 **architects, and homeowners:** E. Gregory McPherson et al., *Chicago's Evolving Urban Forest: Initial Report of the Chicago Urban Forest Climate Project*, USDA Northeastern Forest Experiment Station, General Technical Report NE-169, January 1993, p. 5.

202 **"net environmental benefits":** Ibid., p. 4.

203 **"'in the country'":** McPherson, interview with the author, September 15, 2014.
203 **"measuring climate effects":** Greg McPherson, "Putting Research to Work," *Urban Forests*, October/ November 1992, p. 5.
204 **"14 Norway maple":** E. Gregory McPherson, David J. Nowak, and Rowan A. Rowntree, *Chicago's Urban Forestry Ecosystem: Results of the Chicago Urban Forest Climate Project*, USDA Northeastern Forest Experiment Station, General Technical Report NE-186, June 1994, p. 13.
204 **"and then weighed":** Ibid., p. 6.
204 **"some proto-typical buildings":** McPherson, interview with the author, September 15, 2014.
204 **"a beach picnic":** Makra, interview with the author, October 21, 2014.
204 **increase these benefits:** McPherson et al., *Chicago's Evolving Urban Forest*, p. 4.
205 **"a desired outcome":** McPherson, Nowak, and Rowntree, *Chicago's Urban Forestry Ecosystem*, p. 11.
205 **"a new generation":** Makra, interview with the author, October 21, 2014.
205 **the American elm:** Ibid.
206 **"and cooling bill":** McPherson, Nowak, and Rowntree, *Chicago's Urban Forestry Ecosystem*, p. 111.
206 **"building air conditioning":** Ibid., p. 109.
206 **value of $1 million:** Ibid., p. iv.
207 **fostering cleaner air:** Ibid., p. 77.
207 **"by urban trees":** Ibid., p. 76.
208 **"the long term":** Ibid., p. 115.
208 **"and air resources":** Ibid., p. 135.
208 **"been very different":** Makra, interview with the author, October 21, 2014.
208 **"a big plan":** Rowntree, interview with the author, July 24, 2014.
208 **"a poor light":** Local official who wishes to remain anonymous, interview with the author.
208 **"shared Rowan's disappointment":** Makra, interview with the author, July 24, 2014.
209 **"water the trees":** *Sacramento Tree Foundation: Celebrating 25 Years*, 2007, p. 7.
209 **"got my job":** Ray Tretheway, interview with the author, August 20, 2010.
211 **"tons of media":** Jim Hardie, interview with author in Los Angeles, July 13, 2015.
211 **Sacramento Tree Foundation:** *Sacramento Tree Foundation: Celebrating 25 Years* (pamphlet), 2007, pp. 8–9.
211 **"a $2 million budget":** Melanie Turner, "Green Awards: Ray Tretheway, Sacramento Tree Foundation," *Sacramento Business Journal*, April 13, 2012.
212 **"me to Rowan":** Tretheway, interview with the author, April 10, 2010.
212 **"in dollar terms":** Misha Sarkovich, "SMUD Shade Tree and Cool Roof Programs: Case Study in Mitigating the Urban Heat Island Effects," 2006 unpublished manuscript given to the author.
213 **"of the trees":** Tretheway, interview with the author, August 20, 2010.

CHAPTER FIFTEEN: "We Stand a Great Chance of Seeing a Return of the Stately and Valuable American Elm": Rebirth of an Iconic Tree?

215 **"a scientific experiment":** Denny Townsend, interview with the author, July 27, 2010.
215 **"at the doorway":** Richard Wolkomir, "Racing to Revive Our Embattled Elms," *Smithsonian*, June 1998, p. 44.
215 **"inches of cork":** Les Line, "The Return of an American Classic," *Audubon*, September/October 1997, p. 74.

216 **"wild native tree":** Denny Townsend, interview with the author, July 8, 2014.
216 **"in this test":** Ibid.
216 **"the American elm":** Wolkomir, "Racing to Revive Our Embattled Elms," p. 44.
216 **"they were better":** Denny Townsend, interview with the author, October 6, 2010.
217 **"in 10 years":** "New Varieties of Elm Raise Hope of Rebirth for Devastated Tree," *New York Times,* December 5, 1989, p. C4.
218 **"its desirable qualities":** Charles R. Burnham, "The Restoration of the American Chestnut," *American Scientist* 76 (September 1988): 486.
218 **"early backcross trees":** Susan Freinkel, *American Chestnut: The Life, Death, and Rebirth of an American Tree* (Berkeley: University of California Press, 2007), p. 185.
219 **"science of chestnuts":** Tom Horton, "American Chestnut," *American Forests,* 116 (Winter 2010): 35.
219 **"a lifetime proposition":** Freinkel, *American Chestnut,* p. 182.
219 ***Ophiostoma ulmi* to each:** Denny Townsend, interview with the author, August 12, 2015.
219 **"each inoculation date":** A. M. Townsend et al., "Variation in Response of Selected American Elm Clones to *Ophiostoma ulmi,*" *Journal of Environmental Horticulture* 13, no. 3 (September 1995): 127.
219 **"perhaps 'American Liberty'":** Ibid., p. 128.
220 **"It was eye-opening":** Townsend, interview with the author, August 12, 2015.
220 **"in American elm":** Townsend et al., "Variation in Response of Selected American Elm Clones," p. 128.
220 **"valuable American elm":** Senator Christopher S. Bond, speaking regarding the National Arboretum of the Agricultural Research Service on June 6, 1996, to the Senate, *Congressional Record* 142, no. 82, p. S5947.
221 **"drought, and flooding":** Alden M. Townsend, "USDA Genetic Research on Elm," in *The Elms: Breeding, Conservation, and Disease Management,* ed. Christopher P. Dunn (Boston: Kluwer Academic, 2000), p. 271.
221 **"the old wounds":** Tom Prosser, interview with the author, December 4, 2015.
222 **"elm was sealed":** Keith Warren, interview with the author, July 15, 2015.
222 **"were very rigorous":** Ibid.
223 **"attractive arcade-forming trees":** George Ware, "Elm Breeding and Improvement at the Morton Arboretum," *Morton Arboretum Quarterly* 28, no. 1 (Spring 1992): 2.
224 **"the American Midwest":** Ann Scott Tyson, "Hardy Asian Elm Tree Shows Promise in US," *Christian Science Monitor,* October 20, 1994, p. 15.
224 **"a dozen more":** Keith Warren, "The Return of the Elm: Status of Elms in the Nursery Industry," in Dunn, ed., *The Elms,* p. 341.
224 **"a true tolerance":** A. M. Townsend and L. W. Douglass, "Variation Among American Elm Clones in Longterm Dieback, Growth, and Survival Following *Ophiostoma* Inoculation," *Journal of Environmental Horticulture* 19, no. 2 (June 2001): 103.
225 **"biotypes after inoculation":** Ibid.
225 **"to look for":** "Buying a Disease-Resistant Elm? Check the Advantages of an American Liberty Elm," *Elm Leaves,* Summer 2002, front page.
225 **"claims for resistance":** Ibid.
225 **"about 'American Liberty'":** Townsend, interview with the author, July 8, 2014.
225 **"is no good":** John Hansel, interview with the author in Keene, New Hampshire, September 23, 2010.
225 **"into the landscape":** Townsend, interview with the author, July 8, 2014.

CHAPTER SIXTEEN: "I Never Saw Such a Bug in My Life": Attack of the Asian Long-Horned Beetles

227 **"was going on":** Ingram S. Carner, interview with the author, July 30, 2010.
228 **"calls his office":** Ibid.
228 **"against our collection":** Mary M. Woodson, "Cities Under Seige," *American Forests* 106, no. 2 (Summer 2000): 7–9.
228 **flooding U.S. ports:** Peter Alsop, "Invasion of the Longhorn Beetles," *Smithsonian*, November 2009.
229 **"change of pace":** Joseph P. Gittleman, interview with the author, June 25, 2010.
229 **"your own anything":** Diane Cardwell, "Trying to Catch Up to a Hungry Beetle," *New York Times*, February 21, 2002, p. B-1.
229 **"citrus or wheat":** Gittleman, interview with the author, June 25, 2010.
230 **be better viewed:** Douglas Martin, "Community Work for Tree Killer, But Penalties Will Stiffen Today," *New York Times*, January 12, 1996, p. B-1.
230 **"was a war":** Henry Stern, interview with the author, July 24, 2014.
230 **"in other projects":** Gittleman, interview with the author, June 25, 2010.
231 **"to me anymore":** David Rohde, "Felling Trees Before Beetle Eats Them," *New York Times*, April 6, 1997, sec. 13, p. 8.
231 **"little new tree":** Interview with a resident in the park, June 25, 2010.
232 **"was profoundly distressing":** Jan Hoffman, "Speaking for the Trees in the Concrete Forest," *New York Times*, August 3, 1999, p. B2.
232 **"one per site":** Gittleman, interview with the author, June 25, 2010.
234 **"accept any responsibility":** Joe McCarthy, interview with the author, October 20, 2009.
234 **"the forest monoculture":** American scientist who prefers to remain anonymous, interview with the author.
234 **said one official:** Judy Antipin and Thomas Dilley, *Chicago vs. Asian Longhorn Beetle: A Portrait of Success* (case study), USDA, October 2004, p. 17.
235 **"were the lifesavers":** Ibid.
235 **"it's been machine-gunned":** Laura Gatland, "Polka-Dot Pest Chomps on Windy City's Trees," *Christian Science Monitor*, August 28, 1998, p. 3.
235 **"kept us cool":** Bob Kurson, "Beetles Force Sad Farewell to Family Tree," *Chicago Sun-Times*, August 16, 1998.
236 **"very sad day":** Antipin and Dilley, *Chicago vs. Asian Longhorn Beetle*, p. 21.
236 **"just really hard":** Ibid., p. 22.
236 **"grandeur is gone":** Ibid., p. 34.
236 **"come out later":** Maureen O'Donnell, "Daley Issues Beetle Alert," *Chicago Sun-Times*, May 25, 1999.
237 **"in those cases":** Clint McFarland, interview with the author, September 14, 2010.
237 **"New York City":** Gittleman, interview with the author, June 25, 2010.
238 **"kids are vectors":** Ibid.
238 **"children used it":** Corey Kilgannon, "Destructive Asian Longhorn Beetles Hit Manhattan, and a Park Loses 13 Trees," *New York Times*, August 22, 1999, sec. 14, p. 5.
238 **"end of story":** Gittelman, interview with the author, June 25, 2010.
238 **"like Central Park":** Kilgannon, "Destructive Asian Longhorn Beetles."
238 **"save the forest":** Ibid.
238 **"in North America":** http://www.centralparknyc.org/tree-guide/american-elm.html.

CHAPTER SEVENTEEN: "On That Branch Was a Four-Inch Green Shoot with Leaves": Ground Zero Survivor Trees

240 **"was pulverized concrete":** Rebecca Clough, interview with the author, November 7, 2014.

241 **"just working there":** Michael Browne, interview with the author, November 6, 2014. and subsequent e-mails.

242 **"the Brooklyn Bridge":** Bram Gunther, interview with the author, October 10, 2014.

242 **"Pentagon," she recalls:** Deborah Gangloff, interview with the author, November 24, 2014.

243 **"*thing to revive*":** Ibid.

243 **"a positive response":** "Support Grows for Memorial Trees," *American Forests* 108 (Spring 2002): 14.

243 **"for the sun":** Anne Raver, "Honoring Loss with the Power of Green," *New York Times*, September 5, 2002, p. F-1. See also E. B. White, *Here Is New York* (New York: Little Bookroom, 2000).

244 **"and the nation":** Living Memorials Project, www.livingmemorialsproject.net/about.htm.

244 **"strength and revival":** Matt Arnn, "Urban Forest Helps Lower Manhattan Heal," *Community & Forests: Newsletter of Communities Committee of 7th American Forestry Congress* 6, no. 1 (Spring 2002): 5.

244 **"on a tree":** Browne, interview with the author, November 6, 2014.

244 **"salvaging healthy specimens":** Rebeccca Clough, written memo given to the author, May 22, 2014.

244 **"find an ally":** Michael Browne, e-mail to the author, November 7, 2014.

244 **"on the truck":** Rebecca J. Clough, written statement to correct errors in children's book *The Survivor Tree*, May 22, 2014.

244 **"and a backhoe":** Browne, e-mail to the author, November 7, 2014.

244 **"perform his magic":** Ibid.

245 **"giant sycamore tree":** Valeria B. Girandola, "The Miracle Tree," Spigot Science for Kids and Classrooms.

245 **"the recovery effort":** *A Guide to St. Paul's Chapel*, 2010 brochure.

246 **"*going to survive*":** Gary Marrs, interview with the author, December 9, 2014.

246 **"'save this tree'":** Ibid.

247 **"my favorite spot":** Kari Watkins, interview with the author, December 8, 2014.

247 **"see what happened":** Mark Bays, interview with the author, December 1, 2014.

248 **the entire site:** Oklahoma City National Memorial & Museum, "We Come Here to Remember" (DVD).

248 **"respecting the tree":** Mark Bays, interview with the author, June 17, 2015, at the memorial site in Oklahoma City.

249 **a "wonderful lesson":** Oklahoma City National Memorial & Museum, "We Come Here to Remember."

250 **"ability to heal":** Browne, e-mail to the author, November 7, 2014.

250 **"see the tree":** Richard Cabo, interview with the author, October 29, 2014.

250 **"more peaceful world":** "Eddie Bauer Leads Memorial Groves Effort," *American Forests* 108 (Spring 2002): 11.

250 **"very difficult site":** Deborah Gangloff, interview with the author, April 21, 2015.

251 **"memorial stands still":** "Pentagon Memorial Trees," *American Forests* 115 (Winter 2009): 15.

251 **"generations to come":** "Brooklyn Celebrates Life with the Planting of the City's First Memorial Grove," *Daily Plant*, December 4, 2002, http://www.nycgovparks.org/parks/sunset-park/dailyplant/15287.

251 **"an unselfish act":** Matthew Arnn, e-mail to the author, March 27, 2015.

251 **"look at them":** Arnn found this Warhol quote in *The Philosophy of Andy Warhol (From A to B and Back Again)* (New York: Houghton Mifflin Harcourt, 1977), p. 154.

CHAPTER EIGHTEEN: "I Was Surprised It Was So Aggressive": Waging War on the Emerald Ash Borer

253 **"growth this year":** David L. Roberts, "Ash Decline in Michigan—2002," Ash Trees Decline, October 2001, http://treedoctor.anr.msu.edu/ash/decline.html#decline.

254 **"the genus *Agrilus*":** David L. Smitley, interview with the author, March 30, 2015.

254 **"emerges next spring":** Ibid.

254 **"on weakened trees":** David L. Roberts, interview with the author, March 22, 2015.

255 **"a dozen locations":** Roberts, interview with the author, March 25, 2015.

255 **"the same color":** Roberts, interview with the author, March 22, 2015.

255 **"various geographical areas":** http://treedoctor.anr.msu.edu/.

256 **"it's just huge":** Deborah G. McCullough, interview with the author, March 25, 2015.

257 **"Chinese reference books":** Therese M. Poland and Deborah G. McCullough, "Emerald Ash Borer: Invasion of the Urban Forest and the Threat to North America's Ash Resource," *Journal of Forestry* 104, no. 3 (April/May 2006): 119.

257 **"from infested ash":** Daniel A. Herms and Deborah G. McCullough, "Emerald Ash Borer Invasion of North America: History, Biology, Ecology, Impacts, and Management," *Annual Review of Entomology* 59 (2014): 15.

257 **"unusually aggressive insect":** Jordon D. Marché II, "Fool Me Twice, Shame on Me: The Emerald Ash Borer in Southeastern Michigan," *Forest History Today* 18, no. 2 (Fall 2012): 11.

257 **"officials for advice":** Pat Shellenbarger, "Mysterious Asian Bug Perils State Ash Trees," *Grand Rapids Press*, July 17, 2002, p. A-1.

258 **"and burn it":** Diane Cardwell, "Trees at Risk from Invader with Appetite," *New York Times*, February 10, 2002, p. 33.

258 **"New York and Vermont":** Ibid.

259 **"just don't know":** Neil Calvanese, interview with the author, June 25, 2010.

260 **"its unadorned greatness":** Donald Culross Peattie, *The Natural History of Trees* 2nd ed. (Boston: Houghton Mifflin, 1966), p. 571.

260 **"ash each year":** Deborah G. McCullough, "Will We Kiss Our Ash Good-bye?" *American Forests* 119 (Winter 2013): 21.

261 **"along that firewood":** Sharon Lucik (EAB Program, USDA APHIS, Brighton, Michigan), interview with the author, March 24, 2015.

261 **"in urban areas":** Don Behm and Annysa Johnson, "Ash to Ashes: Voracious Beetles Are Gobbling Hardy Urban Trees," *Milwaukee Journal Sentinel*, November 24, 2002, p. B-1.

262 **"dollars in damage":** Erika Blake, "Beetle Battles Target Pests in Ash Trees," *Toledo Blade*, December 25, 2002, p. B-1.

262 **"beyond southeast Michigan":** David Shepardson, "Beetle Devours Michigan Ash Trees," *Detroit News*, November 19, 2002, p. 1-C.

262 **"beyond that circle":** APHIS official who prefers to remain anonymous, interview with the author.

262 **"a continental scale":** Herms and McCullough, "Emerald Ash Borer Invasion of North America," p. 14.
262 **"day everything changed":** Daniel Herms, "EAB Retrospective Perspective" (PowerPoint presentation), slide 2, Ohio State University, 2008.
263 **"make it through":** Tom Henry, "Beetle Poses Dire Threat," *Toledo Blade*, March 30, 2003, p. A-11.
263 **"in my lifetime":** Ibid.
263 **"a widespread problem":** Ibid.
263 **"alive and well":** Kevin Brown, "Cities Await Report in Fight to Save Ash Trees," *Detroit News*, February 25, 2003, p. C-4.
264 **"did not know":** Craig Kellogg, interview with the author, April 6, 2015.
264 **"single tree left":** Tom Holden, interview with the author, April 21, 2015.
264 **dying, or doomed:** Cooperative Emerald Ash Borer Project 2007, "The Green Menace" (PowerPoint presentation), slide 11 ("Discovery and Response").
265 **"find this thing":** Kellogg, interview with the author, April 6, 2015.
265 **"a dispersal flight":** Robin A. J. Taylor et al., "Emerald Ash Borer Flight Potential," presentation at EAB Research and Technology Development Meeting, October 5–6, 2004, Romulus, Michigan.
266 **"than previously thought":** Poland and McCullough, "Emerald Ash Borer," p. 120.
266 **"in low-light conditions":** Joseph A. Francese et al., "Studies to Develop an EAB Survey Trap," presentation at EAB Research and Technology Development Meeting, October 5–6, 2004, Romulus, Michigan.
266 **"EAB natural enemies":** Leah S. Bauer et al., "Update on Emerald Ash Borer Natural Enemy Surveys in Michigan and China," presentation at EAB Research and Technology Development Meeting, October 5–6, 2004, Romulus, Michigan.
267 **"eradication, but containment":** Kellogg, interview with the author, April 6, 2015.
267 **"years in jail":** Michigan Emerald Ash Borer Response Project, "The REAL Story About Emerald Ash Borer in Michigan" (press release), no date, but early 2006.
268 **"got excited then":** Dirk VanderHart, "Moving Target," *Columbus Dispatch*, June 14, 2005, p. A-6.
268 **"deer and moose":** David Cappaert et al., "Emerald Ash Borer in North America: A Research and Regulatory Challenge," *American Entomologist*, Fall 2005, p. 154.
268 **"just highly visible":** Nate Siegert, interview with the author, May 7, 2015.
269 **"firewood or lumber":** Kellogg, interview with the author, April 7, 2015.

CHAPTER NINETEEN: "Putting in an Urban Forest Instead of a Storm Drain": High-Tech Meets a Million Trees

271 **"those trees provided":** Greg McPherson, interview with the author, March 30, 2010.
272 **"a conveyor belt":** Paula Peper, interview with the author, October 20, 2014.
272 **"oak tree has":** Ibid.
272 **"of urban forestry":** Ibid.
272 **"list of benefits":** Ibid.
273 **"street tree benefits":** Scott Maco, interview with the author, May 25, 2010.
273 **"tenth-of-an-acre circle":** David Nowak, interview with the author, July 29, 2010.
273 **"in the code":** Ibid.
274 **"not even compatible":** Mark Buscaino, "I-Tree Beginnings," personal recollections as written for the author, April 7, 2010.

274 **"'offer customer support'":** Ibid.
274 **"'it out there'":** Mark Buscaino, interview with the author, February 12, 2016.
275 **"the ecosystem benefits":** Deborah Gangloff, interview with the author, February 9, 2016.
275 **"$4.6 million annually":** Charles Seabrook, "Hotlanta Gets Hotter as Trees Fall to Development," *Atlanta Constitution*, March 27, 1996, p. 1.
275 **"and surrounding counties":** Charles Seabrook, "The Ungreening of Atlanta," *Atlanta Constitution*, September 14, 1997, p. E8.
275 **"case for preservation":** Kathleen Wong, "A Pixel Worth 1,000 Words," *U.S. News & World Report*, July 19, 1999, p. 48.
275 **"streets and sidewalks":** Stephen C. Fehr, "Mayor Working to Keep It Green," *Washington Post*, November, 17, 1999.
276 **"robust, public spaces":** Stephen C. Fehr, "Mayor Working to Keep It Green," *Washington Post*, November 17, 1999. https://www.washingtonpost.com/archive/local/1999/11/17/mayor-working-to-keep-it-green/4a5153e6-760d-45dc-aa14-3699311de381/.
276 **former lush glory:** Barbara Shea, "History of the Founding of Casey Trees," written privately and lent to the author, February 18, 2016.
276 **"interpretation and format":** Gangloff, interview with the author, February 9, 2016.
276 **"for our science":** Nowak, interview with the author, February 23, 2016.
277 **"importance of trees":** Greg Ina, interview with the author, March 26, 2015.
277 **"it can be":** Mark Buscaino, ""i-Tree: A Model of Technology Development, Dissemination, Support and Refinement to Benefit the American Public," USDA Forest Service Urban and Community Forestry Program Memo on i-Tree, 2006.
277 **"Economic and Environmental":** Buscaino, "I-Tree Beginnings."
277 **"not have happened":** Ina, interview with the author, March 26, 2015.
277 **"home to Seattle":** Maco, interview with the author, May 25, 2010.
277 **"of the science":** Buscaino, ""i-Tree: A Model of Technology Development, Dissemination, Support and Refinement to Benefit the American Public."
277 **"technology, reinventing ourselves":** Fiona Watt, interview with the author, November 20, 2009.
278 **"in their neighborhood":** Peper, interview with the author, September 29, 2015.
278 **"city in the United States":** "Trees Count! Summary." http://www.milliontreesnyc.org/html/about/urban_forest_census.shtml, 2009.
279 **"good management purposes":** Peper, interview with the author, September 29, 2015.
280 **"the original 8,000":** Paula J. Peper et al., "New York City, New York Municipal Forest Resource Analysis," Center for Urban Forest Research, USDA Forest Service, Pacific Southwest Research Station, April 2007, p. 31.
281 **"about $200 million":** Diane Cardwell, "Bid for a Million Trees Starts with One in Bronx," *New York Times*, October 10, 2007.
281 **"meet our goal":** Ibid.
281 **"have better procurement":** Watt, interview with the author, November 20, 2009.
282 **"out the middleman":** Matthew Stephens, interview with the author, November 20, 2009.
282 **"liners from Schmidt's":** Stephens, interview with the author, January 15, 2010.
283 **"the urban environment":** "Updates on City Ordinances," MillionTreesNYC Special Edition, p. 1A, within *Gooddirt* (NYRP newsletter), Spring/Summer 2008.
284 **"go to hell":** Sara Catania, "It's Not Easy Being Green," *LA Weekly*, July 28–August 3, 2000, p. 1.

284 **"real green jobs":** Andy Lipkis, interview with the author, July 14, 2015.
284 **"insanity to me":** Connie Koenenn, "The Green Team," *Los Angeles Times*, July 20, 1999, p. E-1.
284 **"the way in":** Catania, "It's Not Easy Being Green."
285 **"'as we can'":** Ann Linnea, *Keepers of the Trees* (New York: Skyhorse, 2010), p. 18.
286 **"a storm drain":** Ibid., p. 19.
286 **"watching things grow":** Laurie Kaufman, "The Future Is Green . . . Blue: Tree-People's Urban Watershed Work," *City Trees*, January/February 2006, p. 14.
286 **"as wealthier neighborhoods":** Los Angeles City official requesting anonymity, interview with the author, April 13, 2010.
287 **"and environmental justice":** Lisa Sarno, interview with the author, April 23, 2010.
287 **"and park trees":** Sacramento Tree Foundation, "A State of Trees Report, a Call to Action for the Sacramento Region," 2000, p. 1.
287 **"haphazard and neglectful":** Ibid., p. 6.
288 **"Foundation," said Tretheway:** Sacramento Tree Foundation, Celebrating 25 Years, 2007 report, p. 14.
288 **"a Congressional campaign":** Shea, "History of the Founding of Casey Trees," February 18, 2016.
289 **"board with trees":** Ibid.
289 **"on private lands":** Buscaino, interview with the author, March 24, 2016.
289 **"new 'great streets'":** Barbara Shea, "History of Casey Trees" (PowerPoint presentation), 2013.
290 **"private, and nonprofit":** Barbara Shea, interview with the author, March 21, 2016.
290 **"for invasive species?"** Maco, interview with the author, May 25, 2010.

CHAPTER TWENTY: "Help Restore a Lost Piece of American History": Return of the Elm

292 **"Dutch elm disease":** Dennis Cauchon, "American Elm Makes Slow Return," *USA Today*, May 8, 2007.
292 **"replace elm trees":** Ibid.
293 **"quintessential civic tree":** Adrian Higgins, "The American Elm Stages a Comeback," *Washington Post*, March 31, 2005, p. HO1.
293 **"highest survival rate":** Catherine A. Smith, "Return of the American Elm," *American Forests* 113 (Spring 2007): 44.
294 **"of the landscape":** James L. Sherald, *Elms of the Monumental Core: History and Management Plan*, Natural Resource Report, Draft, National Park Service, National Capital Region, Center for Urban Ecology, Washington, D.C., December 2009, p. 65.
294 **"in the future":** A. M. Townsend, S. E. Bentz, and L. W. Douglass, "Evaluation of 19 American Elm Clones for Tolerance to Dutch Elm Disease," *Journal of Environmental Horticulture* 23, no. 1 (March 2005): 21–24.
296 **"beautiful fountainlike specimen":** Tom Zetterstrom, interview with the author, October 13, 2015.
296 **rest of America:** Eliot Tozer, "The American Elm," *Garden Design*, November/December 2004, p. 18.
297 **"limbs and leaves":** Thomas J. Campanella, *Republic of Shade* (New Haven, CT: Yale University Press, 2003), p. ix.
297 **"and flattering light":** Ibid.
297 **"was the inspiration":** William Jacobi, interview with the author, November 11, 2015.

297 **"any elm tree":** Colorado State University, Department of Bioagricultural Sciences and Pest Management, "National Elm Trial: Executive Summary," 2015, http://bspm.agsci.colostate.edu/people-button/faculty-new/william-jacobi/national-elm-trial/.
297 **"and fall color":** Ibid.
298 **"me any trees":** William Jacobi, interview with author, November 10, 2015.
298 **"in different regions":** Jason Griffin, e-mail to the author, May 10, 2016.
298 **"from indifferent landowners":** http://www.elmpost.org/.
299 **"be highly effective":** Ibid.
299 **"a useful guideline":** Ibid.
299 **"happening to it":** http://www.elmpost.org/proves.htm.
299 **"days went by":** Ibid.
299 **"with elm trees":** Ibid.
300 **"toward their bases":** Ibid.
300 **"in the spring":** Ibid.
300 **"on its own":** Ibid.
301 **"I once did":** Bruce Carley, "Valley Forge Elm Proves Itself in Acton, MA," http://www.elmpost.org/warning/htm/.
301 **"to my neighbors":** Joseph R. Pickens, interview with author, November 22, 2015.
301 **"wonderful cathedral effect":** Ibid.
301 **"it 'The Survivor'":** Ibid.
302 **"gorgeous, tall structure":** John Hansel, interview with the author, September 23, 2010.
302 **"immune to DED":** http://www.elmpost.org/.
303 **"elms as parents":** Bruce Carley, e-mail to the author, October 5, 2015.
303 **"branches are concerned":** Robert DeFeo of National Park Service, e-mail to Tom Zetterstrom, April 20, 2011.
303 **"'cultivars like 'Princeton'":** E-mail dated February 23, 2014, from Tom Zetterstrom to Peter Crane, dean of Yale School of Forestry and Environmental Science, and shared with the author.
304 **"the DED fungus":** Jim Slavicek, interview with the author, May 26, 2015.

CHAPTER TWENTY-ONE: "Oh, My God! They're Really Here": Further Conquests of the Asian Beetles

306 **commissioner Mike Picardi:** "Insect Invasion: Emerald Ash Borer Found on South Side, Officials Say," *Chicago Tribune*, June 20, 2008, p. 10.
306 **"was very motivating":** Edith Makra, interview with the author, March 9, 2015.
306 **kinds of ash:** Ibid.
307 **"to other communities":** "Insect Invasion."
307 **"a great story":** Gerry Smith, "From Insect Casualty to Handmade Luxury: Exhibit at Morton Arboretum," *Chicago Tribune*, August 23, 2008.
307 **"insecticide or not":** Tim De Chant, "Some Turning to Chemicals to Save Ash Trees," *Chicago Tribune*, June 26, 2008, p. C21.
308 **"They're really here":** Linda Bock, "Worcester Woman Honored for Dropping Dime on Beetle Invaders," *Worcester Telegram & Gazette*, August 11, 2009.
308 **"hurt my grandchildren":** Ibid.
308 **"going to change":** Ibid.
309 **"during the year":** Evelyn Herwitz, *Trees at Risk* (Worcester, MA: Chandler House, 2001), p. 163.
309 **"damaged or diseased":** Ibid., p. 175.
309 **"be proven correct":** Evelyn Herwitz, interview with the author, August 15, 2014.

310 **"raining down beetles":** Ken Gooch, interview with the author, June 25, 2010.

310 **Maple Producers Association:** Rodrique Ngowi, "Authorities Hope Beetle Invasion Can Be Ground to a Halt," *Washington Post,* November 5, 2008, p. A18.

311 **in its yard:** Clint McFarland, interview with the author, September 14, 2010.

311 **"years in Brooklyn":** Speaking in Emily Driscoll's 2010 documentary *BUGGED: The Race to Eradicate the Asian Longhorned Beetle.*

312 **"all this cracking":** McFarland, interview with the author, September 14, 2010.

312 **"can't happen again":** Peter Alsop, "Invasion of the Longhorn Beetles," *Smithsonian,* November 2009.

312 **"trees are gone":** Nick Kotsopoulos, "Beetle Battle," *Worcester Telegram,* March 3, 2009. http://www.telegram.com/article/20090303/NEWS/903030353/1101.

313 **"the ALB situation":** E-mail dated December 10, 2009, from Dermot O'Donell to Worcester mayor Joe O'Brien, and given to the author.

314 **"was really crying":** Helen McLaughlin, interview with the author, September 14, 2010.

314 **"every two years":** McFarland, interview with the author, September 14, 2010.

314 **"choice, do we?":** Alsop, "Invasion of the Longhorn Beetles."

315 **"out six maples":** McFarland, interview with the author, August 8, 2014.

315 **"lot of anxiety":** Stephen W. Schneider, interview with the author, September 15, 2014.

316 **"a lure tree":** Schneider, interviews with the author, September 15 and 22, 2014.

316 **"certify the inspections":** Clint McFarland, interview with the author, October 14, 2014.

316 **"few beetles emerged":** McFarland, interview with the author, August 8, 2014.

316 **"hit that hard":** Dru Sabatello, interview with the author, September 10, 2015.

317 **"'of a monoculture'":** Ibid.

317 **"and the wave":** Ibid.

317 **"get people crazy":** Ibid.

317 **"were at risk":** Laurie Taylor, interview with the author, August 24, 2015.

317 **"with neighborhood associations":** Sabatello, interview with the author, September 10, 2015.

318 **"people did call":** Sabatello, interview with the author, August 13, 2015.

318 **most plausible alternative:** Taylor, interview with the author, August 24, 2015.

318 **"to help ourselves":** Laurie Taylor, e-mail to the author, August 25, 2015.

319 **"not the beetle":** Laurie Blake, "Emerald Ash Borer Treatments Costing Less, Working Better," *Star Tribune,* December 10, 2013.

319 **"them healthy trees":** Ibid.

320 **"can believe that":** Taylor, interview with the author, August 24, 2015.

320 **"costs nearly double":** Deborah G. McCullough, "Will We Kiss Our Ash Goodbye?" *American Forests* 19, no. 4 (Winter 2013): 19.

320 **"canopy should be":** R. J. Laverne, interview with the author, May 27, 2015.

321 **"and beneficial shade":** Ibid.

321 **"ability to concentrate":** Ibid.

321 **"all ash trees":** R. J. Laverne, e-mail to the author, September 14, 2015.

322 **"their directed attention":** Laverne, interview with the author, May 27, 2015

322 **"including the soundscape":** Robert J. Laverne, "Loss of Urban Forest Canopy and the Related Effects on Soundscape and Human Directed Attention" (PowerPoint presentation), 2008.

322 **"so to speak":** Geoffrey H. Donovan, interview with the author, September 2, 2015.

323 **"of a house":** Geoffrey H. Donovan and David T. Butry, "Trees in the City: Valuing Street Trees in Portland, Oregon," *Landscape and Urban Planning* 94 (2010): 77–83.

323 **"thinking about health":** Donovan, interview with the author, September 2, 2015.

323 **"for the baby":** Ibid.

323 **to healthier babies:** Geoffrey H. Donovan et al., "Urban Trees and the Risk of Poor Birth Outcomes," *Health & Place* 17 (2011): 390–93.

324 **"of my brain":** Geoffrey H. Donovan, "The Economic Impact of Urban Forests," talk given at Greenprint Summit 2014: Planning for Trees and Public Health, Sacramento Tree Foundation, January 30, 2014. https://www.youtube.com/watch?v=P8UzN-PBLzg&list=PLK1swLtvhR_B-fpcLnWgl8L93mXZDcQ0F&index=5.

324 **"point of view":** Donovan, interview with the author, September 2, 2015.

324 **"public health benefits":** Geoffrey H. Donovan et al., "The Relationship Between Trees and Human Health: Evidence from the Spread of the Emerald Ash Borer," *American Journal of Preventive Medicine* 44, no. 2 (2013): 139–45.

325 **"no relationship there":** Donovan, talk given at Greenprint Summit 2014.

325 **"death for people":** Donovan, interview with the author, September 2, 2015.

CHAPTER TWENTY-TWO: "A Tree Is Shaped by Its Experiences": The Survivor Trees

327 **"become so symbolic":** Richard Cabo, interview with the author, October 29, 2014.

327 **"their healing powers":** Ronaldo Vega, "Survivor Tree Seedlings Link NYC, OKC," *The Memo Blog*, 911Memorial.org, April 19, 2015.

328 **"on its body":** Bernd Heinrich, *The Trees in My Forest* (New York: Ecco, 2003), p. 79.

328 **"and stand strong":** Mark Bays, e-mail to the author, June 25, 2015.

328 **"that tree is":** David W. Dunlap, "A 9/11 Survivor Blossoms in the Bronx," City Room, *New York Times*, April 30, 2009.

328 **"its hardest times":** Ibid.

329 **"lop it off":** Jane Goodall, *Seeds of Hope* (New York: Grand Central, 2014), p. 335.

329 **"has to endure":** Aline Reynolds, "One Survivor from 9/11 Returns Home, for Good," *Downtown Express*, December 29, 2010–January 5, 2011.

329 **"[oak] trees nearby":** Michael Browne, e-mail to the author, July 9, 2015.

330 **"by another one":** http://forums2.gardenweb.com/discussions/171413/cleveland-vs-bradford-pear; then see posting by Rhizo_1(North AL) zone 7, August 19, 2009.

330 **"can be noticed":** Steve Nix, "'Chanticleer' Callery Pear—The Flowering Pears," n.d., forestry.about.com/od/urbanforestry/a/callery_pear.htm.

331 **"not my imagination":** Susannah Breslin, "A Tree That Smells Like . . . Well . . . Um," *The Frisky*, April 12, 2010, http://www.thefrisky.com/2010-04-12/a-tree-that-smells-like-semen/.

331 **"like dead fish":** "It's a Beautiful Tree But It Causes a Stink," *All Things Considered*, NPR, aired April 24, 2015.

331 **"going to grow":** Ibid.

332 **"into the world":** Goodall, *Seeds of Hope*, pp. 333–35.

333 **by Superstorm Sandy:** Associated Press, "9/11 Memorial Unveils Survivor Tree Seedling Program," *Huffington Post*, September 11, 2013.

334 **"the entire place":** Browne, e-mail to the author, July 9, 2015.

334 **"piece of Americana":** Rebecca Clough, interview with the author, November 7, 2014.

AFTERWORD: "The Answer Is Urban Forests"

337 **a brick wall:** R. S. Ulrich, "View Through a Window May Influence Recovery from Surgery," *Science* 224, no. 4647: 420–21.

338 **"my emotional state":** Sara O. Marberry, "A Conversation with Roger Ulrich," *Healthcare Design*, September 1, 2010.

338 **"much like sleep":** Eric Jaffe, "How Urban Parks Enhance Your Brain," Atlantic Cities online, July 16, 2012.

338 **and lower aggression:** Frances E. Kuo and William C. Sullivan, "Aggression and Violence in the Inner City, Effects of Environment via Mental Fatigue," *Environment and Behavior* 22, no. 4 (July 2001): 543–71.

338 **"that was awesome":** Eric Jaffe, "This Side of Paradise," *Observer* 23, no. 5 (May/June 2010).

338 **"endocrine system anxiety":** Carrie Madren, "A Tree-Lined Path to Good Health," *American Forests* 117 (Autumn 2011): 46.

339 **"in cognitive functioning":** Omid Kardan et al., "Neighborhood Greenspace and Health in a Large Urban Center," *Scientific Reports* 5 (July 9, 2015).

339 **"in our cubicles":** Jaffe, "How Urban Parks Enhance Your Brain."

339 **"seven years younger":** Kardan et al., "Neighborhood Greenspace and Health in a Large Urban Center."

339 **"are at play":** Elizabeth Preston, "Trees Make Canadians Feel Healthier," Discover magazine.com, July 10, 2015.

340 **"enjoy the trees":** "ACTrees Interview: TD's Craig Alexander on the Economics of Urban Trees," ACTrees.org, June 16, 2014.

340 **"and climate change":** Biography of William (Ned) Friedman on Arnold Arboretum Web site.

340 **"a botanical garden":** Alvin Powell, "Wild Ambition at the Arboretum," *Harvard Gazette*, November 30, 2015.

341 **"for species identification":** CBOL Plant Working Group, "A DNA Barcode for Land Plant," *Proceedings of the National Academy of Sciences* 106, no. 31 (August 4, 2009): 12794.

341 **"experiments are working":** William Powell, "The American Chestnut's Genetic Rebirth," *Scientific American* 310 v., issue 3 (March 2014): 68–73.

342 **genuine American chestnuts:** Bart Chezar, 2015 Annual Chestnut Report, February 14, 2016 (typescript manuscript provided to the author).

342 **"a humbling experience":** Ibid., p 1.

342 **"can be unpredictable":** Bernd Heinrich, "Revitalizing Our Forests," *New York Times*, December 20, 2013, op-ed page.

342 **"to climate change":** Gary M. Lovett et al., "Nonnative Forest Insects and Pathogens in the United States: Impacts and Policy Options," *Ecological Applications*, doi:10.1890/15-1176, pp 1–19.

343 **"the real world":** Center for Tree Science on the Morton Arboretum Web site.

343 **"the year 2040":** Chicagorti.org website.

344 **"on the increase":** David J. Nowak and Eric J. Greenfield, "Tree and Impervious Cover Change in U.S. Cities," *Urban Forestry & Urban Greening* 11 (2012): 29.

344 **"world" observed Midler:** "Mayor De Blasio Celebrates One Millionth Tree," Press Release from Office of the Mayor, New York City, November 20, 2015.

344 **"canopy has declined":** Mark Buscaino, interview with the author, May 2, 2016.

344 **"to keep planting":** Ibid.

345 **"and public health":** Ray Tretheway, interview with the author, May 13, 2016.

345 **"by their shade":** John Melvin, interview with the author, September 29, 2015.

345 **"into public greenspaces":** Scott Steen, "Looking Backwards and Forwards," *American Forests* 121 (Spring/ Summer 2015): 6.

346 **"to be together":** Dan Lambe, interview with the author, April 22, 2016.

346 **"survives the treatment":** Andrew Jackson Downing, "Pity the Town Trees," *Horticulturalist and Journal of Rural Art and Rural Taste* 1, no. 4 (October 1846): 198.

346 **"for urban life":** Jad Daley, "Trees: Helping Cities Solve Climate Change," *Huffington Post,* January 12, 2016.

346 **"answer is there":** Justin Gillis, "Restored Forests Breathe Life into Efforts Against Climate Change," *New York Times,* December 24, 2014, p. A8.

Index

INDEX

ALSO AVAILABLE

Eiffel's Tower

The Thrilling Story Behind Paris's Beloved Monument and the Extraordinary World's Fair That Introduced It

In this history of the Eiffel Tower, Jill Jonnes offers an eye-opening look not only at the construction of one of the world's most iconic structures, but also at the epochal event that surrounded its arrival as a wonder of the world–the 1889 Exposition Universelle and the celebrities that made it an affair to remember, from Buffalo Bill, sharpshooter Annie Oakley, and Thomas Edison to Whistler, Gaugin, and Van Gogh.

"In splendid detail, Jonnes examines the importance of the tower in its own historical moment." –Caroline Weber, *The New York Times Book Review* (Editors' Choice)

PENGUIN BOOKS

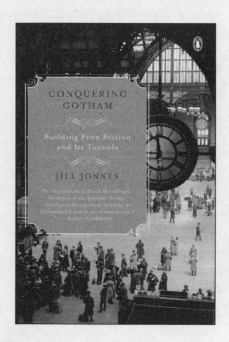